GLENCOE MATH

AUTHORS
Carter • Cuevas • Day • Malloy
Kersaint • Reynosa • Silbey • Vielhaber

Mc
Graw
Hill
Education

MHEonline.com

Send all inquiries to:
McGraw-Hill Education
8787 Orion Place
Columbus, OH 43240

ISBN: 978-0-07-667964-5 (*Volume 1*)
MHID: 0-07-667964-0

Printed in the United States of America.

4 5 6 7 LMN 20 19 18 17

CONTENTS IN BRIEF

Units organized by domain

Glencoe Math is organized into units based on groups of related standards called domains. The Standards for **MP** Mathematical Practices are embedded throughout the course.

MATHEMATICAL PRACTICES HANDBOOK

©Blend Images/Alamy

MP Mathematical Practices

Mathematical Practices Handbook

UNIT 1

Corbis/PunchStock

Ratios and Proportional Relationships

Chapter 1 **Ratios and Rates**

Chapter 2 **Fractions, Decimals, and Percents**

UNIT 2

LatinStock Collection/Alamy

The Number System

Chapter 3 **Compute with Multi-Digit Numbers**

Chapter 4 **Multiply and Divide Fractions**

Chapter 5 **Integers and the Coordinate Plane**

UNIT 3

VICTOR HABBICK VISIONS/Getty Images

Expressions and Equations

Chapter 6 **Expressions**

Chapter 7 **Equations**

Chapter 8 **Functions and Inequalities**

UNIT 4

©Comstock/PunchStock

Geometry

Chapter 9 **Area**

Chapter 10 **Volume and Surface Area**

UNIT 5

L. Mouton/PhotoAlto

Statistics and Probability

Chapter 11 **Statistical Measures**

Chapter 12 **Statistical Displays**

Everything you need,

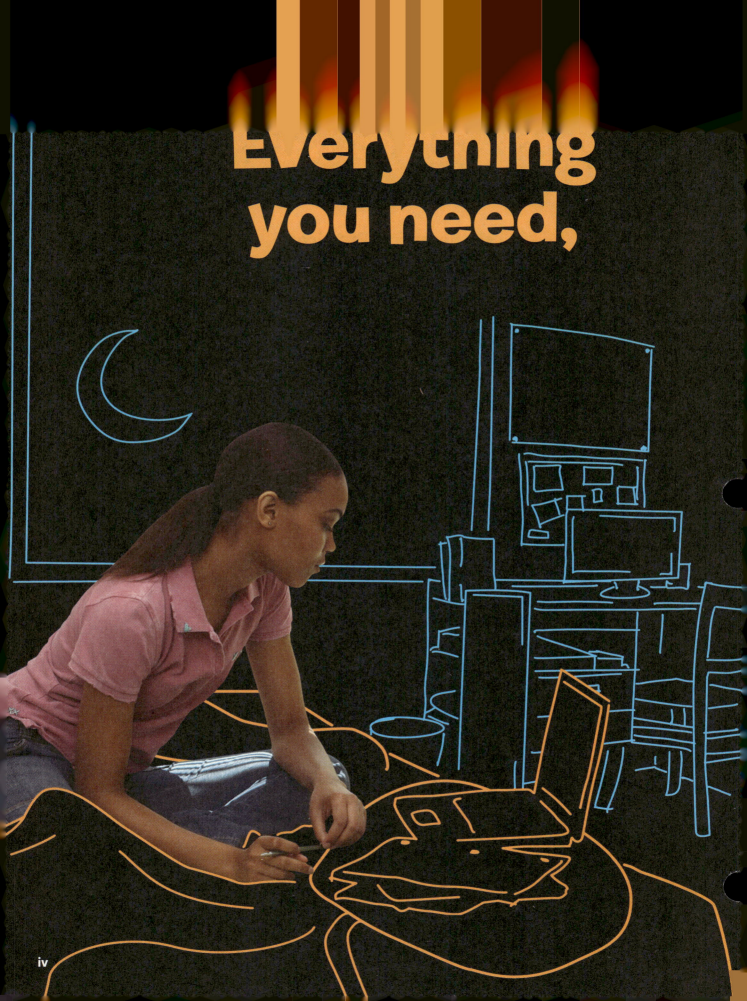

anytime, anywhere.

With ConnectED, you have instant access to all of your study materials—anytime, anywhere. From homework materials to study guides—it's all in one place and just a click away. ConnectED even allows you to collaborate with your classmates and use mobile apps to make studying easy.

Resources built for you—available 24/7:

- Your eBook available wherever you are
- Personal Tutors and Self-Check Quizzes to help your learning
- An Online Calendar with all of your due dates
- eFlashcard App to make studying easy
- A message center to stay in touch

Go Mobile!

Visit mheonline.com/apps to get entertainment, instruction, and education on the go with ConnectED Mobile and our other apps available for your device.

Chapter 1
Ratios and Rates

Real World
p. 31

Essential Question

HOW do you use equivalent rates in the real world?

Chapter 2
Fractions, Decimals, and Percents

Essential Question

WHEN is it better to use a fraction, a decimal or a percent?

Real World
p. 155

UNIT PROJECT 169

People Everywhere

UNIT PROJECT PREVIEW
page 172

Chapter 3
Compute with Multi-Digit Numbers

Essential Question

HOW can estimating be helpful?

Real World
p. 215

Chapter 4
Multiply and Divide Fractions

Essential Question

WHAT does it mean to multiply and divide fractions?

Real World
p. 257

Chapter 5
Integers and the Coordinate Plane

Real World
p. 387

Essential Question

HOW are integers and absolute value used in real-world situations?

UNIT PROJECT 421

Get Out the Map!

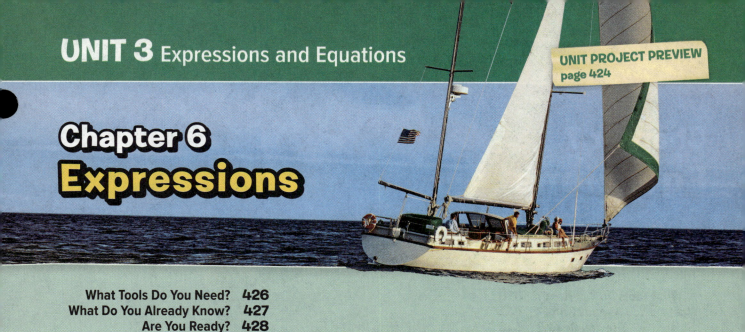

UNIT 3 Expressions and Equations

UNIT PROJECT PREVIEW
page 424

Chapter 6
Expressions

ⓔ Essential Question

HOW is it helpful to write
numbers in different ways?

Real World
p. 495

xi

Chapter 7
Equations

 Essential Question

HOW do you determine if two numbers or expressions are equal?

Real World
p. 535

Chapter 8
Functions and Inequalities

Essential Question

HOW are symbols, such as <, >, and =, useful?

Real World
p. 595

It's Out of This World

UNIT 4 Geometry

UNIT PROJECT PREVIEW
page 652

Chapter 9
Area

Essential Question

HOW does measurement help you solve problems in everyday life?

Real World
p. 661

xiv

Chapter 10
Volume and Surface Area

Essential Question

HOW is shape important when measuring a figure?

Real World
p. 783

UNIT PROJECT 797

A New Zoo

UNIT 5 Statistics and Probability

UNIT PROJECT PREVIEW
page 800

Chapter 11
Statistical Measures

Essential Question

HOW are the mean, median, and mode helpful in describing data?

p. 809

Chapter 12
Statistical Displays

Real World
p. 891

Essential Question

WHY is it important to carefully evaluate graphs?

UNIT PROJECT **925**

Let's Exercise

State Standards for MATHEMATICS, Grade 6

Glencoe Math, Course 1, focuses on four critical areas: (1) using concepts of ratio and rate to solve problems; (2) understanding division of fractions; (3) using expressions and equations; and (4) understanding of statistical reasoning.

Content Standards

Ratios and Proportional Relationships

- Understand ratio concepts and use ratio reasoning to solve problems.

The Number System

- Apply and extend previous understandings of multiplication and division to divide fractions by fractions.
- Compute fluently with multi-digit numbers and find common factors and multiples.
- Apply and extend previous understandings of numbers to the system of rational numbers.

Expressions and Equations

- Apply and extend previous understandings of arithmetic to algebraic expressions.
- Reason about and solve one-variable equations and inequalities.
- Represent and analyze quantitative relationships between dependent and independent variables.

Geometry

- Solve real-world and mathematical problems involving area, surface area, and volume.

Statistics and Probability

- Develop understanding of statistical variability.
- Summarize and describe distributions.

(MP) Mathematical Practices

1. Make sense of problems and persevere in solving them.
2. Reason abstractly and quantitatively.
3. Construct viable arguments and critique the reasoning of others.
4. Model with mathematics.
5. Use appropriate tools strategically.
6. Attend to precision.
7. Look for and make use of structure.
8. Look for and express regularity in repeated reasoning.

Track Your Progress

These pages list the key ideas that you should be able to understand by the end of the year. You will rate how much you know about each one. Don't worry if you have no clue **before** you learn about them. Watch how your knowledge grows as the year progresses!

 I have no clue.   I've heard of it.   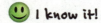 I know it!

	Before			After		
6.RP Ratios and Proportional Relationships	😞	😐	🙂	😞	😐	🙂
Understand ratio concepts and use ratio reasoning to solve problems.						
6.RP.1 Understand the concept of a ratio and use ratio language to describe a ratio relationship between two quantities.						
6.RP.2 Understand the concept of a unit rate a/b associated with a ratio $a:b$ with $b \neq 0$, and use rate language in the context of a ratio relationship.						

6.RP.3 Use ratio and rate reasoning to solve real-world and mathematical problems, e.g., by reasoning about tables of equivalent ratios, tape diagrams, double number line diagrams, or equations.
  a. Make tables of equivalent ratios relating quantities with whole number measurements, find missing values in the tables, and plot the pairs of values on the coordinate plane. Use tables to compare ratios.
  b. Solve unit rate problems including those involving unit pricing and constant speed.
  c. Find a percent of a quantity as a rate per 100 (e.g., 30% of a quantity means 30/100 times the quantity); solve problems involving finding the whole, given a part and the percent.
  d. Use ratio reasoning to convert measurement units; manipulate and transform units appropriately when multiplying or dividing quantities.						

6.NS The Number System	😞	😐	🙂	😞	😐	🙂
Apply and extend previous understandings of multiplication and division to divide fractions by fractions.						
6.NS.1 Interpret and compute quotients of fractions, and solve word problems involving division of fractions by fractions, e.g., by using visual fraction models and equations to represent the problem.						
Compute fluently with multi-digit numbers and find common factors and multiples.						
6.NS.2 Fluently divide multi-digit numbers using the standard algorithm.						
6.NS.3 Fluently add, subtract, multiply, and divide multi-digit decimals using the standard algorithm for each operation.						
6.NS.4 Find the greatest common factor of two whole numbers less than or equal to 100 and the least common multiple of two whole numbers less than or equal to 12. Use the distributive property to express a sum of two whole numbers 1–100 with a common factor as a multiple of a sum of two whole numbers with no common factor.						

	Before			After		
6.NS The Number System *continued*	☹	😐	🙂	☹	😐	🙂
Apply and extend previous understandings of numbers to the system of rational numbers.						
6.NS.5 Understand that positive and negative numbers are used together to describe quantities having opposite directions or values (e.g., temperature above/below zero, elevation above/below sea level, credits/debits, positive/negative electric charge); use positive and negative numbers to represent quantities in real-world contexts, explaining the meaning of 0 in each situation.						
6.NS.6 Understand a rational number as a point on the number line. Extend number line diagrams and coordinate axes familiar from previous grades to represent points on the line and in the plane with negative number coordinates. a. Recognize opposite signs of numbers as indicating locations on opposite sides of 0 on the number line; recognize that the opposite of the opposite of a number is the number itself, e.g., $-(-3) = 3$, and that 0 is its own opposite. b. Understand signs of numbers in ordered pairs as indicating locations in quadrants of the coordinate plane; recognize that when two ordered pairs differ only by signs, the locations of the points are related by reflections across one or both axes. c. Find and position integers and other rational numbers on a horizontal or vertical number line diagram; find and position pairs of integers and other rational numbers on a coordinate plane.						
6.NS.7 Understand ordering and absolute value of rational numbers. a. Interpret statements of inequality as statements about the relative position of two numbers on a number line diagram. b. Write, interpret, and explain statements of order for rational numbers in real-world contexts. c. Understand the absolute value of a rational number as its distance from 0 on the number line; interpret absolute value as magnitude for a positive or negative quantity in a real-world situation. d. Distinguish comparisons of absolute value from statements about order.						
6.NS.8 Solve real-world and mathematical problems by graphing points in all four quadrants of the coordinate plane. Include use of coordinates and absolute value to find distances between points with the same first coordinate or the same second coordinate.						
6.EE Expressions and Equations	☹	😐	🙂	☹	😐	🙂
Apply and extend previous understandings of arithmetic to algebraic expressions.						
6.EE.1 Write and evaluate numerical expressions involving whole-number exponents.						

	Before			After		
6.EE Expressions and Equations *continued*	☹	😐	😊	☹	😐	😊

6.EE.2 Write, read, and evaluate expressions in which letters stand for numbers.

    a. Write expressions that record operations with numbers and with letters standing for numbers.

    b. Identify parts of an expression using mathematical terms (sum, term, product, factor, quotient, coefficient); view one or more parts of an expression as a single entity.

    c. Evaluate expressions at specific values of their variables. Include expressions that arise from formulas used in real-world problems. Perform arithmetic operations, including those involving whole number exponents, in the conventional order when there are no parentheses to specify a particular order (Order of Operations).

6.EE.3 Apply the properties of operations to generate equivalent expressions.

6.EE.4 Identify when two expressions are equivalent (i.e., when the two expressions name the same number regardless of which value is substituted into them).

Reason about and solve one-variable equations or inequalities.

6.EE.5 Understand solving an equation or inequality as a process of answering a question: which values from a specified set, if any, make the equation or inequality true? Use substitution to determine whether a given number in a specified set makes an equation or inequality true.

6.EE.6 Use variables to represent numbers and write expressions when solving a real-world or mathematical problem; understand that a variable can represent an unknown number, or, depending on the purpose at hand, any number in a specified set.

6.EE.7 Solve real-world and mathematical problems by writing and solving equations of the form $x + p = q$ and $px = q$ for cases in which p, q and x are all nonnegative rational numbers.

6.EE.8 Write an inequality of the form $x > c$ or $x < c$ to represent a constraint or condition in a real-world or mathematical problem. Recognize that inequalities of the form $x > c$ or $x < c$ have infinitely many solutions; represent solutions of such inequalities on number line diagrams.

Represent and analyze quantitative relationships between dependent and independent variables.

6.EE.9 Use variables to represent two quantities in a real-world problem that change in relationship to one another; write an equation to express one quantity, thought of as the dependent variable, in terms of the other quantity, thought of as the independent variable. Analyze the relationship between the dependent and independent variables using graphs and tables, and relate these to the equation.

	Before			After		
6.G Geometry	🙁	😐	🙂	🙁	😐	🙂
Solve real-world and mathematical problems involving area, surface area, and volume.						
6.G.1 Find the area of right triangles, other triangles, special quadrilaterals, and polygons by composing into rectangles or decomposing into triangles and other shapes; apply these techniques in the context of solving real-world and mathematical problems.						
6.G.2 Find the volume of a right rectangular prism with fractional edge lengths by packing it with unit cubes of the appropriate unit fraction edge lengths, and show that the volume is the same as would be found by multiplying the edge lengths of the prism. Apply the formulas $V = l\,w\,h$ and $V = b\,h$ to find volumes of right rectangular prisms with fractional edge lengths in the context of solving real-world and mathematical problems.						
6.G.3 Draw polygons in the coordinate plane given coordinates for the vertices; use coordinates to find the length of a side joining points with the same first coordinate or the same second coordinate. Apply these techniques in the context of solving real-world and mathematical problems.						
6.G.4 Represent three-dimensional figures using nets made up of rectangles and triangles, and use the nets to find the surface area of these figures. Apply these techniques in the context of solving real-world and mathematical problems.						

	Before			After		
6.SP Statistics and Probability	🙁	😐	🙂	🙁	😐	🙂
Develop understanding of statistical variability.						
6.SP.1 Recognize a statistical question as one that anticipates variability in the data related to the question and accounts for it in the answers.						
6.SP.2 Understand that a set of data collected to answer a statistical question has a distribution which can be described by its center, spread, and overall shape.						
6.SP.3 Recognize that a measure of center for a numerical data set summarizes all of its values with a single number, while a measure of variation describes how its values vary with a single number.						
Summarize and describe distributions.						
6.SP.4 Display numerical data in plots on a number line, including dot plots, histograms, and box plots.						
6.SP.5 Summarize numerical data sets in relation to their context, such as by: a. Reporting the number of observations. b. Describing the nature of the attribute under investigation, including how it was measured and its units of measurement. c. Giving quantitative measures of center (median and/or mean) and variability (interquartile range and/or mean absolute deviation), as well as describing any overall pattern and any striking deviations from the overall pattern with reference to the context in which the data were gathered. d. Relating the choice of measures of center and variability to the shape of the data distribution and the context in which the data were gathered.						

Mathematical Practices Handbook

Essential Question

WHAT practices help me learn and talk about math?

Mathematical Practices

The standards for mathematical practice will help you become a successful problem solver and to use math effectively in your daily life.

What You'll Learn

MP Throughout this handbook, you will learn about each of these mathematical practices and how they are integrated in the chapters and lessons of this book.

① **Focus on Mathematical Practice**
Persevere with Problems

② **Focus on Mathematical Practice**
Reason Abstractly and Quantitatively

③ **Focus on Mathematical Practice**
Construct an Argument

④ **Focus on Mathematical Practice**
Model with Mathematics

⑤ **Focus on Mathematical Practice**
Use Math Tools

⑥ **Focus on Mathematical Practice**
Attend to Precision

⑦ **Focus on Mathematical Practice**
Make Use of Structure

⑧ **Focus on Mathematical Practice**
Use Repeated Reasoning

Place a checkmark below the face that expresses how much you know about each Mathematical Practice. Then explain in your own words what it means to you.

☹ I have no clue.   😐 I've heard of it.   😊 I know it!

Mathematical Practices

Mathematical Practice	☹	😐	😊	What it Means to Me
①				
②				
③				
④				
⑤				
⑥				
⑦				
⑧				

Persevere with Problems

How do I make sense of problems?

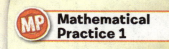

Mathematical Practice 1

Make sense of problems and persevere in solving them.

Yolanda will put 72 photos in a scrapbook. She will put the same number of photos on each of 6 pages. Four photos will be in each row. How many rows will be on each page?

As you work on a problem, you may find you need to change how you are solving it. This is called persevering.

You can use a four-step plan to solve problems. Let's study each step.

1. **Understand** Read the problem. Circle the information that you are given and underline what you are trying to find.

2. **Plan** Decide on an appropriate strategy to use. Some strategies are listed below.

Make a Table

Guess, Check, and Revise

Estimate or Exact Answer

Draw a Diagram

Look for a Pattern

Use Logical Reasoning

What strategy will you use to solve the problem above?

_____

3. **Solve** Apply your strategy to solve the problem.

_____

4. **Check** Determine if your solution is accurate and makes sense. What should you do if your solution doesn't make sense?

_____

_____

Solve each problem by using two different strategies. Then explain which strategy is more effective. Some sample strategies include Guess, Check, and Revise, Make a Table, Draw a Diagram, and Solve a Simpler Problem.

5. Tina bought some nail polish and lip gloss at the store. The nail polish cost $4.50 per bottle and lip gloss cost $7 per tube. Tina bought 9 items and spent $50.50. How many of each item did she buy?

Strategy 1

Strategy 2

6. Of 80 state parks, many have fishing, biking, and caving opportunities. Based on the results in the table, how many parks allow biking or caving but not fishing?

Fishing	68	Biking and Caving	2
Biking	46	Fishing and Caving	0
Caving	4	Biking and Fishing	36

Strategy 1

Strategy 2

ind it in Your Book!

MP **Persevere with Problems**

Give an example where Mathematical Practice 1 is used in Chapter 1. Then explain why your example represents this practice.

Reason Abstractly and Quantitatively

How can I use numbers to describe situations?

Suppose you receive gift cards to an online store for your birthday. Aunt Hilda sends you a $10 gift card, Uncle Norman sends you a $25 gift card, and your grandparents also send you a gift card. You've had your eye on five MP3 downloads, three games, and a wireless game controller. How much does the gift card from your grandparents need to be worth to cover the total cost?

Item	Cost ($)
MP3 download	0.99
Game 1	47.91
Game 2	17.61
Game 3	27.95
Game controller	60.00

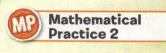

MP **Mathematical Practice 2**

Reason abstractly and quantitatively.

You can represent this situation quantitatively, or by using numbers.

1. What numerical expression represents the total cost of all of the items?

_____

2. If your grandparents gave you a $100 gift card, will you have enough money to buy everything? Explain.

_____

_____

_____

3. List three real-world situations that you could represent quantitatively.

Situation 1	Situation 2	Situation 3

Represent each problem quantitatively.

4. A certain rollercoaster can accommodate 1,600 riders per hour. A digital sign states there are 345 people in line. If you get into line, about how many minutes will it take you to get to the front of the line?

_____

_____

5. Whew! Was this year the hottest year on record? Research and graph the record monthly high temperatures for your town. Then describe how this year's temperatures compare to the record temperatures.

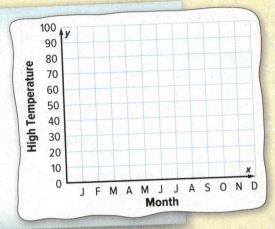

_____

_____

_____

_____

6. You have just arrived at summer camp. In your cabin, the counselor has each camper perform the super-secret handshake with every other person in the cabin. You and another person are in the cabin, so there is one handshake. Another person arrives, and another, until there are 14 people in the cabin. How many handshakes will there be in all?

_____

_____

Find it in Your Book!

MP **Reason Abstractly**

Give an example where Mathematical Practice 2 is used in Chapter 7. Then explain why your example represents this practice.

Construct an Argument

Have you ever had to explain to someone why you were late?

Chances are that saying "I don't know" is not going to go over very well. You need to give a good reason why you were late. In math, you often need to construct a viable, or practical, argument or examine the reasoning of others. When you use rules, definitions, or properties, you are using *deductive reasoning*. When you use examples or patterns, you are using *inductive reasoning*.

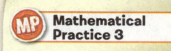

MP Mathematical Practice 3

Construct viable arguments and critique the reasoning of others.

1. The table shows some of the properties of math. Complete the table by completing the example for each property.

Commutative Property of Addition	Commutative Property of Multiplication
$5 + 8 = \boxed{} + 5$	$5 \times 8 = 8 \times \boxed{}$
Associative Property of Addition	**Associative Property of Multiplication**
$6 + (4 + 3) = (\boxed{} + 4) + 3$	$6 \times (4 \times 3) = (6 \times 4) \times \boxed{}$
Identity Property of Addition	**Identity Property of Multiplication**
$9 + 0 = \boxed{}$	$9 \times 1 = \boxed{}$

2. Complete each step to construct an argument for how you can find $3.5 + 1.7 + 1.5$ mentally.

Step	Property
$3.5 + 1.7 + 1.5 = 1.7 + 3.5 + 1.5$	_____
	_____
$= 1.7 + (3.5 + 1.5)$	_____
	_____
$= 1.7 + \underline{}$	Add mentally.
$= \underline{}$	Add mentally.

It's Your Turn!

Solve each problem. Construct a viable argument for your answer.

3. About 80% of the egg rolls sold in U.S. grocery stores are made in Houston, Texas. A skilled wrapper can make 4,000 egg rolls in an 8-hour day. Bettie said this is about one egg roll a minute. Is her claim reasonable? Explain.

_____

_____

4. Not all products labeled cheese are all cheese! A cheese product is less than half actual cheese. Cheese food is more than half, but less than all cheese. Arturo said a 1 pound block that contained 9 ounces of cheese should be called a cheese food. Is his claim reasonable? Explain.

_____

_____

5. A department store advertised that for every $25 you spend at the store, you get $2 off your total bill. Nathanial's bill came to $154 before the discount. He figured he would get a $12 discount. Is his claim reasonable? Explain.

_____

Find it in Your Book!

 MP **Construct an Argument**

Give an example where Mathematical Practice 3 is used in Chapter 1. Then explain why your example represents this practice.

_____

_____

_____

_____

Model with Mathematics

How many different names do you have?

MP **Mathematical Practice 4**

Model with mathematics

Your family, friends, and teachers may call you by different names. You might be Susanna, Suzie, Sue, or Miss Wilson. All of these names still represent *you*. In math, we also use different ways to model the same idea. We can model situations using words, graphs, tables, numbers, symbols, or diagrams.

1. How can you model the number 0.25 using words?

   _____

2. How can you model the number 0.25 using a fraction?

   _____

3. Suppose you and your friends ordered four pizzas. Together, a total of 3 pizzas were eaten. Complete the graphic organizer to model the fraction of the total pizza that was eaten in different ways.

Diagram	Numbers	Words
$\frac{1}{4}$ $\frac{1}{4}$ $\frac{1}{4}$ ☐	$\dfrac{\Box}{\Box}$, 0.75, ☐ %	three-_____ _____ out of four _____-five percent

4. Draw a fraction circle to show the relationship between the amount of pizza you and your friends ate and the total amount of pizza ordered. Explain why the fraction circle is a good model.

   _____

   _____

Use the models shown to solve each problem.

5. You are downloading MP3 singles onto your computer. The singles cost $0.99 each.

a. Tables Complete the table to show the cost of 1, 2, 3, 4, and 5 singles.

Number of Singles, s	Total Cost ($), c

b. Graphs Graph the ordered pairs (number of singles, total cost) on the coordinate plane.

6. Luna's family is driving along the highway. The table shows the time, in hours, traveled and the total distance, in miles, traveled.

Time, t (h)	Distance, d (mi)
1	68
2	136
3	204
4	272
5	340

a. Tables Use the table to find the total distance traveled after 7 hours. _____

b. Words Describe the relationship between time and distance traveled.

_____

_____

Find it in Your Book!

  Model with Mathematics

Give an example of where Mathematical Practice 4 is used in Chapter 8. Then explain why your example represents this practice.

_____

_____

_____

_____

Use Math Tools

How likely is it that a baseball player will hit a homerun using a hockey stick?

Since he is using a hockey stick as a bat, he's not very likely to make any hit, much less a homerun! Similar to using the correct equipment in sports, you also need to use the correct tools in mathematics. So, let's investigate how to choose and use the appropriate tools to solve math problems.

1. To build a hockey goal, you might use a hammer or a drill. While you wouldn't use a hammer or a drill to solve a math problem, you might use some of the tools and strategies listed below. Complete the graphic organizer by adding some of your own tools and strategies.

Math Tools

Tools	
counters	algebra tiles
ruler	computer

Strategies	
estimation	choose the correct unit
draw a diagram	mental math

2. Suppose you want to build a hockey goal and need to determine how much the materials will cost. List three tools or techniques that you could use to determine the total cost of the materials.

_____

3. Describe a situation in which you might use estimation.

_____

_____

_____

It's Your Turn!

For Exercises 4–7, select the tools and/or strategies you could use to solve each problem. Then write the corresponding letter(s) of each tool or strategy.

Tools	Strategies
A. paper and pencil	**F.** draw a diagram
B. calculator	**G.** estimation
C. measuring tape/ruler	**H.** mental math
D. Internet	**I.** number sense
E. virtual manipulatives	**J.** make a prediction

4. Yikes, the supply list for school is huge! You need to buy everything on the list. How much will it all cost?

_____

5 spiral notebooks	10 pens
3-ring binder	3 pencils
notebook paper	dictionary
4 highlighters	10 folders
scientific calculator	flash drive

5. You want to bake a cake for your best friend's birthday. How much will it cost to buy all of the ingredients?

_____

6. The table shows the different admission prices to an amusement park. You and your family want to spend the day at the amusement park, have lunch, and buy a souvenir. How much money should you take?

_____

Wonderama!	
Ticket Prices	
Adult	$22.95
Children (4–11)	$18.95
Seniors	$21.95

Ⓕind it in Your Book!

MP **Use Math Tools**

Give an example of where Mathematical Practice 5 is used in Chapter 1. Then explain why your example represents this practice.

_____

_____

_____

_____

Attend to Precision

How can the everyday meaning of a word help you remember the math meaning?

MP Mathematical Practice 6

Attend to precision.

Mathematics is not just about numbers. You also use words, symbols, and drawings to communicate mathematics. Being precise in your communication means using mathematical words accurately. Some words are used in everyday life as well as mathematics.

1. Look up the word "annex" in the dictionary. What does it mean to "annex"?

   _____

   _____

2. Give two examples of how the word "annex" is used in everyday life.

Example 1	Example 2

3. How is "annex" used in mathematics?

   _____

   _____

4. Explain how you would solve 24 – 9.34. Use correct mathematical vocabulary in your explanation.

   _____

   _____

   _____

Solve. Then use correct mathematical vocabulary to explain how you solved each problem.

5. The table shows the amount of time Manny exercised this weekend. How long did Manny exercise altogether?

_____

_____

_____

Day	Part of Hour Exercised
Saturday	$\frac{2}{3}$
Sunday	$\frac{5}{6}$

6. Catie is making bracelets for her friends using small rubber bands. She has 364 rubber bands and she wants to make 6 bracelets. How many rubber bands will she use for each bracelet? How many rubber bands will be left over?

_____

_____

_____

7. When baking, it is important to use exact measures of ingredients. Kevin wants to double the recipe for carrot cake. A single batch requires $1\frac{1}{4}$ cup oil. He estimates he will need about 2 cups of oil. How much oil will he need? Why is estimation not a good strategy when baking?

_____

_____

ind it in Your Book!

MP Be Precise

Give an example of where Mathematical Practice 6 is used in Chapter 1. Then explain why your example represents this practice.

_____

_____

_____

Make Use of Structure

Can you think of an easier way to find a solution?

Mathematics is based on patterns and structure. You can look for those patterns to help solve problems. If you know how numbers work together, you can find easier ways to solve problems.

MP Mathematical Practice 7

Look for and make use of structure.

1. The restaurant receipt shown does not have a total. Use paper and pencil to find the total spent, not including tip.

   _____

Mimi's Diner

two drinks	$2.50
appetizer	$5.50
entrée	$11.75
dessert	$4.25
Total	

2. What are some other strategies for finding a solution?

   _____

3. Use the terms below to describe how you would solve a problem using mental math or estimation.

   compatible numbers   rounding   properties of numbers

Mental Math

Estimation

4. Using one of the methods above, explain how you found the total of the restaurant receipt from Exercise 1.

   _____

   _____

   _____

   _____

Explain how you would solve each problem. Then solve.

5. Six of the 24 students in a homeroom have an electronic tablet. Predict how many students out of 200 have an electronic tablet.

_____

_____

6. Tucker ran 2.25 kilometers each day for 5 days. How many kilometers did he run altogether?

_____

_____

_____

7. The perimeter of the triangle can be found by adding the lengths of the sides. Find the perimeter of the triangle.

_____

_____

_____

$4\frac{5}{8}$ in. $3\frac{3}{8}$ in.

5 in.

Find it in Your Book!

 Make Use of Structure

Give an example of where Mathematical Practice 7 is used in Chapter 1. Then explain why your example represents this practice.

_____

_____

_____

_____

How can repeated reasoning help you find shortcuts?

MP **Mathematical Practice 8**

Look for and express regularity in repeated reasoning.

In mathematics, if you recognize a pattern, it can help you to find shortcuts to solve the rest of the problem or similar problems. When you look for and use patterns, you are using repeated reasoning.

The Taco Hut likes to give away free food on a regular basis to encourage customers to return. The table shows which items they give away for free and how often.

Food	Frequency
Taco	every 4 days
Nachos	every 5 days
Burrito	every 6 days

1. In the first 60 days of the promotion, on which days will a taco be given away for free?

   _____

2. In the first 60 days, on which days will nachos be given away for free?

   _____

3. In the first 60 days, on which days will a burrito be given away for free?

   _____

4. Are there any days in which more than one food item is given away? If so, which days?

   _____

   _____

   _____

5. If all three items were given away today, how many days until all three are given away on the same day? Explain your reasoning.

   _____

   _____

Solve each problem by using repeated reasoning.

6. Elyse is saving up to buy a new digital camera. She begins with $25 in her savings account and deposits $10 each week. How many more weeks must she save to purchase model X250 than model X150? Explain.

Camera Model	Cost ($)
X150	99
X250	129

_____

_____

_____

7. Anya tracked the amount of snowfall during two different snowstorms. During the first snowstorm, it snowed 1 inch per hour. The second storm produced an inch of snow every two hours. How many more inches of snow did the first storm produce in 6 hours than the second storm? Graph the situation on the coordinate plane.

_____

_____

Inches of Snow — y
9 8 7 6 5 4 3 2 1
0 1 2 3 4 5 6 7 8 9 x
Number of Hours

8. The drama club is building a wall for their next production. The wall is 12 feet long. They are placing the studs every 18 inches. If there are studs on each end of the wall, how many studs will they place?

_____

_____

Find it in Your Book!

MP **Use Repeated Reasoning**

Give an example of where Mathematical Practice 8 is used in Chapter 1. Then explain why your example represents this practice.

_____

_____

_____

Use the Mathematical Practices

Solve.

During the summer, you earn money to buy school clothes. The table shows how much money you earn for each job. This summer, you babysat for 32 hours, mowed 10 lawns, and worked at the pool for 20 hours. You spent $256.78 on school clothes.

Job	Money Earned
babysitting	$5.00/hr
lawn mowing	$15.00/lawn
pool	$6.20/hr

a. Do you have any money left? If so, how much? _____

b. You want to buy a tablet that costs $399. Do you have enough money left? If not, how many more hours would you need to babysit

   to have enough to buy the tablet? _____

Determine which mathematical practices you used to determine the solution. Shade the circles that apply.

Which **MP** **Mathematical Practices** did you use?
Shade the circle(s) that applies.

① Persevere with Problems   ⑤ Use Math Tools

② Reason Abstractly   ⑥ Attend to Precision

③ Construct an Argument   ⑦ Make Use of Structure

④ Model with Mathematics   ⑧ Use Repeated Reasoning

Reflect

 Answering the Essential Question

Use what you learned about the mathematical practices to complete the graphic organizer. Write 3 examples that you could use for each category.

Select Tools and Strategies

1. _____
2. _____
3. _____

Organize Ideas

1. _____
2. _____
3. _____

Essential Question

WHAT practices help me learn and talk about math?

1. _____
2. _____
3. _____

Use Multiple Representations

1. _____
2. _____
3. _____
4. _____

Use a Problem-Solving Model
(List the four steps)

Answer the Essential Question. WHAT practices help me learn and talk about math?

_____

_____

_____

UNIT 1
Ratios and Proportional Relationships

 Essential Question

HOW can you use mathematics to describe change and model real-world situations?

Chapter 1
Ratios and Rates

A ratio is a comparison of two quantities. In this chapter, you will explore ratio concepts and use ratio reasoning to solve rate problems.

Chapter 2
Fractions, Decimals, and Percents

Equivalent forms of fractions, decimals, and percents can be written and used to solve problems. In this chapter, you will apply these relationships to solve percent problems.

Collaborate

Watch

People Everywhere You can use the U.S. Census to compare different characteristics of states' populations, such as the average number of people per square mile, the percent of youth compared to the percent of adults, and the ratio of women to men.

Choose two populations at your school, such as boys and girls, or sixth graders and seventh graders. Then choose a characteristic, such as sports played or number of siblings. Compare the characteristics of the two populations. Report your findings in a double bar graph.

At the end of Chapter 2, you'll complete a project to compare characteristics of the population of different states. Time to get in the right *state* of mind!

Comparing _____

Chapter 1
Ratios and Rates

 Essential Question

HOW do you use equivalent rates in the real world?

Content Standards
6.RP.1, 6.RP.2, 6.RP.3, 6.RP.3a, 6.RP.3b, 6.NS.4

MP **Mathematical Practices**
1, 3, 4, 5, 6, 7, 8

 Math in the Real World

Cheetahs are the fastest land animals. They can chase prey by running at speeds of 60 miles per hour.

A cheetah can only maintain top speeds for a short time. If a cheetah runs 1 mile in 60 seconds, fill in the diagram to show how far the cheetah will run in 210 seconds.

210 s

60 s

1 mile

FOLDABLES
Study Organizer

 1 Cut out the Foldable on page FL3 of this book.

2 Place your Foldable on page 82.

 3 Use the Foldable throughout this chapter to help you learn about ratios and rates.

Vocabulary

coordinate plane	origin	unit price
equivalent ratio	prime factorization	unit rate
graph	rate	*x*-axis
greatest common factor	ratio	*x*-coordinate
least common multiple	ratio table	*y*-axis
ordered pair	scaling	*y*-coordinate

Study Skill: Studying Math

New Vocabulary New vocabulary terms are clues about important concepts. Learning new vocabulary words is more than just memorizing the definition. Whenever you see a new vocabulary word, ask yourself:

• How does this fit with what I already know?

• How is this alike or different from something I learned earlier?

Organize your answers in a word map like the one shown.

Make a word map for *proper fraction*.

Definition from Text	In Your Own Words
A rate is a comparison by division of two quantities with different kinds of units.	A rate compares two amounts with different units using division.

rate

Examples	Nonexamples
• 45 miles per hour	• 5 black cats out of 15 cats
• 16 books for 8 students	• 3 sugar cookies to 9 cookies

Definition	In Your Own Words

proper fraction

Examples	Nonexamples
•	•
	•

What Do You Already Know?

Place a checkmark below the face that expresses how much you know about each concept. Then scan the chapter to find a definition or example of it.

😞 I have no clue. 😐 I've heard of it. 😊 I know it!

Concept	😞	😐	😊	Definition or Example
greatest common factor				
ordered pairs				
prime factorization				
ratios				
the coordinate plane				
unit rates				

(Table title: Ratios and Rates)

When Will You Use This?

Here are a few examples of how unit rates are used in the real world.

Activity 1 Use a newspaper to find an ad for two different grocery stores. See if you can find an offer for the same item. How do the prices compare? Which one do you think is offering a better deal? Explain your reasoning.

_____

_____

_____

Activity 2 Go online at **connectED.mcgraw-hill.com** to read the graphic novel **Squeezing Lemons**. According to the advertisements, what is the advertised price of lemonade at each store?

_____

Mei, Pilar, and David in
Squeezing Lemons

Next item on the agenda is buying lemonade for the school dance.

Are You Ready?

Try the Quick Check below.
Or, take the Online Readiness Quiz.

Check ✓

Quick Review

5.NBT.6, 5.NF.5b

Example 1

Find $6\overline{)348}$.

$$
\begin{array}{r}
58 \\
6\overline{)348} \\
-30 \\
\hline
48 \\
-48 \\
\hline
0
\end{array}
$$

Divide each place-value position from left to right.

Since $48 - 48 = 0$, there is no remainder.

Example 2

Write $\frac{40}{64}$ in simplest form.

$$\frac{40}{64} = \frac{5}{8}$$

÷8 ÷8

Divide the numerator and denominator by the greatest common factor (GCF), 8.

Since the GCF of 5 and 8 is 1, the fraction $\frac{5}{8}$ is in simplest form.

Quick Check

Divide Whole Numbers Find each quotient.

1. $3\overline{)87}$

2. $8\overline{)584}$

3. $52\overline{)312}$

Show your work.

Simplify Fractions Write each fraction in simplest form.

4. $\frac{32}{48} =$ _____

5. $\frac{7}{28} =$ _____

6. $\frac{15}{25} =$ _____

7. An airplane has flown 260 miles out of a total trip of 500 miles. What fraction, in simplest form, of the trip has been completed?

_____

How Did You Do?

Which problems did you answer correctly in the Quick Check?
Shade those exercise numbers below.

① ② ③ ④ ⑤ ⑥ ⑦

Factors and Multiples

Vocabulary Start-Up

A *common factor* is a number that is a factor of two or more numbers. The greatest of the common factors of two or more numbers is called the **greatest common factor** (GCF).

The least nonzero number that is a multiple of two or more whole numbers is the **least common multiple** (LCM) of the numbers.

Fill in the charts below.

GCF
• stands for: _____
Define: • Greatest   _____
• Common   _____
• Factor   _____

LCM
• stands for: _____
Define: • Least   _____
• Common   _____
• Multiple   _____

 ## Real-World Link

Bryan is making balloon arrangements. He has 8 blue and 12 green balloons. What is the greatest amount of arrangements he

can make if he wants them to be identical? _____

Which **Mathematical Practices** did you use?
Shade the circle(s) that applies.

① Persevere with Problems

② Reason Abstractly

③ Construct an Argument

④ Model with Mathematics

⑤ Use Math Tools

⑥ Attend to Precision

⑦ Make Use of Structure

⑧ Use Repeated Reasoning

 Essential Question

HOW do you use equivalent rates in the real world?

Vocabulary

greatest common factor
least common multiple

Content Standards
6.NS.4

MP Mathematical Practices
1, 3, 4, 8

Find the Greatest Common Factor

You can use common factors or prime factors to find the GCF.

 Example

1. There are one-slice servings of three types of cake on a table. Each row has an equal number of servings and only one type of cake. What is the greatest number of servings in each row?

Cakes	
Type	**Number of Servings**
marble	10
red velvet	15
chocolate	20

To solve this problem, use common factors.

factors of 10: **1**, 2, **5**, 10

factors of 15: **1**, 3, **5**, 15

factors of 20: **1**, 2, 4, **5**, 10, 20   The common factors are 1 and 5.

The GCF of 10, 15, and 20 is 5. So, the greatest number of pieces of cake that can be placed in each row is 5.

Got it? Do this problem to find out.

a. Lana earned $49 on Friday, $42 on Saturday, and $21 on Sunday selling bracelets. She sold each bracelet for the same amount. What is the most she could have charged for each bracelet?

a. _____

Example

2. Find the GCF of 12 and 18.

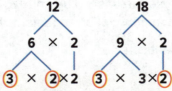

2 and 3 are the common prime factors.

So, the GCF of 12 and 18 is 2×3, or 6.

Got it? Do these problems to find out.

b. _____

c. _____

Find the greatest common factor of each set of numbers.

b. 12, 66   **c.** 18, 30   **d.** 32, 48

d. _____

Find the Least Common Multiple

You can find the least common multiple (LCM) by using a number line, making a list, or by using prime factors.

Examples

 Tutor

3. **Find the LCM of 2 and 3.**

Method 1 **Use a number line.**

Put a red **X** above each nonzero multiple of 2 and a blue **X** above each nonzero multiple of 3.

The least number with both a red and a blue X is 6.

So, 6 is the least common multiple of 2 and 3.

Method 2 **Use an organized list.**

List the nonzero multiples of 2 and 3.

multiples of 2: 2, 4, **6**, 8, 10, **12**,... $1 \times 2, 2 \times 2, 3 \times 2,...$

multiples of 3: 3, **6**, 9, **12**, 15,... $1 \times 3, 2 \times 3, 3 \times 3,...$

Notice that 6 and 12 are common multiples.

So, the least common multiple of 2 and 3 is 6.

4. **Find the LCM of 14 and 21 using prime factorization.**

Write the prime factorization of each number.

14   **21**

⑦ × 2   ⑦ × 3   7 is the only common prime factor.

Multiply using each common prime factor only once.

So, the LCM is $7 \times 2 \times 3$ or 42.

Got it? **Do these problems to find out.**

Find the least common multiple of each set of numbers.

   e. 2, 6   **f.** 4, 5, 10   **g.** 3, 5, 7

 Show your work.

e. _____

f. _____

g. _____

Example

5. Ernesto has painting class every 2 weeks. Kamala has a pottery class every 5 weeks. Ernesto and Kamala met at the art building for class this week. How many weeks will it be until they see each other again?

multiples of 2: 2, 4, 6, 8, **10**, 12, 14,...

multiples of 5: 5, **10**, 15, 20, 25, 30,...

The least common multiple of 2 and 5 is 10. So, Ernesto and Kamala will see each other again in 10 weeks.

Guided Practice

 Check ✓

Find the greatest common factor of each set of numbers. (Example 1 and 2)

1. 8, 32 _____

Show your work.

2. 24, 60 _____

3. 3, 12, 18 _____

Find the least common multiple of each set of numbers. (Examples 3 and 4)

4. 7, 9 _____

5. 6, 15 _____

6. 9, 12, 15 _____

7. The Movie House gives away a $5 coupon for every 4 movies purchased. They give away a bag of popcorn for every 3 movies purchased. How many movies would you have to purchase in all before receiving both a $5 coupon and a bag of popcorn at the same purchase? (Example 5)

_____

8. @ **Building on the Essential Question** How does finding the greatest common factor help you to solve real-world problems? _____

_____

_____

Rate Yourself!

Are you ready to move on? Shade the section that applies.

YES ? NO

For more help, go online to access a Personal Tutor.

Tutor

Independent Practice

Go online for Step-by-Step Solutions
eHelp

Find the greatest common factor of each set of numbers. (Example 2)

1. 8, 14 _____

Show your work.

2. 21, 24, 27 _____

3. 21, 35, 49 _____

4. 12, 18, 26 _____

Find the least common multiple of each set of numbers. (Examples 3 and 4)

5. 5 and 6 _____

6. 6 and 9 _____

7 6, 12, and 15 _____

8. 3, 9, and 15 _____

9 A gardener has 27 pansies and 36 daisies. He plants an equal number of each type of flower in each row. What is the greatest possible number of pansies in each row? (Example 1)

_____

_____

10. Fourteen boys and 21 girls will be equally divided into groups. Find the greatest number of groups that can be created if no one is left out. (Example 1)

_____

_____

11. Inez waters her plants every two days. She trims them every 15 days. She did both today. When will she do both again? (Example 5) _____

12. **Identify Repeated Reasoning** An airport offers two shuttles that run on different schedules. If both shuttles leave the airport at 4:00 P.M., at what time will they next leave the airport together?

Shuttle Schedule	
Shuttle	**Departs**
A	every 6 minutes
B	every 9 minutes

_____

_____

🔥 H.O.T. Problems Higher Order Thinking

13. MP **Model with Mathematics** Write and solve a real-world problem that can be solved using the greatest common factor of two numbers.

_____

_____

_____

14. MP **Identify Repeated Reasoning** How can you use number patterns to find the least common multiple of 120 and 360?

_____

_____

_____

15. MP **Persevere with Problems** If the GCF of two numbers is 1, they are called *relatively prime*. Find three sets of relatively prime numbers.

_____

16. MP **Use a Counterexample** Determine whether each statement is *true* or *false*. If *true*, explain why. If *false*, give a counterexample.

a. The GCF of any two even numbers is always even.

_____

_____

b. The GCF of any two odd numbers is always odd.

_____

c. The GCF of an odd number and an even number is always even.

_____

_____

Extra Practice

Find the greatest common factor of each set of numbers.

17. 15, 20 _5_____

Homework Help

factors of 15: ①3⑤15

factors of 20: ①2, 4⑤10, 20

The common factors are 1 and 5.

The GCF is 5.

18. 30, 48, 60 _____

19. 24, 30, 42 _____

20. 24, 40, 56 _____

Find the least common multiple of each set of numbers.

21. 3 and 5 _15_____

multiples of 3: 3, 6, 9, 12,⑮18, 21, 24, 27,㉚

multiples of 5: 5, 10,⑮20, 25,㉚

The common multiples are 15 and 30.

The LCM is 15.

22. 12 and 18 _____

23. 5, 10, and 15 _____

24. 9, 12, and 18 _____

25. A grocery store clerk has 16 oranges, 20 apples, and 24 pears. The clerk needs to put an equal number of apples, oranges, and pears into each basket. What is the greatest number of baskets that can be made so that no fruit is left?

_____

_____

26. 🔴**MP** **Identify Repeated Reasoning** The science department buys the equipment shown in the table. They bought all three items this year. In how many years will they have to buy all three items again?

_____

Item	Time Bought
Microscopes	every 5 years
Safety goggles	every 4 years
Test tubes	every 2 years

27. Drusilla replaces the light bulb in the hall closet every 9 months and replaces the air filter every 6 months. She just replaced both items this month. After how many months will she replace both the light bulb and the air filter? Select all that apply.

☐ 12 months   ☑ 18 months   ☐ 36 months   ☐ 48 months

28. Macy is painting a design that contains two repeating patterns. One pattern repeats every 8 inches. The other repeats every 12 inches. The design is 19 feet long. Both patterns begin at the same place. Use the pattern pieces to create a sample of the design. Use the sample to determine the number of times the patterns begin in the same place. ⬚

⬚◁▷⬚ 8 in.   ⬚◁•▷⬚ 12 in.

8-inch pattern: ⬚

12-inch pattern: ⬚

Spiral Review

Write each fraction in simplest form. 5.NF.5b

29. $\dfrac{9}{18} =$ _____

30. $\dfrac{21}{35} =$ _____

31. $\dfrac{36}{48} =$ _____

32. Josiah ran $\dfrac{4}{5}$ mile. How many tenths are equal to $\dfrac{4}{5}$ mile? Use bar diagrams to find the answer. 5.NF.5b

$\frac{1}{5}$	$\frac{1}{5}$	$\frac{1}{5}$	$\frac{1}{5}$	$\frac{1}{5}$

_____

_____

33. Pizza Palace cuts a medium pizza into 8 slices. The same size pizza at Pizza Pioneers is cut into 16 slices. Jasmine ate 4 slices of a medium pizza from Pizza Pioneers. What fraction of the pizza from Pizza Palace is equal to $\dfrac{4}{16}$? Explain. 5.NF.5b

_____

Inquiry Lab
Ratios

 HOW can you use tables to relate quantities?

Content Standards
6.RP.1, 6.RP.3, 6.RP.3a

MP **Mathematical Practices**
1, 3, 4

Max has 3 fiction books and 6 nonfiction books to donate to the community center. He wants to package them so that there are an equal number of fiction and nonfiction books in each group. He also wants to have as many packages as possible. How many books are in each group?

What do you know? _____

_____

_____

What do you need to find? _____

_____

Hands-On Activity 1

Step 1 Use 3 red counters to represent the fiction books. Use 6 yellow counters to represent the nonfiction books.

Step 2 Determine the smallest possible equal-size groups. Use mats to divide the counters into the groups.

Each group has an equal number of fiction books and an equal number of nonfiction books.

Each group has [ ] fiction book and [ ] nonfiction books.

Hands-On Activity 2

Maria is also collecting books. She wants to make packages that have 3 fiction books and 4 nonfiction books. She already has 9 fiction books. How many nonfiction books will she need?

Use a multiplication table to compare the numbers.

Step 1 Complete the rows for 3 and 4 on a multiplication table.

fiction →	3	6								
nonfiction →	4	8								

Step 2 Read across the top until you reach 9. Find the corresponding number in the bottom row and circle the 2 numbers.

Maria needs ⬚ nonfiction books.

Hands-On Activity 3

Sanjay has 27 jerseys. Divide them into two groups so that for every 4 red jerseys, there are 5 blue jerseys.

Step 1 Complete the rows for 4 and 5 on a multiplication table.

red →	4	8								
blue →	5	10								

Step 2 Read across both rows until you find two numbers with a sum of 27.

There are ⬚ red jerseys and ⬚ blue jerseys.

Check Draw a picture to check your answer.

Investigate

Work with a partner. Determine the number of pieces of fruit that should be put in each group. Make as many equal-size groups as possible using all the fruit. Use and draw counters to represent the fruit.

1. 3 apples and 9 pears

2. 4 peaches and 6 oranges

Show your work.

3. 4 plums and 7 bananas

4. 6 apricots and 9 mangos

Work with a partner. Use a multiplication table to solve the following problems.

5. Evie wants groups of 3 notebooks and 5 pens. She already has 12 notebooks. How many pens will she need?

6. Louis wants groups of 6 daisies and 8 tulips for flower arrangements. He already has 24 daisies. How many tulips will he need?

7. Selma has 77 strawberries. Divide them into two groups so that for every 4 strawberries in Group 1 there are 7 strawberries in Group 2.

Analyze and Reflect

8. **MP Identify Repeated Reasoning** Describe the patterns used in the tables in Activities 2 and 3.

_____

_____

_____

9. **MP Reason Inductively** How would finding the least common multiple help you when dividing items into equal groups?

_____

_____

Create

10. **MP Model with Mathematics** Write and solve a word problem in which there are 3 yellow beads for every 2 blue beads.

_____

_____

_____

11. **MP Model with Mathematics** Write and solve a real-world word problem in which there are 3 tables for every 8 chairs.

_____

_____

_____

12. **MP Model with Mathematics** Write and solve a real-world word problem in which there are 3 pancakes for every person.

_____

_____

_____

13. **Inquiry** HOW can you use tables to relate quantities?

_____

_____

Ratios

Dogs In her dog walking business, Mrs. DeCarbo walks 2 large dogs and 8 small dogs.

Compare the number of small dogs to large dogs. Use yellow counters to represent the large dogs. Use red counters to represent the small dogs. Draw the counters in the box.

Essential Question

HOW do you use equivalent rates in the real world?

Vocabulary

ratio

Content Standards
6.RP.1, 6.RP.3

MP Mathematical Practices
1, 3, 4, 5

1. $2 + \boxed{} = 8$ There are $\boxed{}$ *more* small dogs than large dogs.

2. $2 \times \boxed{} = 8$ There are $\boxed{}$ *times* as many small dogs as large dogs.

3. $8 - \boxed{} = 2$ There are $\boxed{}$ *fewer* large dogs than small dogs.

4. $8 \div \boxed{} = 2$ The number of large dogs is $\dfrac{\boxed{}}{\boxed{}}$ the number of small dogs.

Which MP Mathematical Practices did you use?
Shade the circle(s) that applies.

① Persevere with Problems
② Reason Abstractly
③ Construct an Argument
④ Model with Mathematics
⑤ Use Math Tools
⑥ Attend to Precision
⑦ Make Use of Structure
⑧ Use Repeated Reasoning

Write a Ratio in Simplest Form

There are many different ways to compare amounts or *quantities*. A **ratio** is a comparison of two quantities by division. A ratio of 2 red paper clips to 6 blue paper clips can be written in three ways.

2 **to** **6**   **2**:**6** $\dfrac{2}{6}$

As with fractions, ratios are often expressed in simplest form.

Example

 Tutor

1. Write the ratio in simplest form that compares the number of red paper clips to the number of blue paper clips. Then explain its meaning.

Write the ratio as a fraction. Then simplify.

red paper clips ····▸ $\dfrac{2}{6}$ = $\dfrac{1}{3}$ ◂···· The GCF of 2 and 6 is 2.
blue paper clips ····▸

The ratio of red to blue paper clips is $\dfrac{1}{3}$, 1 to 3, or 1:3. This means that for every 1 red paper clip there are 3 blue paper clips.

 Show your work.

Got it? Do this problem to find out.

a. Write the ratio in simplest form that compares the number of suns to the number of moons. Then explain its meaning.

a. _____

Use Ratios to Compare Categorical Data

Each piece of categorical data can only be assigned to one group. Bar diagrams (or tape diagrams) and frequency tables can be used to represent categorical data. Ratios can be used to compare the data.

Examples

2. **Several students named their favorite flavor of gum. Write the ratio that compares the number who chose fruit to the total number of students.**

Favorite Flavor of Gum	
Flavor	**Number of Responses**
Peppermint	9
Cinnamon	8
Fruit	3
Spearmint	1

Fruit: 3

Total: $9 + 8 + 3 + 1$, or 21

fruit flavor responses $\cdots\blacktriangleright$ $\dfrac{3}{21} = \dfrac{1}{7}$ $\cdots$ The GCF of 3 and 21 is 3.
total responses $\cdots\blacktriangleright$

The ratio is $\dfrac{1}{7}$, 1 to 7, or 1:7.

So, 1 out of every 7 students preferred fruit-flavored gum.

3. **Monday's yogurt sales are recorded in the table. Write the ratio that compares the sales of strawberry yogurt to the total sales. Then explain its meaning.**

Flavor	Number Sold
Peach	3
Blueberry	6
Vanilla	7
Strawberry	8

Strawberry: ☐

Total: ☐ + ☐ + ☐ + ☐, or ☐

strawberry yogurt sold $\cdots\blacktriangleright$ $\dfrac{☐}{☐} \div \dfrac{☐}{☐} = \dfrac{☐}{☐}$ or ☐ to ☐
total sold $\cdots\blacktriangleright$

So, ☐ out of every ☐ yogurt cups sold were strawberry.

Got it? Do this problem to find out.

b. A pet store sold the animals listed in the table in one week. Write the ratio of cats to pets sold that week. Then explain its meaning.

Pet	Number Sold
Birds	10
Dogs	14
Cats	8

b. _____

Accuracy

It is important to read the entire problem so that an accurate answer can be determined.

Example

4. Katy wants to divide her 30 flowers into two groups, so that the ratio is **2 to 3**.

Step 1 Use a bar diagram to show a ratio of 2 to 3.

6	6

6	6	6

⎫ 30 flowers

Step 2 There are 5 equal sections. So, each section represents 30 ÷ 5 or 6 flowers.

There are 12 flowers in one group and 18 in the other.

 Check ✓

Guided Practice

Write each ratio as a fraction in simplest form. Then explain its meaning. (Example 1)

1. Show your work.

pens to pencils

2.

pennies:dimes

3. Last month, Ed ate 9 apples, 5 bananas, 4 peaches, and 7 oranges. Find the ratio of bananas to the total number of fruit. Then explain its meaning. (Examples 2 and 3)

_____

_____

4. Divide 28 cans of soda into two groups so the ratio is 3 to 4. (Example 4) _____

_____

5. ⓔ **Building on the Essential Question** How can you use mental math to determine if a ratio is simplified?

_____

_____

Rate Yourself!

How confident are you about ratios? Shade the ring on the target.

I'm on target.

I need help.

For more help, go online to access a Personal Tutor.

FOLDABLES Time to update your Foldable!

Independent Practice

Go online for Step-by-Step Solutions

Write each ratio as a fraction in simplest form. Then explain its meaning. (Example 1)

1. _____   2. _____

_____   _____

flutes:drums

sandwiches to milk cartons

3 A class has 6 boys and 15 girls. What is the ratio of boys to girls?

(Example 2) _____

4. The table shows the number of books Salvador has read. Find the ratio of mystery books to the total. Explain its meaning. (Example 3)

_____

Type	Number of Books
Mystery	10
Nonfiction	7
Science Fiction	5
Western	2

5. Divide 33 photos into two groups so the ratio is 4 to 7. (Example 4)

_____

6. **MP** **Model with Mathematics** Refer to the graphic novel frame below for Exercises a–b.

a. For each store, what is the ratio of the number of cans to the price?

_____

b. What would be the ratio of the number of cans to the price at Super Saver and Price Busters if a coupon for $1 off the total purchase is

used? _____

7 (MP) **Use Math Tools** The graph shows the number of Wimbledon championships of several countries.

a. Write the ratio that compares the championships won by Australia to the total number won by the United States in simplest form. Then explain its meaning.

_____

_____

_____

b. Write the ratio that compares the championships won by Australia to the total number of championships. Then explain its meaning.

_____

_____

_____

H.O.T. Problems Higher Order Thinking

8. (MP) **Model with Mathematics** Create three different drawings showing a number of rectangles and circles in which the ratio of rectangles to circles is 3:1.

Show your work.

9. (MP) **Persevere with Problems** Find the missing number in the following pattern. Explain your reasoning.

12, 24, 72, 288, [ ]

_____

10. (MP) **Persevere with Problems** The table shows how Levon spends his time at the gym. Over the course of a week, he wants to spend 600 minutes at the gym. How much more time will he spend lifting weights than on the treadmill? Explain your reasoning.

Activity	Time (min)
Treadmill	25
Lifting weights	35

_____

_____

_____

Extra Practice

Write each ratio as a fraction in simplest form. Then explain its meaning.

11. $\frac{1}{4}$; *for every 1 triangle there are 4 rectangles.*

12. _____

triangles to rectangles

There are 2 triangles and 8 rectangles.

The ratio is $\frac{2}{8}$. $\frac{2}{8} \div \frac{2}{2} = \frac{1}{4}$

soccer balls:footballs

13. An animal shelter has 36 kittens and 12 puppies available for adoption. What is the ratio of puppies to kittens?

   _____

14. Find the ratio of black cell phone covers sold to the total number of cell phone covers sold last week. Then explain its meaning.

   _____

Color	Number of Cell Phone Covers Sold
Green	5
Silver	6
Red	3
Black	4

15. On the first day of the food drive, Mrs. Teasley's classes brought in 6 cans of fruit, 4 cans of beans, 7 boxes of noodles, and 4 cans of soup. Find the ratio of cans of fruit to the total number of food items collected. Then explain its meaning.

   _____

   _____

16. The rise and span for a roof are shown. The pitch of a roof is the ratio of the rise to the half-span. If the rise is 8 feet and the span is 30 feet, what is the pitch in simplest form?

   _____

   _____

rise

span

17. **MP** **Justify Conclusions** Debra found that 6 of the 24 students in her class own a cell phone. What is the ratio of students that own a cell phone to students that do not? Explain your reasoning to a classmate.

   _____

   _____

18. At a putt-putt course there are 50 yellow golf balls, 45 red golf balls, 65 blue golf balls, 40 orange golf balls, and 60 green golf balls. Select the correct ratio to complete the table.

8:13	4:5
5:6	3:4
10:9	12:13

Comparison	Ratio
yellow to red	
yellow to green	
green to blue	

Comparison	Ratio
red to green	
orange to yellow	
orange to blue	

19. The table shows the number of each type of sports card that Jaclyn has collected.

baseball	basketball	football	soccer
45	14	20	21

Write a ratio in simplest form that compares the number of basketball cards to the total number of cards. ☐

Spiral Review

Find the equivalent fraction. 5.NF.5b

20. $\dfrac{3}{7} = \dfrac{\boxed{}}{21}$

21. $\dfrac{1}{6} = \dfrac{\boxed{}}{24}$

22. $\dfrac{4}{5} = \dfrac{28}{\boxed{}}$

23. The Sanchez family is going on vacation. If they drive for 3 hours at the posted speed, how many miles will they travel? 5.NBT.5

_____

SPEED LIMIT **65**

24. Everett made $\dfrac{3}{5}$ of the baskets he shot. Suppose he shot 60 baskets. How many did he make? 5.NF.4 _____

25. There are 36 students in Mrs. Keaton's sixth grade class. If $\dfrac{5}{12}$ of her students are girls, how many girls are in the class? 5.NF.4 _____

Inquiry Lab
Unit Rates

 Inquiry **HOW can you use bar diagrams to compare quantities in real-world situations?**

Content Standards
6.RP.2, 6.RP.3, 6.RP.3b
MP **Mathematical Practices**
1, 3, 8

Jamila and Anica were rollerblading. They skated 14 miles in 2 hours. If they skated at a constant rate, how many miles did they skate in 1 hour?

What do you know? _____

_____

What do you need to find? _____

_____

Hands-On Activity 1

Step 1 Use a bar diagram to represent 14 miles. The box is separated into two equal sections to represent 2 hours.

```
|------------------- 14 miles -------------------|
|      1 hour      |      1 hour      |
```

Step 2 Each section represents one hour. Determine the number of miles skated in one hour.

```
|------------------- 14 miles -------------------|
|      1 hour      |      1 hour      |
|---- 7 miles -----|
```

So, they skated ☐ miles in one hour.

Hands-On Activity 2

A package of 5 crackers contains 205 Calories. How many Calories are in one cracker?

Step 1 Draw a bar diagram to represent 205 Calories. Divide the bar diagram into 5 equal sections to represent 5 crackers.

|-------------- 205 Calories --------------|

|-------|

☐ Calories

Step 2 Label the first section "1 cracker." Determine the number of Calories in 1 cracker.

So, one cracker contains ☐ Calories.

Hands-On Activity 3

A bottle of body wash costs $2.88 and contains 12 ounces. How much does it cost per ounce?

Step 1 Draw a bar diagram to represent _____. Divide the bar diagram into ☐ equal sections to represent ☐ ounces.

|-------------------------------- ☐ --------------------------------|

|-----|

☐

Step 2 Label the first section "_____." Determine the cost for 1 ounce of body wash.

So, one ounce of body wash costs $ ☐ .

Investigate

Work with a partner to solve. Use a bar diagram.

1. Travis drove 129 miles in 3 hours. He drove at a constant speed. How many miles did he drive in 1 hour? _____

2. Six oranges cost $5.34. How much does 1 orange cost? _____

3. Doug read 231 pages in 7 hours. He read the same number of pages each hour. How many pages did he read in 1 hour? _____

pages

4. Mariah has 72 flowers in 4 vases. She put the same number of flowers in each vase. How many flowers are in 1 vase? _____

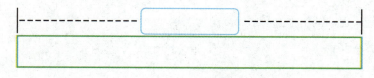

flowers

Work with a partner to complete the problem.

5. In the bakery, a container of cookies is $4.55 and contains 13 servings. The coins below equal $4.55. Divide the coins into 13 equal groups

   to determine the cost per serving. Circle each group. _____

6. **MP Reason Inductively** How does dividing the coins into equal groups help solve the problem?

   _____

   _____

7. **MP Justify Conclusions** The comparison of miles to hours in Activity 1 is 14:2, which can be reduced to 7:1. How is simplifying similar to division?

   _____

   _____

Create

On Your Own

8. **MP Identify Repeated Reasoning** Write a rule for how to compare two quantities so that the second quantity has a value of 1 without using a diagram.

   _____

   _____

9. **MP Model with Mathematics** Write a real-world word problem in which the unit rate is 6 miles per hour.

   _____

   _____

10. **Inquiry** HOW can you use bar diagrams to compare quantities in real-world

    situations? _____

    _____

Rates

Vocabulary Start-Up

Use your glossary, which starts on page GL1, to complete the definitions of the vocabulary words in the table.

Definition	Examples
fraction: A number that represents part of a _____ or part of a _____ .	$\frac{1}{2}, \frac{3}{4}, \frac{9}{12}, \frac{45}{3}$
ratio: A comparison of two _____ by _____ .	2 out of 3, 2 to 3, 2:3, $\frac{2}{3}$
rate: A _____ comparing two _____ with different kinds of _____ .	$\frac{36\ \text{miles}}{3\ \text{hours}}$ 36 miles for every 3 hours $26 for 5 bags 19 songs in 5 minutes
unit rate: A _____ that is _____ so that it has a denominator of _____ .	$\frac{12\ \text{miles}}{1\ \text{hour}}$, 12 miles per hour $5.20 for 1 bag 3.8 songs in 1 minute

 ## Real-World Link

Desiree typed a 15-character text message in 5 seconds.

1. Write the rate Desiree typed as a fraction. $\dfrac{\boxed{}\ \text{characters}}{\boxed{}\ \text{seconds}}$

2. What operation would you use to write the fraction in simplest form? _____

Which **Mathematical Practices** did you use?
Shade the circle(s) that applies.

① Persevere with Problems   ⑤ Use Math Tools

② Reason Abstractly   ⑥ Attend to Precision

③ Construct an Argument   ⑦ Make Use of Structure

④ Model with Mathematics   ⑧ Use Repeated Reasoning

 Essential Question

HOW do you use equivalent rates in the real world?

Vocabulary

rate
unit rate
unit price

Content Standards
6.RP.2, 6.RP.3, 6.RP.3b

MP Mathematical Practices
1, 3, 4

Find a Unit Rate

A **rate** is a ratio comparing two quantities of different kinds of units. A **unit rate** has a denominator of 1 unit when the rate is written as a fraction. To write a rate as a unit rate, divide the numerator and the denominator of the rate by the denominator.

Ratio	Rate	Unit Rate
$15:5 =$	$\dfrac{15 \text{ characters}}{5 \text{ seconds}} =$	$\dfrac{3 \text{ characters}}{1 \text{ second}}$

Examples

1. **Samantha picked 45 oranges in 5 minutes. Write this rate as a unit rate.**

Write the rate as a fraction. Compare the number of oranges to the number of minutes. Then divide.

$$\frac{45 \text{ oranges}}{5 \text{ minutes}} = \frac{9 \text{ oranges}}{1 \text{ minute}}$$

So, the unit rate is $\dfrac{9 \text{ oranges}}{1 \text{ minute}}$, or 9 oranges per minute.

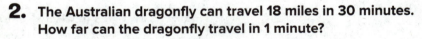

2. **The Australian dragonfly can travel 18 miles in 30 minutes. How far can the dragonfly travel in 1 minute?**

Write the rate as a fraction. Compare the distance to the number of minutes. Then divide.

$$\frac{18 \text{ miles}}{30 \text{ minutes}} = \frac{3 \text{ miles}}{5 \text{ minutes}}$$

The ratio 3 to 5 cannot be simplified to a whole number rate. It can be written as $\dfrac{3 \text{ miles}}{5 \text{ minutes}}$ or as a unit rate of $\frac{3}{5}$ mile to 1 minute.

The dragonfly can travel $\frac{3}{5}$ mile every minute.

> Show your work.

Got it? **Do these problems to find out.**

a. Ama downloaded 35 songs in 5 minutes. How many songs did she download per minute?

b. Jonathan is baking several loaves of bread to sell in his bakery. He used 9 cups of water and 12 cups of whole wheat flour. How much water was used per cup of flour?

Simplifying Ratios

The least common factor of 3 and 5 is 1. To find the unit rate of the ratio $\frac{3 \text{ miles}}{5 \text{ minutes}}$, divide both the numerator and denominator by 5. So, the unit rate in fration form is $\frac{3}{5}$ mile per minute.

a. _____

b. _____

Example

3. An adult's heart beats about 2,100 times every 30 minutes. A baby's heart beats about 2,600 times every 20 minutes. How many more beats does a baby's heart beat in 60 minutes than an adult's heart?

Step 1 Find the unit rates.

Adult: $\dfrac{2,100 \text{ beats}}{30 \text{ minutes}}$ or $\dfrac{70 \text{ beats}}{1 \text{ minute}}$

Baby: $\dfrac{2,600 \text{ beats}}{20 \text{ minutes}}$ or $\dfrac{130 \text{ beats}}{1 \text{ minute}}$

Step 2 Using the unit rate for each, determine the number of beats in 60 minutes.

Adult: $70 \times 60 = 4,200$ beats
Baby: $130 \times 60 = 7,800$ beats

Step 3 Find the difference.

$7,800 - 4,200 = 3,600$

So, a baby's heart beats 3,600 more times in 60 minutes than an adult's heart.

Got it? Do this problem to find out.

c. A hummingbird's heart rate while resting is about 7,500 beats every 30 minutes. How many more beats does a hummingbird's heart beat in 60 minutes than a human baby's heart?

Show your work.

c. _____

Key Phrases

Key phrases such as *per, in,* and *for every* are often used to describe unit rates.

Find a Unit Price

You can use what you know about unit rates to find a unit price. The **unit price** is the cost per unit. To write a price as a unit price, divide the numerator and the denominator of the rate by the denominator.

$$\dfrac{\$36}{4 \text{ tickets}} = \dfrac{\$9}{1 \text{ ticket}}$$

For example, it costs $36 for 4 movie tickets. So, the cost per unit, or per ticket, is $9.

Tutor

4. **Financial Literacy** Four potted plants cost **$88. What is the price per plant?**

Write the rate as a fraction. Compare the total cost to the number of plants. Then divide.

$$\overset{\div 4}{\frac{\$88}{4 \text{ plants}}} = \frac{\$22}{1 \text{ plant}}$$
$$\div 4$$

So, the price per potted plant is $22.00.

Guided Practice

Check

Write each rate as a unit rate. (Examples 1 and 2)

1. 44 points in 4 quarters = _____

Show your work.

2. 125 feet in 5 seconds = _____

3. 360 miles traveled on 12 gallons of gasoline = _____

4. 12 meters in 28 seconds = _____

5. Molly shot 20 baskets in 4 minutes. Nico shot 42 baskets in 6 minutes. How many more baskets did Nico shoot per minute? (Example 3) _____

6. For Carolina's birthday, her mom took her and 4 friends to a water park. Carolina's mom paid $40 for 5 student tickets. What was the price for one student ticket? (Example 4)

_____

7. **ℚ** **Building on the Essential Question** How are rates and ratios related? _____

_____

_____

_____

Independent Practice

Go online for Step-by-Step Solutions eHelp

Write each rate as a unit rate. (Examples 1 and 2)

1. 72 ounces in 6 steaks = _____

2. 162 water bottles in 9 cases = _____

Show your work.

3 Marcella divided 40.8 gallons of paint among 8 containers. How much paint is in each container? (Example 1) _____

4. Central Subs made 27 sandwiches using 12 pounds of turkey. How much turkey was used per sandwich? (Example 2)

_____

5. The results of a car race are shown. Determine who drove the fastest. Explain.

(Example 3) _____

_____

_____

Drivers' Times		
Driver	Laps	Time (min)
Cutwright	35	84
Evans	42	96.6
Loza	38	102.6

6. Theo's mom bought an eight-pack of juice boxes at the store for $4. Find the unit rate for the juice boxes. (Example 4) _____

7. Joshua's cousin pledged $12 for a charity walk. If Joshua walked 3 miles, how much did his cousin pay per mile? (Example 4)

_____

8. **MP Justify Conclusions** The Lovin' Lemon Company sells a 4-gallon jug of lemonade for $24. The Sweet and Sour Company sells an eight-pack of 1-quart bottles of lemonade for $16.00. Which company has a higher unit price? Explain your answer. _____

_____

9 The Shanghai Maglev Train is one of the fastest trains in the world, traveling about 2,144 miles in 8 hours.

a. How many miles does it travel in one hour? _____

b. The distance between Columbus, Ohio, and New York City is about 560 miles. How many hours would it take the train to travel between the cities? _____

10. **MP** **Multiple Representations** The table shows the approximate population and areas of five states. *Population density* is the number of people per square unit of an area.

State	Population Estimate (as of July 2007)	Area (square miles)
California	36,500,000	163,707
Florida	18,300,000	65,758
Iowa	2,990,000	56,276
New Jersey	8,690,000	8,722
Wyoming	522,000	97,818

a. **Numbers** Find the population density of each state. Round to the nearest tenth.

_____

_____

_____

b. **Graph** Make a bar graph of the five population densities.

c. **Words** Connecticut has about the same population as Iowa, but its area is 4,875 square miles. Without calculating, compare Connecticut's population density to Iowa's. Justify your answer.

_____

_____

_____

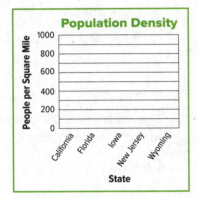

Population Density

People per Square Mile: 0, 200, 400, 600, 800, 1000

States: California, Florida, Iowa, New Jersey, Wyoming

State

🔥 H.O.T. Problems Higher Order Thinking

11. **MP** **Find the Error** Julie wrote the rate $108 in 6 weeks as a unit rate. Find her mistake and correct it.

_____

_____

_____

$$\frac{\$108}{6 \text{ weeks}} = \frac{\$54}{3 \text{ weeks}}$$

12. **MP** **Persevere with Problems** The ratio of red jelly beans to yellow jelly beans in a dish is 3:4. If Greg eats 3 red jelly beans and 6 yellow ones, the ratio is 4:5. How many yellow jelly beans were originally in the dish?

_____

13. **MP** **Justify Conclusions** If you travel at a rate of 45 miles per hour, how many minutes will it take you to travel 1 mile? Justify your response.

_____

_____

Extra Practice

Write each rate as a unit rate.

14. Davis printed 24 photos in 8 minutes. How many photos did he print per minute?

3 photos per minute _____

15. Carrie planted 48 tulips in 12 minutes. How many tulips did she plant per minute?

_____

16. Vinnie decorated 72 cookies in 36 minutes. How many cookies did he decorate per minute?

_____

17. Alana biked 45 miles in 3 hours. How many miles did she bike per hour?

_____

18. A Ruby Throated Hummingbird beats its wings 159 times in 3 seconds. How many times does the Ruby Throated Hummingbird beat its wings

per second? _____

19. The Reyes family bought four concert tickets for $252. What was the price

per ticket? _____

20. An adult blinks about 450 times in 30 minutes. A 12-year-old blinks about 150 times in 15 minutes. How many more times does an adult blink in

60 minutes than a 12-year-old? _____

21. Find the number of meters each record holder ran in one second of each event. Round to the nearest tenth.

a. 200 meters, 19.30 seconds, Usain Bolt, Jamaica _____

b. 400 meters, 43.18 seconds, Michael Johnson, USA _____

c. 100 meters, 9.69 seconds, Usain Bolt, Jamaica _____

22. **MP** **Justify Conclusions** The 24 students in Mr. Brown's homeroom sold 72 magazine subscriptions. The 28 students in Mrs. Garcia's homeroom sold 98 magazine subscriptions. Whose homeroom sold more magazine

subscriptions per student? Explain your reasoning. _____

_____

_____

23. Boxes of fruit snacks are on sale at the grocery. The boxes are the same size. A family needs to purchase 24 boxes.

3 boxes for $5   4 boxes for $6

Which snacks should they purchase if they want to spend a lesser amount?

How much money will they save?

24. A runner is training for a half marathon. Her training schedule is shown in the table.

Graph and label the unit rates for each day on the number line.

Day	Distance (mi)	Time (min)
Tuesday	4	38
Wednesday	6	72
Thursday	4	37
Saturday	3	31.5
Sunday	12	138

Which day did she run the fastest?

Spiral Review

Simplify each fraction. **5.NF.5b**

25. $\dfrac{16}{80} = \dfrac{\Box}{\Box}$

26. $\dfrac{4}{10} = \dfrac{\Box}{\Box}$

27. $\dfrac{48}{200} = \dfrac{\Box}{\Box}$

28. Josephine wants to put a wallpaper border around the ceiling of her room. The dimensions are shown at the right. How many feet of border does she need? **5.NF.2**

$10\frac{9}{12}$ ft

$8\frac{1}{2}$ ft

Ratio Tables

 Real-World Link Tools

Refreshments A punch recipe uses one container of soda and three containers of juice to make one batch of punch.

1. Draw red counters to show the number of containers of soda and draw yellow counters to show the number of containers of juice needed to make 2 batches of punch.

$$\text{soda} \rightarrow \boxed{}$$
$$\text{juice} \rightarrow \boxed{}$$

2. Draw red counters to show the number of containers of soda and draw yellow counters to show the number of containers of juice needed to make 3 batches of punch.

$$\text{soda} \rightarrow \boxed{}$$
$$\text{juice} \rightarrow \boxed{}$$

3. Find the ratio in simplest form of soda to juice needed for 1, 2, and 3 batches. What do you notice?

_____

Which **MP** **Mathematical Practices** did you use?
Shade the circle(s) that applies.

① Persevere with Problems
② Reason Abstractly
③ Construct an Argument
④ Model with Mathematics

⑤ Use Math Tools
⑥ Attend to Precision
⑦ Make Use of Structure
⑧ Use Repeated Reasoning

Equivalent Ratios

The quantities in the opening activity can be organized into a table. This table is called a **ratio table** because the columns are filled with pairs of numbers that have the same ratio.

Soda	1	2	3
Juice	3	6	9

The ratios $\frac{1}{3}$, $\frac{2}{6}$, and $\frac{3}{9}$ are equivalent, since each simplifies to a ratio of $\frac{1}{3}$.

Equivalent ratios express the same relationship between quantities.

Examples

1. To make yellow icing, you mix 6 drops of yellow food coloring with 1 cup of white icing. How much yellow food coloring should you mix with 5 cups of white icing to get the same shade?

Use a ratio table. Since $1 \times 5 = 5$, multiply each quantity by 5.

So, add 30 drops of yellow food coloring to 5 cups of icing.

$\times 5$

Drops of Yellow	6	30
Cups of Icing	1	5

$\times 5$

2. In a recent year, Joey Chestnut won a hot dog eating contest by eating nearly 66 hot dogs in 12 minutes. If he ate at a constant rate, determine about how many hot dogs he ate every 2 minutes.

Divide each quantity by one or more common factors until you reach a quantity of 2 minutes.

So, he ate about 11 hot dogs every 2 minutes.

$\div 2 \quad \div 3$

Hot Dogs	66	33	11
Time (min)	12	6	2

$\div 2 \quad \div 3$

Show your work.

Got it? Do these problems to find out.

a. A patient receives 1 liter of IV fluids every 8 hours. At that rate, find how many hours it will take to receive 4 liters of IV fluids.

IV Fluids (L)	1	4
Time (h)	8	

a. _____

b. To make cranberry jam, you need 12 cups of sugar for every 16 cups of cranberries. Find the amount of sugar needed for 4 cups of cranberries.

Sugar (c)	12	
Cranberries (c)	16	4

b. _____

Use Scaling

Multiplying or dividing two related quantities by the same number is called **scaling**. Sometimes you may need to *scale back* and then *scale forward* to find an equivalent ratio.

Examples

Tutor

3. **Cans of corn are on sale at 10 for $4. Find the cost of 15 cans.**

Cans of Corn	10		15
Cost in Dollars	4		■

There is no whole number by which you can multiply 10 to get 15. So, scale back to 5 and then scale forward to 15.

÷2 ×3

Cans of Corn	10	**5**	15
Cost in Dollars	4	**2**	**6**

÷2 ×3

Divide each quantity by a common factor, 2.

Then, since 5 × 3 = 15, multiply each quantity by 3.

So, 15 cans of corn would cost $6.

- -

4. **Joe mows lawns during his summer vacation to earn money. He took 14 hours last week to mow 8 lawns. At this rate, how many lawns could he mow in 49 hours?**

Is there a whole number by which you can multiply 14 to get 49? _____

Scale back to _____, and then scale forward to _____.

÷2 ×7

Number of Hours	14	**7**	49
Number of Lawns	8	**4**	**28**

÷2 ×7

So, Joe can mow _____ lawns in 49 hours.

Got it? Do this problem to find out.

c. A child's height measures 105 centimeters. Estimate her height in inches.

Height (cm)	25		105
Height (in.)	10		

Show your work.

c. _____

Example

5. On her vacation, Leya exchanged $50 American and received $60 Canadian. Use a ratio table to find how many Canadian dollars she would receive for $20 American.

Set up a ratio table. Use scaling to find the desired quantity.

÷10 ×4

Canadian Dollars	60	**6**	**24**
American Dollars	50	**5**	20

÷10 ×4

Divide each quantity by a common factor, 10.

Then, since 5 × 4 = 20, multiply each quantity by 4.

Leya would receive $24 Canadian for $20 American.

Guided Practice

Complete each ratio table to solve each problem.

1. Santiago receives an allowance of $7 every week. How much total does he receive after 4 weeks? (Example 1)

Allowance ($)	7			
Number of Weeks	1			4

_____

2. Tonya runs 8 kilometers in 60 minutes. At this rate, how long would it take her to run 2 kilometers? (Example 2)

Distance Run (km)	8		2
Time (min)	60		

_____

3. Lamika buys 12 packs of juice boxes that are on sale and pays a total of $48. Use a ratio table to determine how much Lamika will pay to buy 8 more packs of juice boxes at the same store. (Example 5)

Number of Juice Boxes			
Price ($)			

_____

4. **Building on the Essential Question** How can you determine if two ratios are equivalent?

_____

_____

_____

Rate Yourself!

How well do you understand ratio tables? Circle the image that applies.

Clear   Somewhat Clear   Not So Clear

For more help, go online to access a Personal Tutor.

Independent Practice

Go online for Step-by-Step Solutions

Complete each ratio table to solve each problem.

1 To make 5 apple pies, you need about 2 pounds of apples. How many pounds of apples do you need to make 20 apple pies? (Example 1)

Number of Pies	5		20
Pounds of Apples	2		

_____

2. Four balls of wool will make 8 knitted caps. How many balls of wool will Malcolm need if he wants to make 6 caps? (Examples 3 and 4)

Balls of Wool	4		
Number of Caps	8		6

_____

3 Before leaving to visit Mexico, Levant traded 270 American dollars and received 3,000 Mexican pesos. When he returned from Mexico, he had 100 pesos left. How much will he receive when he exchanges these pesos for dollars? (Example 2)

American Dollars	270		
Mexican Pesos	3,000		100

_____

4. On a bike trip across the United States, Rodney notes that he covers about 190 miles every 4 days. If he continues at this rate, use a ratio table to determine about how many miles he could bike in 6 days. (Example 5)

Miles Biked			
Days			

_____

5. **MP** **Identify Repeated Reasoning** A punch recipe that serves 24 people calls for 4 liters of lemon-lime soda, 2 pints of sherbet, and 6 cups of ice.

a. Complete a ratio table to represent this situation.

b. How much of each ingredient would you need to make an identical recipe that serves 12 people? 36 people?

People Served	
Liters of Soda	
Pints of Sherbet	
Cups of Ice	

_____

_____

c. How much of each ingredient would you need to make an identical recipe that serves 18 people? Explain your reasoning.

_____

_____

_____

6. On a typical day, flights at a local airport arrive at a rate of 10 every 15 minutes. At this rate, how many flights would you expect to arrive in 1 hour?

Number of Flights			
Minutes			

_____

7. MP **Identify Structure** Complete the graphic organizer to explain how equivalent ratios are used to find larger quantities and smaller quantities.

Equivalent Ratios

Larger Quantity

See Example 1.
Operation used: ☐
Real-World Example:

_____

Smaller Quantity

See Example 2.
Operation used: ☐
Real-World Example:

_____

🔥 H.O.T. Problems Higher Order Thinking

8. MP **Find the Error** Melody used the ratio table at the right to find the number of people served with 15 pounds of ground turkey. Find her error and correct it.

Pounds of Ground Turkey	2	1	15
People Served	6	5	19

_____

_____

_____

9. MP **Justify Conclusions** There are 18 bulls and 45 cows on a ranch. If 4 more bulls and 4 more cows were added, will the ratio of bulls to cows remain the same? Justify your answer using a ratio table.

Bulls			
Cows			

_____

_____

_____

10. MP **Use Math Tools** Complete the ratio table to illustrate a real-world relationship among two quantities in which the scale factor is 4.

Extra Practice

Complete each ratio table to solve each problem.

11. A zoo requires that 1 adult accompany every 7 students that visit the zoo. How many adults must accompany 28 students? _4 adults_

		+1	+1	+1	
Number of Adults	1	2	3	4
Number of Students	7	14	21	28

+7 +7 +7

12. Valentina purchased 200 beads for $48 to make necklaces. If she needs to buy 25 more beads, how much will she pay if she is charged the same rate?

Number of Beads	200		25
Cost in Dollars	48		

13. If a hummingbird were to get all of its food from a feeder, then a 16-ounce nectar feeder could feed about 80 hummingbirds a day. How many hummingbirds would you expect to be able to feed with a 12-ounce feeder?

Ounces of Nectar	16		12
Number of Birds Fed	80		

14. When a photo is reduced or enlarged, its length to width ratio usually remains the same. Aurelia wants to enlarge a 4-inch by 6-inch photo so that it has a width of 15 inches. Use a ratio table to determine the new length of the photo.

Length (in.)	4			
Width (in.)	6			

15. Landon owns a hybrid SUV that can travel 400 miles on a 15-gallon tank of gas. Determine how many miles he can travel on 6 gallons.

16. **MP Justify Conclusions** A veterinarian needs to know an animal's weight in kilograms. If 20 pounds is about 9 kilograms and a dog weighs 30 pounds, use a ratio table to find the dog's weight in kilograms. Explain your reasoning.

Pounds			
Kilograms			

17. Jaylen is making biscuits using the recipe.

How many cups of flour will he need to make

30 biscuits? [   ]

Whole Wheat Biscuits

2 c	Whole wheat flour
4 tsp	Baking powder
$\frac{1}{2}$ tsp	Salt
2 tbsp	Shortening
1 c	Milk
1	Small egg

Makes 8 biscuits

18. Beth walked 3 blocks in 15 minutes. Based on this rate, determine if each statement is true or false.

a. Beth walks 9 blocks in 45 minutes. ☐ True ☐ False

b. Beth walks 4 blocks in 25 minutes. ☐ True ☐ False

c. Beth walks 8 blocks in 40 minutes. ☐ True ☐ False

d. Beth walks 12 blocks in 60 minutes. ☐ True ☐ False

Spiral Review

Identify each point shown on the graph. 5.G.1

19. (_____ , _____)

20. (_____ , _____)

21. Liam is drawing a map. He needs to plot four points to identify four places on his map. Plot and label the following points. 5.G.2

a. the library at (3, 2)

b. the school at (6, 4)

c. the park at (8, 1)

d. Liam's house at (2, 8)

Graph Ratio Tables

Vocabulary Start-Up

The **coordinate plane** is formed when two perpendicular number lines intersect at their zero points. This point is called the **origin**. The horizontal number line is called the **x-axis** and the vertical number line is called the **y-axis**. An **ordered pair**, such as (2, 3), is a pair of numbers used to locate a point on the coordinate plane.

Fill in the blanks with the hightlighted words from above.

_____

_____

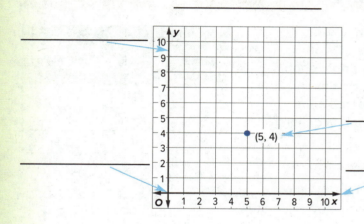

_____

_____

Essential Question

HOW do you use equivalent rates in the real world?

Vocabulary

coordinate plane
origin
x-axis
y-axis
ordered pair
x-coordinate
y-coordinate
graph

Content Standards
6.RP.3, 6.RP.3a

MP Mathematical Practices
1, 3, 4

 ## Real-World Link

In 3 minutes, a North American wood turtle can travel about 17 yards. If the x-axis represents minutes and the y-axis represents yards, write an ordered pair to represent this situation.

(,   )
minutes yards

 Which MP Mathematical Practices did you use?
Shade the circle(s) that applies.

① Persevere with Problems
② Reason Abstractly
③ Construct an Argument
④ Model with Mathematics

⑤ Use Math Tools
⑥ Attend to Precision
⑦ Make Use of Structure
⑧ Use Repeated Reasoning

Graph Ordered Pairs

You can use an ordered pair to name any point on the coordinate plane. The first number in an ordered pair is the **x-coordinate**, and the second number is the **y-coordinate**.

The x-coordinate corresponds to a number on the x-axis. ⟶ **(3, 6)** ⟵ The y-coordinate corresponds to a number on the y-axis.

You can express information in a table as a set of ordered pairs. To see patterns, **graph** the ordered pairs on the coordinate plane.

 ## Examples

The table shows the cost in dollars to create CDs of digital photos at a photo shop. The table also shows this information as ordered pairs (number of CDs, cost in dollars).

Cost to Create CDs		
Number of CDs, x	Cost in Dollars, y	Ordered Pair (x, y)
1	3	(1, 3)
2	6	(2, 6)
3	9	(3, 9)

1. Graph the ordered pairs.

Start at the origin. Use the x-coordinate and move along the x-axis. Then use the y-coordinate and move along the y-axis. Draw a dot at each point.

2. Describe the pattern in the graph.

The points appear in a line. Each point is one unit to the right and three units up from the previous point.

So, the cost increases by $3 for every CD created.

Got it? Do these problems to find out.

The table shows Gloria's earnings for 1, 2, and 3 hours. The table also lists this information as ordered pairs (hours, earnings).

a. Graph the ordered pairs.

b. Describe the pattern in the graph.

Gloria's Earnings		
Hours, x	Dollars Earned, y	Ordered Pair (x, y)
1	5	(1, 5)
2	10	(2, 10)
3	15	(3, 15)

Show your work.

a.

b. _____

Compare Ratios

You can use tables and graphs to compare ratios. The greater the ratio, the steeper the line will appear.

 Real World

Examples Tutor

Two friends are making scrapbooks. Renée places 4 photos on each page of her scrapbook. Gina places 6 photos on each page of her scrapbook.

3. Make a table for each scrapbook that shows the total number of photos placed, if each book has 1, 2, 3, or 4 pages. List the information as ordered pairs (pages, photos).

Renée's Scrapbook		
Pages, x	Photos, y	(x, y)
1	4	(1, 4)
2	8	(2, 8)
3	12	(3, 12)
4	16	(4, 16)

Gina's Scrapbook		
Pages, x	Photos, y	(x, y)
1	6	(1, 6)
2	12	(2, 12)
3	18	(3, 18)
4	24	(4, 24)

4. Graph the ordered pairs for each friend on the same coordinate plane.

Graph the ordered pairs for Renée's scrapbook in blue.

Graph the ordered pairs for Gina's scrapbook in red.

5. How does the ratio of photos to each page compare for each person? How is this shown on the graph?

The ratio of photos to pages for Renée's scrapbook is 4:1 while the ratio for Gina's scrapbook is 6:1. On the graph, both sets of points appear to be in a straight line, but the line for Gina is steeper than the line for Renée.

> **STOP and Reflect**
>
> Marta is also making a scrapbook. She places 5 photos on each page. How does the ratio of photos to each page compare for her book, Gina's book, and Renée's book?

Two friends are each saving money in their bank accounts. Marcus saves $10 each week while David saves $15 each week. (Examples 1–5)

1. Make a table for each friend that shows the total amount saved for 1, 2, 3, and 4 weeks. List the information as ordered pairs (weeks, total dollars saved).

Show your work.

Marcus				David		
Weeks, x	Total Saved ($), y	(x, y)		Weeks, x	Total Saved ($), y	(x, y)
1				1		
2				2		
3				3		
4				4		

2. Graph the ordered pairs for each friend on the same coordinate plane.

3. How do the ratios of Marcus's savings and David's savings compare? How is this shown on the graph?

_____

_____

_____

_____

4. **ⓔ Building on the Essential Question** How can graphing help solve a problem involving ratios?

_____

_____

Rate Yourself!

How confident are you about graphing ratios? Check the box that applies.

For more help, go online to access a Personal Tutor.

Tutor

FOLDABLES Time to update your Foldable!

Independent Practice

Go online for Step-by-Step Solutions

The table shows the total time it took Samir to read 0, 1, 2, and 3 pages of the book. The table also lists this information as ordered pairs (number of pages, total minutes). (Examples 1–2)

Samir's Reading		
Number of Pages, x	Total Minutes, y	Ordered Pair (x, y)
0	0	(0, 0)
1	4	(1, 4)
2	8	(2, 8)
3	12	(3, 12)

1 Graph the ordered pairs.

2. Describe the pattern in the graph.

_____

_____

Ken's Home Supply charges $5 for each foot of fencing. Wayne's Warehouse charges $6 for each foot of fencing. (Examples 3–5)

3. Make a table for each store that shows the total cost for 1, 2, 3, or 4 feet of fencing. List the information as ordered pairs (feet of fencing, total cost).

Ken's Home Supply		
Fencing (ft), x	Cost ($), y	(x, y)
1		
2		
3		
4		

Wayne's Warehouse		
Fencing (ft), x	Cost ($), y	(x, y)
1		
2		
3		
4		

4. Graph the ordered pairs for each store on the same coordinate plane.

5 Using the tables and graphs, write a few sentences comparing the ratios of amount charged per foot of fencing for each store. How is this shown on the graph?

_____

_____

_____

6. **MP Justify Conclusions** Patty's Pies made 2 peach pies using 10 cups of peaches. They made 3 pies using 15 cups of peaches and 4 pies using 20 cups of peaches. Predict how many cups of peaches would be needed to make 9 peach pies. Explain. _____

_____

_____

7. **MP Multiple Representations** The *golden rectangle* is a rectangle in which the ratio of the length to the width is approximately 1.618 to 1. This ratio is called the *golden ratio*.

Length, x	Width, y	(x, y)

a. **Table** Make a ratio table to show the approximate lengths of golden rectangles given widths that are 1, 2, 3, and 4 units. List the information as ordered pairs (length, width).

b. **Graph** Graph the ordered pairs on the coordinate plane.

c. **Analyze** How does the area of each rectangle change as the dimensions change?

_____

_____

_____

H.O.T. Problems Higher Order Thinking

8. **MP Model with Mathematics** Write a real-world problem using ratios or rates that could be represented on the coordinate plane. _____

_____

_____

9. **MP Persevere with Problems** Give the coordinates of the point located halfway between (2, 1) and (2, 4). _____

10. **MP Persevere with Problems** The graph shows the cost of purchasing pencils from the school office. The graph is missing a point to indicate the cost of 12 pencils. Complete the graph by plotting the missing information. Explain your answer.

_____

_____

_____

Extra Practice

The table shows the total number of miles Ariel runs for several days. The table also lists this information as ordered pairs (number of days, total miles).

11. Graph the ordered pairs.

Ariel's Running Record		
Days, x	Miles, y	(x, y)
1	3	(1, 3)
2	6	(2, 6)
3	9	(3, 9)
4	12	(4, 12)

12. Describe the pattern in the graph. _The graph shows that as the number of days increases by 1, the number of miles run increases by 3._

There are two employees for every tiger in the tiger exhibit at a local zoo. For every elephant in the elephant exhibit, there are four employees.

13. Make a table for each animal that shows the total number of employees for 1, 2, 3, or 4 animals. List the information as ordered pairs (number of animals, number of employees).

Tiger Exhibit		
Animals, x	Employees, y	(x, y)
1		
2		
3		
4		

Elephant Exhibit		
Animals, x	Employees, y	(x, y)
1		
2		
3		
4		

14. Graph the ordered pairs for each exhibit on the same coordinate plane.

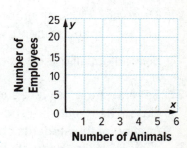

15. **MP** **Justify Conclusions** Using the tables and graphs, write a few sentences comparing the ratios of the number of employees per animal. How is this shown on the graph?

_____

_____

_____

16. The table gives the ratio of teachers to students at Jefferson Middle School.

At Hamilton Middle School, the ratio of teachers to students is 12 to 312. Which statement corretly compares the ratio of the teachers to students at the two schools?

Jefferson Middle School	
Students, x	Teachers, y
24	1
48	2
72	3
96	4

- ☐ There are more students per teacher at Hamilton Middle School than at Jefferson Middle School.
- ☐ Both schools have an equivalent ratio of students to teachers.
- ☐ There are more students at Hamilton Middle School than at Jefferson Middle School.
- ☐ There are more students per teacher at Jefferson Middle School than at Hamilton Middle School.

17. Nina earns $15 for each yard she mows. She wants to buy a dress that costs $109. How many yards will she need to mow to earn the money for the dress? Explain.

Spiral Review

Simplify each fraction. 5.NF.5b

18. $\frac{13}{78} = \frac{\square}{\square}$

19. $\frac{26}{130} = \frac{\square}{\square}$

20. $\frac{20}{240} = \frac{\square}{\square}$

21. There are 270 sixth grade students and 45 chaperones going on a field trip. How many students will be with each chaperone if the groups are divided equally? 5.NBT.5 _____

22. Several students were surveyed about their favorite class. The results are shown in the table. What fraction of the students chose music as their favorite subject? Write the fraction in simplest form. 5.NF.3

Favorite Class	
Art	26
English	19
Math	21
Music	16
Science	32

The Four-Step Plan

Content Standards
6.RP.3

MP **Mathematical Practices**
1, 3, 4

Case #1 Cabin Fever

At a summer camp, the ratio of cabins to campers is 15 to 180. An equal number of campers are staying in each cabin.

How many campers are in each cabin?

1 Understand *What are the facts?*

- You know there are 15 cabins for 180 campers.
- You need to find how many campers are in each cabin.

2 Plan *What is your strategy to solve this problem?*

Divide 180 by 15. Before you calculate, estimate.

Estimate $200 \div 20 =$ ☐

3 Solve *How can you apply the strategy?*

Use long division to find the number of campers in each cabin.

$$
\begin{array}{r}
\boxed{} \\
15 \overline{)180} \\
-\boxed{} \\
\hline
30 \\
-30 \\
\hline
0
\end{array}
$$

There are ☐ campers in each cabin.

4 Check *Does the answer make sense?*

Check by multiplying. Since $12 \times 15 =$ ☐ , the answer is correct.

Analyze the Strategy

MP **Justify Conclusions** How many campers would be in each cabin if the ratio of cabins to campers was 15 to 225? Explain.

_____

_____

Show Me the Money

The table shows Kaylee's weekly allowance.

Age		10	11	12	13
Weekly Allowance ($)		2	4	6	■

If the pattern continues, how much allowance will Kaylee earn when she is 13 years old?

1 Understand

Read the problem. What are you being asked to find?

I need to find _____

Underline key words and values in the problem. What information do you know?

The top row shows an increase of ☐ year. The bottom row shows an increase of $ ☐ per year.

2 Plan

Choose an operation.

I will use _____ to solve this problem.

3 Solve

Describe the pattern in the table. Then complete it using your problem-solving strategy.

_____

$$+1 \quad +1 \quad +1$$

Age		10	11	12	13
Weekly Allowance ($)		2	4	6	

$$+2 \quad +2 \quad \boxed{}$$

6 + ☐ = ☐ So, Kaylee will earn $ ☐ when she is 13 years old.

4 Check

Use information from the problem to check your answer.

Use subtraction to check your answer. ☐ − ☐ = 6

Work with a small group to solve the following cases.
Show your work on a separate piece of paper.

Case #3 Walking

Megan uses a pedometer to find how many steps she takes each school day. She took 32,410 steps over the course of 5 days. She took the same number of steps each day and each step is 28 inches.

How many miles did she walk on Monday? Round to the nearest hundredth. (Hint: There are 5,280 feet in one mile.)

Case #4 Savings

James is earning money to buy a $100 bicycle. For each dollar James earns, his mother has agreed to give him $1. So far, he has earned $14 mowing lawns and $7 washing cars.

How much more must James earn in order to buy the bicycle?

Case #5 Money

Mrs. Eddington is buying a new big-screen television. She made an initial payment of $50 and paid a total of $890 over 12 months.

How much did she pay each month?

Use any strategy!

Case #6 Sports Equipment

Mrs. Dimas has $130 to buy basketballs for Edison Middle School.

How many can she buy at $15 each? Interpret the remainder.

Mid-Chapter Check

Vocabulary Check

1. Fill in the blank in the sentence below with the correct term. (Lesson 2)

A _____ is a comparison of two quantities by division.

Skills Check and Problem Solving

Find the greatest common factor or least common multiple of each set of numbers. (Lesson 1)

2. 24 and 18

GCF = _____

3. 12 and 20

LCM = _____

4. 16 and 32

GCF = _____

5. Write *15 cookies to 40 brownies* as a ratio in simplest form. (Lesson 2) _____

6. Write 171 miles in 3 hours as a unit rate. (Lesson 3) _____

7. **MP Use Math Tools** The table below shows the amount in Josiah's account each week. List the information as ordered pairs and then graph the ordered pairs. Describe the pattern in the graph. (Lesson 5)

Josiah's Savings		
Week, *x*	Savings ($), *y*	Ordered Pair (*x*, *y*)
1	5	
2	10	
3	15	
4	20	
5	25	

_____

_____

8. **MP Persevere with Problems** An artist is using three different colors in a mosaic. The ratio of green to blue to yellow color tiles in the mosaic is 4:6:9. She has 42 blue tiles to use. How many green and yellow tiles does she need? (Lesson 4)

_____

Equivalent Ratios

 Real-World Link

Photography Andrea spent $2 to make 10 prints from a photo booth. Later, she spent $6 to make 30 prints.

Number of Prints	Cost ($)
10	2
30	6

1. Express the relationship between the number of prints she made and the total cost for each situation as a rate in fraction form.

   ◻/◻ prints   and   ◻/◻ prints

2. Compare the relationship between the numerators of each rate in Exercise 1. Compare the relationship between the denominators of these rates.

   _____

   _____

   _____

3. What is the unit rate for 10 prints? _____

4. What is the unit rate for 30 prints? _____

5. Are the rates in Exercise 1 equivalent? Explain.

   _____

   _____

Which MP Mathematical Practices did you use?
Shade the circle(s) that applies.

① Persevere with Problems   ⑤ Use Math Tools

② Reason Abstractly   ⑥ Attend to Precision

③ Construct an Argument   ⑦ Make Use of Structure

④ Model with Mathematics   ⑧ Use Repeated Reasoning

Use Unit Rates

There are different ways to determine if two ratios or rates are equivalent. One way is by examining unit rates. By comparing quantities as rates in simplest form, you can determine if the relationship between the two quantities stays the same.

$$\frac{10 \text{ prints}}{\$2} = \frac{5 \text{ prints}}{\$1} \quad \text{and} \quad \frac{30 \text{ prints}}{\$6} = \frac{5 \text{ prints}}{\$1}$$

÷2 ÷2 ÷6 ÷6

Since the rates have the same unit rate, they are equivalent ratios.

Examples

Tutor

Determine if each pair of rates is equivalent. Explain your reasoning.

1. 20 miles in 5 hours; 45 miles in 9 hours

Write each rate as a fraction. Then find its unit rate.

$$\frac{20 \text{ miles}}{5 \text{ hours}} = \frac{4 \text{ miles}}{1 \text{ hour}} \qquad \frac{45 \text{ miles}}{9 \text{ hours}} = \frac{5 \text{ miles}}{1 \text{ hour}}$$

÷5 ÷5 ÷9 ÷9

Since the rates do not have the same unit rate, they are not equivalent.

2. 3 T-shirts for $21; 5 T-shirts for $35

$$\frac{\$21}{3 \text{ T-shirts}} = \frac{\$7}{1 \text{ T-shirt}} \qquad \frac{\$35}{5 \text{ T-shirts}} = \frac{\$7}{1 \text{ T-shirt}}$$

÷3 ÷3 ÷5 ÷5

Since the rates have the same unit rate, they are equivalent.

Got it? Do these problems to find out.

Determine if each pair of rates is equivalent. Explain your reasoning.

    a. 36 T-shirts in 3 boxes; 60 T-shirts in 6 boxes

    b. 42 flowers in 7 vases; 54 flowers in 9 vases

Unit Rates

The unit rate in Example 2, $\frac{\$7}{1 \text{ T-shirt}}$, is called the unit price since it gives the cost per unit.

a. _____

b. _____

Show your work.

Example

 Tutor

3. Felisa read the first **60 pages** of a book in **3 days**. She read the last **90 pages** in **6 days**. Are these reading rates equivalent? Explain your reasoning.

$$\frac{60 \text{ pages}}{3 \text{ days}} = \frac{20 \text{ pages}}{1 \text{ day}} \qquad (\div 3)$$

$$\frac{90 \text{ pages}}{6 \text{ days}} = \frac{15 \text{ pages}}{1 \text{ day}} \qquad (\div 6)$$

Since the rates do not have the same unit rate, they are not equivalent. So, Felisa's reading rates are not equivalent.

Got it? Do these problems to find out.

c. Marcia made 10 bracelets for 5 friends. Jen made 12 bracelets for 4 friends. Are these rates equivalent? Explain your reasoning.

d. Club A raised $168 by washing 42 cars. Club B raised $152 by washing 38 cars. Are these fundraising rates equivalent? Explain your reasoning.

Use Equivalent Fractions

If a unit rate is not easily found, use equivalent fractions to decide whether the ratios or rates are equivalent.

 Examples

 Tutor

Determine if the pair of ratios or rates is equivalent. Explain your reasoning.

4. 3 free throws made out of 7 attempts;
9 free throws made out of 14 attempts

Write each ratio as a fraction.

$$\frac{3 \text{ free throws}}{7 \text{ attempts}} \overset{?}{=} \frac{9 \text{ free throws}}{14 \text{ attempts}} \qquad \overset{\times 3}{\underset{\times 2}{}}$$

The numerator and the denominator are not multiplied by the same number. So, the fractions are not equivalent.

Since the fractions are *not* equivalent, the ratios are not equivalent.

Proportion

A proportion is an equation stating that two ratios or rates are equivalent.

c. _____

show your work.

d. _____

5. Selena is comparing the cost of two packages of DVDs. A package of 6 DVDs costs $90 and a package of 3 DVDs costs $45. Are the rates equivalent? Explain your reasoning.

$$\frac{6 \text{ DVDs}}{\$90} \overset{\div 2}{\underset{\div 2}{=}} \frac{3 \text{ DVDs}}{\$45}$$

The numerator and the denominator are divided by the same number. So, the fractions are equivalent.

Since the fractions are equivalent, the ratios are equivalent.

Got it? Do this problem to find out.

Show your work.

e. _____

e. Mrs. Jeffries has 12 girls out of 16 students on the Student Council. The Earth Day Committee has 4 girls out of 8 students. Are the ratios equivalent? Explain your reasoning.

Guided Practice

Check ✓

Determine if each pair of ratios or rates is equivalent. Explain your reasoning.

1. $24 saved after 3 weeks; $52 saved after 7 weeks (Examples 1 and 2)

Show your work.

_____

_____

_____

2. 270 Calories in 3 servings; 450 Calories in 5 servings (Examples 1 and 2)

_____

_____

_____

3. Micah can do 75 push-ups in 3 minutes. Eduardo can do 130 push-ups in 5 minutes. Are these rates equivalent?

Explain. (Example 3)

_____

_____

4. A human adult takes about 16 breaths in 60 seconds. A puppy takes about 8 breaths in 15 seconds. Are these rates equivalent? Explain your reasoning. (Examples 4 and 5)

_____

_____

5. ⒺBuildiBuilding on the Essential Question** How can you determine if two ratios are equivalent?

_____

_____

Rate Yourself!

Are you ready to move on? Shade the section that applies.

I have a few questions.

I'm ready to move on.

I have a lot of questions.

For more help, go online to access a Personal Tutor.

Tutor

FOLDABLES Time to update your Foldable!

Independent Practice

Go online for Step-by-Step Solutions eHelp

Determine if each pair of ratios or rates is equivalent. Explain your reasoning. (Examples 1–2, 4–5)

1 $3 for 6 bagels; $9 for 24 bagels

_____

_____

2. $12 for 3 paperback books; $28 for 7 paperback books

_____

_____

3 3 hours worked for $12; 9 hours worked for $36

_____

_____

4. 12 minutes to drive 30 laps; 48 minutes to drive 120 laps

_____

_____

5. Jenny is comparing the cost of two packages of socks. One package has 8 pairs of socks for $12. Another package has 3 pairs of socks for $6. Are the rates equivalent? Explain your reasoning.

_____

_____

6. Jade enlarged the photograph at the right to a poster. The size of the poster is 60 inches by 100 inches. Is the ratio of the poster's length and width equivalent to the ratio of the photograph's length and width? Explain your reasoning. (Example 3)

3 in.

5 in.

_____

_____

7 **MP** **Justify Conclusions** On a math test, it took Kiera 30 minutes to do 6 problems. Heath finished 18 problems in 40 minutes. Did the students work at the same rate? Explain your reasoning.

_____

_____

8. **MP** **Be Precise** Refer to the graphic novel frame below for Exercises a–b.

a. What is the unit price for the cans of lemonade at each of the stores?

_____

_____

b. From which store should Mei, Pilar, and David purchase the cans of lemonade? Explain.

_____

_____

H.O.T. Problems Higher Order Thinking

9. **MP** **Which One Doesn't Belong?** Identify the rate that does not belong with the other three. Justify your response.

| 4.5 feet per second | 112.5 feet in 25 seconds | 86.4 feet in 18 seconds | 54 feet in 12 seconds |

_____

_____

10. **MP** **Identify Structure** Write two ratios that are equivalent to $\frac{5}{7}$.

_____

11. **MP** **Persevere with Problems** The ratio of girls to boys in the junior high band is 5 to 7. At the beginning of the year, there were 72 students in the band. By the end of the year, the ratio of girls to boys was 3 to 4. If there are now 48 boys in the band, how many girls joined the band during the school year? _____

Extra Practice

Determine if each pair of ratios or rates is equivalent. Explain your reasoning.

12. 16 points scored in 4 games; 48 points scored in 8 games

No; $\dfrac{16 \text{ points}}{4 \text{ games}} = \dfrac{4 \text{ points}}{1 \text{ game}}$ and $\dfrac{48 \text{ points}}{8 \text{ games}} = \dfrac{6 \text{ points}}{1 \text{ game}}$; Since the unit rates are not the same, the rates are not equivalent.

13. 96 words typed in 3 minutes; 160 words typed in 5 minutes

14. 15 computers for 45 students; 45 computers for 135 students

15. 16 out of 28 students own pets; 240 out of 560 students own pets

16. 288 miles on 12 gallons of fuel; 240 miles on 10 gallons of fuel

17. Fenton is building a model of a living room. The model sofa is 16 inches long and 7 inches deep. The real sofa's dimensions are 80 inches long and 35 inches deep. Is the ratio of the model's dimensions equivalent to the ratio of the real sofa's dimensions? Explain your reasoning.

18. Store A sells 12 juice bottles for $4 and store B sells 18 juice bottles for $6. Are the rates equivalent? Explain your reasoning.

19. **MP Justify Conclusions** Rosalinda saved $35 in 5 weeks. Her sister saved $56 in 56 days. Are the rates at which each sister saved equivalent? Explain your reasoning.

20. Sort the ratios listed at the right into bins so that equivalent ratios are grouped together.

40:64	$\frac{5}{75}$	14:35	24:64
$\frac{6}{15}$	$\frac{65}{104}$	$\frac{66}{176}$	$\frac{12}{30}$
15:225	$\frac{6}{16}$	15:24	48:128

Equivalent Ratios

21. A chef buys 9 cucumbers, 18 peppers, and 21 tomatoes at the farmer's market. How much does she spend?

HOME-GROWN VEGETABLES
Cucumbers   6 for $2
Peppers   12 for $9
Tomatoes   6 for $4

Spiral Review

Write an equivalent fraction. 5.NF.5b

22. $\frac{11}{50} = \frac{33}{\boxed{}}$

23. $\frac{4}{5} = \frac{\boxed{}}{80}$

24. $\frac{2}{9} = \frac{28}{\boxed{}}$

25. Socks are on sale 4 pairs for $5. How much would you pay for 8 pairs of socks? 5.NBT.7

_____

26. Sasha bought 3 pens. Malachi bought 1 pen. How much more did Sasha spend than Malachi? 4.OA.3

_____

Pens! 3 for $6!

Inquiry Lab
Ratio and Rate Problems

 Inquiry HOW can you use unit rates and multiplication to solve for missing measures in equivalent ratio problems?

Content Standards
6.RP.3, 6.RP.3b

MP Mathematical Practices
1, 3, 4, 5, 8

Jill and Sammy are racing go-karts. Jill completed 6 laps in 12 minutes. If Sammy raced at the same rate, how many minutes did it take her to complete 3 laps?

What do you know? _____

_____

What do you need to find? _____

_____

Hands-On Activity 1

Step 1 Use a bar diagram to represent the number of laps Jill completed. The time to travel 6 laps is 12 minutes.

Jill's race:
— 12 min —
| 1 lap | 1 lap | 1 lap | 1 lap | 1 lap | 1 lap |

Step 2 Each section represents 1 lap. Determine the number of minutes it took Jill to complete one lap.

Jill completed each lap in 12 ÷ 6, or ☐ minutes.

Step 3 Determine the number of minutes it took Sammy to complete 3 laps.

Jill's race:
— 12 min —
| 1 lap | 1 lap | 1 lap | 1 lap | 1 lap | 1 lap |

Sammy's race:
— ? min —
| 1 lap | 1 lap | 1 lap |

Each lap was completed in ☐ minutes.

So, Sammy's time was 3 × ☐, or ☐ minutes.

Hands-On Activity 2

There are 184 goldfish at a pet store. The goldfish are in 4 tanks, each with the same number of fish. Determine how many fish are in 3 tanks.

Step 1   Use a bar diagram to represent the total number of goldfish.

|------------------ 184 goldfish ----------------|

[ ] goldfish

Step 2   Label each section "1 tank." There are

184 ÷ 4, or [ ] goldfish in each tank.

So, there are 46 × [ ], or [ ] goldfish in three tanks.

Hands-On Activity 3

Devon drove 171 miles in 3 hours. Logan drove 177 miles in 3 hours. At these rates, how many more miles can Logan drive in 7 hours than Devon?

Step 1   Use bar diagrams to represent the number of miles Devon and Logan drove.

Devon

|----------- [ ] -----------|

[ ] miles

Logan

|----------- [ ] -----------|

[ ] miles

Step 2   Label each section "1 hour". In one hour, Devon drove 171 ÷ 3, or

[ ] miles and Logan drove 177 ÷ 3, or [ ] miles.

Devon will drive 7 × [ ], or [ ] miles in 7 hours. Logan will drive 7 × [ ],

or [ ] miles in 7 hours. So, Logan will drive [ ] — [ ] or [ ] more miles

in 7 hours than Devon.

Investigate

Work with a partner. Use a bar diagram to help solve each problem.

1. Determine the miles traveled in 5 hours at a rate of 189 miles in 3 hours.

Show your work.

2. Determine the number of ice cubes in 32 glasses at a rate of 20 ice cubes in 5 glasses.

3. Determine the cost of 5 pounds of bananas if 2 pounds cost $1.16.

4. Determine the time needed to deliver 72 papers at a rate of 9 papers in 18 minutes.

5. Predict the number of blue squares in a quilt with 11 green squares if there are 4 green squares in a quilt with 68 blue squares.

6. Predict the number of roses in a garden with 16 sunflowers if there are 3 sunflowers in a garden with 81 roses.

7. How does using a bar diagram help you in predicting the solution to ratio and rate problems?

8. How can you use a bar diagram to check the accuracy of the solution to a ratio or rate problem?

Work with a partner. Refer to Exercise 4 on the previous page.

9. Suppose Marcus delivers papers at a rate of 9 papers in 18 minutes. How much longer would it take him to deliver 100 papers than 72 papers? Justify your response.

10. How can you determine the time it takes to deliver one paper without drawing a bar diagram?

11. Without using a bar diagram, explain how you would solve the following comparison problem. Then solve the problem. *Ronaldo delivers papers at the rate of 6 papers in 24 minutes. How much longer would it take him to deliver 56 papers than 41 papers?*

Create

12. **MP** **Model with Mathematics** Lee can read at a rate of 1,100 words in 5 minutes. Write and solve a word problem that uses this information.

13. **MP** **Model with Mathematics** Elyse uses 42 gallons of water for a 10-minute shower. Write and solve a prediction problem that uses this information.

14. **Inquiry** HOW can you use unit rates and multiplication to solve for missing measures in equivalent ratio problems?

Ratio and Rate Problems

 Real-World Link

Games An arcade sells game tokens individually or in packages. They are having a sale on token packages, as shown below.

Number of Packages	Price ($)
1	5
2	10
3	15

Essential Question

HOW do you use equivalent rates in the real world?

Content Standards
6.RP.3, 6.RP.3b

MP Mathematical Practices
1, 3, 4, 5, 7

1. How many token packages can you buy with $20? ☐ $25? ☐ Explain.

2. What is the unit price?

3. How much would it cost to buy 6 token packages?

4. The arcade sells individual tokens for $0.25 each. If a token package contains 25 tokens, how much would you save by buying a package of 25 tokens instead of 25 individual tokens? Explain.

Which MP Mathematical Practices did you use?
Shade the circle(s) that applies.

① Persevere with Problems

② Reason Abstractly

③ Construct an Argument

④ Model with Mathematics

⑤ Use Math Tools

⑥ Attend to Precision

⑦ Make Use of Structure

⑧ Use Repeated Reasoning

Solve Ratio Problems

You can use bar diagrams or equations with equivalent ratios to solve ratio and rate problems.

 Real World **Examples** Tutor

1. Heritage Middle School has 150 students. Two out of three students in Mrs. Mason's class prefer gel toothpaste. Use this ratio to predict how many students in the entire middle school prefer gel toothpaste.

Method 1 Use a bar diagram.

Step 1 Draw a bar diagram.

Step 2 Determine how many students are in each section.

150 students		
50	50	50

gel ———————— not gel

Method 2 Use equivalent fractions.

Write an equivalent ratio.

likes gel → $\dfrac{2}{3} = \dfrac{\blacksquare}{150}$ ← likes gel
total →   ← total

$\dfrac{2}{3} = \dfrac{100}{150}$ Since $3 \times 50 = 150$, multiply 2 by 50.

So, 100 students would prefer gel toothpaste.

2. The ratio of the number of text messages sent by Lucas to the number of text messages sent by his sister is 3 to 4. Lucas sent 18 text messages. How many text messages did his sister send?

Method 1 Use a bar diagram.

Step 1 Draw a bar diagram.

Step 2 Determine how many text messages are in each section.

18 texts			24 texts			
6	6	6	6	6	6	6

———— Lucas ————   ———— Sister ————

Method 2 Use equivalent fractions.

Write an equivalent ratio.

Lucas → $\frac{3}{4}$ = $\frac{18}{\blacksquare}$ ← Lucas
his sister → ← his sister

$$\frac{3}{4} = \frac{18}{24}$$

×6

Since 6 × 3 = 18, multiply 4 by 6.

×6

So, Lucas' sister sent 24 text messages.

Got it? Do these problems to find out.

a. In a survey, four out of five people preferred creamy over chunky peanut butter. There are 120 people shopping at the grocery store. Use the survey to predict how many people in the store would prefer creamy peanut butter.

b. A survey found that 12 out of every 15 people in the United States prefer eating at a restaurant over cooking at home. If 400 people selected eating at a restaurant on the survey, how many people took the survey?

Solve Rate Problems

You can use double number lines or equations to solve rate problems.

 Example

 Tutor

3. The Millers drove 105 miles on 4 gallons of gas. At this rate, how many miles can they drive on 6 gallons of gas?

Draw a double number line.

105 ÷ 4 = 26.25   Find the unit rate.

26.25 × 6 = 157.5   Multiply.

miles   0 [ ]   105   [ ]

gallons   0   1   2   3   4   5 [ ]

So, the Millers can drive 157.5 miles on 6 gallons of gas.

Got it? Do this problem to find out.

c. There are 810 Calories in 3 scoops of vanilla ice cream. How many Calories are there in 7 scoops of ice cream?

STOP and Reflect

What is the relationship between ratios and fractions?

Show your work.

a. _____

b. _____

c. _____

Example

4. **Jeremy drove his motorcycle 120 miles in 3 hours. At this rate, how many miles can he drive in 5 hours? At what rate did he drive his motorcycle?**

$$\frac{120 \text{ miles}}{3 \text{ hours}} = \frac{\blacksquare \text{ miles}}{1 \text{ hour}} \qquad \frac{120 \text{ miles}}{3 \text{ hours}} = \frac{40 \text{ miles}}{1 \text{ hour}}$$ Find the unit rate.

$$\frac{40 \text{ miles}}{1 \text{ hour}} \times 5 \text{ hours} = 200 \text{ miles}$$ Multiply.

So, Jeremy can drive 200 miles in 5 hours driving at a rate of 40 miles per hour.

Got it? Do this problem to find out.

Show your work.

d. **STEM** While resting, a human takes in about 5 liters of air in 30 seconds. At this rate, how many liters of air does he take in during 150 seconds?

d. _____

Guided Practice

 Check ✓

Show your work.

1. Out of 30 students surveyed, 17 have a dog. Based on these results, predict how many of the 300 students in the school have a dog? (Example 1)

_____

2. If one out of 12 students at a school share a locker, how many share a locker in a school of 456 students? (Example 2) _____

_____

3. Sybrina jogged 2 miles in 30 minutes. At this rate, how far would she jog in 90 minutes? At what rate did she jog each hour? (Examples 3 and 4)

_____

4. Ⓔ **Building on the Essential Question** How can you use diagrams and equations to solve ratio and rate problems?

_____

_____

_____

Rate Yourself!

Are you ready to move on? Shade the section that applies.

YES   ?   NO

For more help, go online to access a Personal Tutor.   Tutor

FOLDABLES Time to update your Foldable!

Independent Practice

Go online for Step-by-Step Solutions

1. If 45 cookies will serve 15 students, how many cookies are needed for 30 students? (Examples 1 and 2)

_____

_____

2. Four students spent $12 on school lunch. At this rate, find the amount 10 students would spend on the same school lunch. (Example 3)

_____

3. A Clydesdale drinks about 120 gallons of water every 4 days. At this rate, about how many gallons of water does a Clydesdale drink in 28 days? (Example 3)

_____

_____

4. **STEM** In 10 minutes, a heart can beat 700 times. At this rate, in how many minutes will a heart beat 140 times? At what rate can a heart beat? (Example 4)

_____

_____

5. **MP Make a Prediction** The table shows which school subjects are favored by a group of students. Predict the number of students out of 400 that would pick science as their favorite subject.

_____

_____

Favorite Subject	
Subject	**Number of Responses**
Math	6
Science	3
English	4
History	7

6. Liliana takes 4 breaths per 10 seconds during yoga. At this rate, about how many breaths would Liliana take in 2 minutes of yoga?

_____

_____

7. **MP Use Math Tools** Find a report in a newspaper or magazine, or on the Internet that uses results from a survey. Evaluate how the survey uses

ratios to reach conclusions. _____

_____

_____

8. **MP** **Identify Structure** One rate of an equivalent ratio is $\frac{9}{n}$. Select two other rates, one that can be solved using equivalent fractions and the other that can be solved with unit rates. _____

_____

9. **MP** **Find the Error** Elisa's mom teaches at a preschool. There is 1 teacher for every 12 students at the preschool. There are 276 students at the preschool. Elisa is setting up equivalent ratios to find the number of teachers at the preschool. Find her mistake and correct it.

$$\frac{12}{1} = \frac{\blacksquare}{276}$$

10. **MP** **Reason Inductively** Tell whether the following statement is *always, sometimes,* or *never* true for numbers greater than zero. Explain.

> *In equivalent ratios, if the numerator of the first ratio is greater than the denominator of the first ratio, then the numerator of the second ratio is greater than the denominator of the second ratio.*

_____

_____

_____

11. **MP** **Persevere with Problems** Suppose 25 out of 175 people said they like to play disc golf and 5 out of every 12 of the players have a personalized flying disc. At the same rates, in a group of 252 people, predict how many you would expect to have a personalized flying disc. Explain.

_____

_____

_____

_____

12. **MP** **Persevere with Problems** A car traveling at a certain speed will travel 76 feet per second. How many miles will the car travel in 3.1 hours if it maintains the same speed? Round to the nearest tenth. *(Hint: There are 5,280 feet in one mile.)* _____

_____

Extra Practice

13. A survey reported that out of 50 teenagers, 9 said they get their news from a newspaper. At this rate, how many out of 300 teenagers would you expect to get their news from a newspaper? <u>54 teenagers</u>

$$\frac{50}{9} = \frac{300}{54}$$

14. Nata spent $28 on 2 DVDs. At this rate, how much would 5 DVDs cost? At what rate did she spend her money? _____

15. If 15 baseballs weigh 75 ounces, how many baseballs weigh 15 ounces?

_____

16. **MP** **Make a Prediction** Suppose 8 out of every 20 students are absent from school less than five days a year. Predict how many students would be absent from school less than five days a year out of 40,000 students.

_____

17. For a store contest, 4 out of every 65 people who visit the store will receive a free DVD. If 455 people visit the store, how many DVDs were given away? _____

18. There were 340,000 cattle placed on feed. Write an equivalent ratio that could be used to find how many of these cattle were between 700 and 799 pounds. How many of the 340,000 cattle placed on feed were between 700 and 799 pounds?

_____

Cattle Placed on Feed	
Weight Group	**Fraction of Total Cattle**
Less than 600 pounds	$\frac{1}{5}$
600–699 pounds	$\frac{11}{50}$
700–799 pounds	$\frac{2}{5}$
800 pounds or more	$\frac{9}{50}$

19. At a bus station, buses depart at a rate of 3 every 10 minutes. At this rate,

[   ] buses would depart in one hour.

20. Student Council sells bottled water at the cheerleading competition as shown in the table. Determine if each statement is true or false.

Cases Sold	3	6
Time (min)	20	40

  a. They sell 27 cases in 3 hours.   ☐ True   ☐ False

  b. They sell 12 cases in 1 hour 20 minutes.   ☐ True   ☐ False

  c. They sell 18 cases in 2 hours.   ☐ True   ☐ False

  d. They sell 24 cases in 2 hours 40 minutes.   ☐ True   ☐ False

  e. They sell 36 cases in 3 hours 20 minutes.   ☐ True   ☐ False

Spiral Review

Write each fraction as a unit fraction. 5.NF.5b

21. $\dfrac{12}{84} = $ _____

22. $\dfrac{13}{143} = $ _____

23. $\dfrac{23}{138} = $ _____

24. Skylar gained 64 yards on 16 carries during a recent football game. Find the ratio of yards per carry. 5.NBT.5

_____

25. The drama club is washing cars for a fundraiser. If the rate continues, how many cars will they wash in 4 hours? 4.OA.5

_____

_____

Hours	Cars Washed
1	8
2	16
3	24

26. Follow the rule to find the next three numbers in the pattern. Describe the pattern using the terms *even* and *odd*. 4.OA.5

      Add 5: 1, 6, 11, _____ , _____ , _____ ...

_____

21ST CENTURY CAREER

in Chemistry

Cosmetic Chemist

Are you naturally curious and analytical? Do you like discovering new things? If so, a career as a cosmetics chemist might be a good choice for you. Cosmetics chemists spend time researching, mixing, and testing new formulas that will make cosmetic products both effective and safe. A cosmetics chemist explained, "When you're developing a product, you play with chemicals and balance ratios to get it to feel right. Basically, it's trial and error."

College & Career
READINESS

Is This the Career for You?

Are you interested in a career as a cosmetics chemist? Take some of the following courses in high school.
◆ Algebra
◆ Biology
◆ Chemical Science
◆ Chemistry
◆ Statistics
Find out how math relates to a career in Chemisty.

MP Beauty is Only Science Deep

Use the information in the recipes below to solve each problem.

1. Using the soap recipe, write a ratio comparing the amount of palm kernel oil to the amount of rose hydrosol as a fraction in simplest form. _____

2. Write a ratio to compare the amount of jojoba oil to the total amount of the ingredients in the lip balm recipe. _____

3. The lip balm costs about $16 to make. What is the cost per ounce? _____

4. The soap recipe makes 4 bars of soap. What is the weight per bar? _____

5. The lip balm recipe is increased so that 10 ounces of candelilla wax is needed. Complete the ratio table to find the amount of shea butter that is needed. _____

Candelilla wax	2				10
Shea butter	6				

6. The soap recipe is increased so that 75 grams of shea butter are needed. Complete the ratio table to find the amount of sodium hydroxide that is needed. _____

Shea butter	30		75
Sodium hydroxide	42		

Lip Balm

4 oz beeswax

2 oz candelilla wax

5 oz jojoba oil

3 oz olive oil

6 oz shea butter

Yield: 20 oz

Shea Butter Soap

110 g rose hydrosol

42 g sodium hydroxide

30 g shea butter

66 g coconut oil

150 g olive oil

66 g palm kernel oil

3 tsp calendula CO_2

$\frac{3}{4}$ tsp rose essential oil

Yield: 15 oz

MP Career Project

It's time to update your career portfolio! There are many different types of jobs in cosmetic chemistry. Research one of these jobs and write a two- or three-sentence job description.

_____

_____

_____

_____

_____

List other careers that someone with an interest in chemistry could pursue.

- _____
- _____
- _____
- _____
- _____

Chapter Review

Vocabulary Check

Complete the crossword puzzle using the vocabulary list at the beginning of the chapter.

Across

2. the horizontal line on a coordinate plane

3. used to locate a point on the coordinate plane

6. the cost per unit

8. a comparison of two quantities by division

9. to place a dot a the point named by an ordered pair

11. the second number of an ordered pair

Down

1. the vertical line on a coordinate plane

2. the first number of an ordered pair

4. a ratio comparing two quantities with different kinds of units

5. multiply or divide two quantities by the same number

6. a rate simplified so that it has a denominator of 1

7. columns filled with pairs of numbers that have the same ratio

10. (0, 0)

Use Your FOLDABLES

Use your Foldable to help review the chapter.

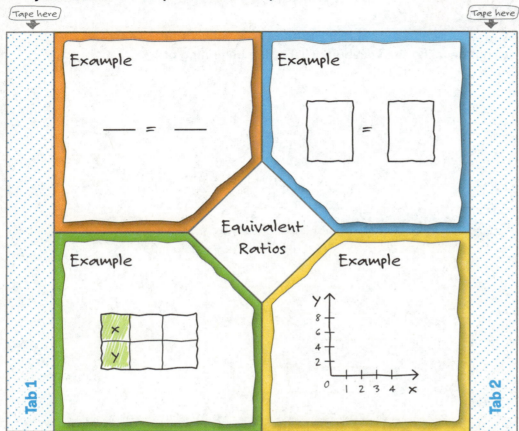

Got it?

Match each ratio with an equivalent ratio.

1. 65:390

2. $\frac{64}{256}$

3. 156:390

4. $\frac{204}{306}$

5. 56:84

6. $\frac{87}{174}$

a. $\frac{2}{5}$

b. $\frac{2}{3}$

c. $\frac{1}{3}$

d. $\frac{1}{6}$

e. $\frac{1}{4}$

f. $\frac{1}{2}$

Power Up! Performance Task

Community Party

A local club is hosting a party at the school gym. Seeking to establish a connection with the community, the club will be providing a fun event and a meal for the children. The doors open at 6:00 P.M. and the games begin at 6:30 P.M.

Ages	Number of Children That Arrived at 6:00 P.M.	Number of Children That Arrived at 6:30 P.M.
6–10	18	6
11–14	12	4

Write your answers on another piece of paper. Show all of your work to receive full credit.

Part A

As the assistant to the party director, you are instructed to split the children into groups so there is an equal number of children in each group. Each group must also have the same number of children from each age range. At 6:00 P.M., what is the greatest number of groups that can be created? How many children of each age range are represented in each group?

Part B

Additional children arrive at 6:30 P.M. What is the greatest number of groups that can be created using the same guidelines in Part A?

Part C

The students begin to play games. Each game takes 10 minutes to complete. There are five game stations, so five games can be played at once. Based on your answer to Part B, how long will it take all the groups to play every game? Explain your answer.

Part D

Each child receives a meal, which consists of a hamburger, a bag of pretzels, and a drink for the children ages 6–10 and a hamburger, two bags of pretzels, and a drink for the children that are older than 10. It costs the club $0.80 for each hamburger, $0.35 for each bag of pretzels, and $0.60 for each drink. Calculate the cost to feed each age group, as well as the total cost to feed all the children.

Reflect

 Answering the Essential Question

Use what you learned about ratios and rates to complete the graphic organizer.

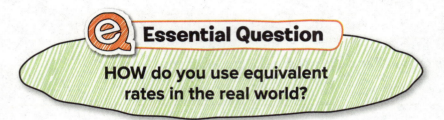 **Essential Question**

HOW do you use equivalent rates in the real world?

Ratio		Rate	
What is it?		What is it?	
Examples		Examples	
Nonexamples		Nonexamples	

How are rates and ratios the same?

How are rates and ratios different?

 Answer the Essential Question. HOW do you use equivalent rates in the real world?

Chapter 2

Fractions, Decimals, and Percents

 Essential Question

WHEN is it better to use a fraction, a decimal, or a percent?

Content Standards
6.RP.3, 6.RP.3c

 Mathematical Practices
1, 2, 3, 4, 5, 6, 7

Math in the Real World

Outer Space Due to the pull of gravity, an astronaut who weighs 180 pounds on Earth would weigh $\frac{1}{6}$ of that on the moon.

Write the astronaut's weight on the box below.

Study Organizer

 Cut out the Foldable on page FL5 of this book.

 Place your Foldable on page 166.

 Use the Foldable throughout this chapter to help you learn about fractions, decimals, and percents.

Vocabulary

least common denominator

percent

percent proportion

proportion

rational number

Study Skill: Reading Math

Everyday Meaning The key to understanding word problems is to understand the meaning of the mathematical terms in the problem.

You will use the terms *factor* and *multiple* in this chapter. Here are two sentences that show their everyday meanings.

- Weather was a *factor* in their decision to postpone the picnic.
- The star quarterback won *multiple* post-season awards.

The table shows how the everyday meaning is connected to the mathematical meaning.

Term	Everyday Meaning	Mathematical Meaning	Connection
Factor	something that actively contributes to a decision or result	2 and 3 are factors of 6.	A factor helps to make a decision. In mathematics, factors "make up" a product.
Multiple	consisting of more than one or shared by many	The multiples of 2 are 0, 2, 4, 6, …	Multiple means many. In mathematics, a number has infinitely many multiples.

Practice Make a list of other words that have the prefixes *fact-* or *multi-*. Determine what the words in each list have in common.

Word	Meaning	Connection

What Do You Already Know?

List three things you already know about fractions, decimals, and percents in the first section. Then list three things you would like to learn about fractions, decimals, and percents in the second section.

Fractions, Decimals, and Percents

What I know	What I want to find out

When Will You Use This?

Here are a few examples of how fractions, decimals, and percents are used in the real world.

Activity 1 Suppose one-half of the students in your class are boys. How would you write one half as a fraction? a decimal? a percent? Which form would you use to represent one-half of the class? Explain your reasoning.

Activity 2 Go online at connectED.mcgraw-hill.com to read the graphic novel *Ring Toss Challenge*. How many rings does each person receive to play the game?

Try the Quick Check below.
Or, take the Online Readiness Quiz.

Quick Review

6.NS.4

Example 1

Find the GCF of 30 and 54.

First, make an organized list of the factors for each number. Then circle the common factors.

30: 1, 2, 3, 5, 6, 10, 15, 30
54: 1, 2, 3, 6, 9, 18, 27, 54

So, the greatest common factor, or GCF, is 6.

Example 2

Find the LCM of 15 and 40.

Write the prime factorization

$15 = 3 \times 5$
$40 = 2 \times 2 \times 2 \times 5$

Find the product of the prime factors. Use the common prime factor, 5, only once.

The least common multiple, or LCM, is $2 \times 2 \times 2 \times 3 \times 5$ or 120.

Quick Check

Greatest Common Factor Find the GCF of each set of numbers.

1. 32 and 52 _____

2. 48 and 60 _____

3. 18, 54, and 72 _____

Least Common Multiple Find the LCM for each set of numbers.

4. 5 and 7 _____

5. 12 and 30 _____

6. 6, 2, 22 _____

7. The front gear of a bicycle has 54 teeth. The back gear has 18 teeth. How many complete rotations must the smaller gear make for both gears to be aligned in the original starting positions?

How Did You Do?

Which problems did you answer correctly in the Quick Check?
Shade those exercise numbers below.

① ② ③ ④ ⑤ ⑥ ⑦

Decimals and Fractions

Real-World Link

Music The instruments below show the part of students in the school orchestra that play each type of instrument.

Brass 0.25

1. Write 0.25 in word form: _____

2. Write 0.25 as a fraction: $\dfrac{\boxed{}}{100}$

Percussion 0.15

3. Write 0.15 in word form: _____

4. Write 0.15 as a fraction: $\dfrac{\boxed{}}{100}$

Strings 0.31

5. Write 0.31 in word form: _____

6. Write 0.31 as a fraction: $\dfrac{\boxed{}}{100}$

Woodwind 0.29

7. Write 0.29 in word form: _____

8. Write 0.29 as a fraction: $\dfrac{\boxed{}}{100}$

 Essential Question

WHEN is it better to use a fraction, a decimal, or a percent?

Vocab **Vocabulary**

rational number

Content Standards
Preparation for 6.RP.3c

MP **Mathematical Practices**
1, 3, 4, 5

Which **MP** **Mathematical Practices** did you use?
Shade the circle(s) that applies.

① Persevere with Problems

② Reason Abstractly

③ Construct an Argument

④ Model with Mathematics

⑤ Use Math Tools

⑥ Attend to Precision

⑦ Make Use of Structure

⑧ Use Repeated Reasoning

Write Decimals as Fractions and Mixed Numbers

Decimals like 0.25, 0.15, 0.31, and 0.29 can be written as fractions with denominators of 10, 100, 1,000, and so on. Any number that can be written as a fraction is a **rational number**.

Decimals like 3.25, 26.82, and 125.54 can be written as mixed numbers in simplest form.

Tutor

Examples

Write each decimal as a fraction in simplest form.

1. 0.6

The place-value chart shows that the place value of the last decimal place is tenths.

$0.6 = \dfrac{6}{10}$ Say *six tenths.*

$= \dfrac{\overset{3}{\cancel{6}}}{\underset{5}{\cancel{10}}}$ Simplify. Divide the numerator and denominator by the GCF, 2.

$= \dfrac{3}{5}$

1,000	100	10	1	0.1	0.01	0.001
thousands	hundreds	tens	ones	tenths	hundredths	thousandths
○	○	○	○	6	○	○

2. 0.45

$0.45 = \dfrac{45}{100}$ Say *forty-five hundredths.*

$= \dfrac{\overset{9}{\cancel{45}}}{\underset{20}{\cancel{100}}}$ Simplify.

$= \dfrac{9}{20}$

1,000	100	10	1	0.1	0.01	0.001
thousands	hundreds	tens	ones	tenths	hundredths	thousandths
○	○	○	○	4	5	○

3. 0.375

$0.375 = \dfrac{375}{1,000}$ Say *three hundred seventy-five thousandths.*

$= \dfrac{\overset{3}{\cancel{375}}}{\underset{8}{\cancel{1,000}}}$ Simplify.

$= \dfrac{3}{8}$

Show your work.

1,000	100	10	1	0.1	0.01	0.001
thousands	hundreds	tens	ones	tenths	hundredths	thousandths
○	○	○	○	3	7	5

Got it? Do these problems to find out.

a. 0.8   **b.** 0.28   **c.** 0.125

Mental Math

Here are some commonly used decimal-fraction equivalencies:

$0.1 = \dfrac{1}{10}$ $0.2 = \dfrac{1}{5}$

$0.25 = \dfrac{1}{4}$ $0.5 = \dfrac{1}{2}$

$0.75 = \dfrac{3}{4}$

It is helpful to memorize these.

a. _____

b. _____

c. _____

Example

Tutor

4. The average length of a conch shell is **9.85 inches. Express 9.85 as a mixed number in simplest form.**

$$9.85 = 9\frac{85}{100}$$ *Say nine and eighty-five hundredths.*

$$= 9\frac{\overset{17}{\cancel{85}}}{\underset{20}{\cancel{100}}} \text{ or } 9\frac{17}{20} \text{ in.}$$ *Simplify.*

Got it? Do this problem to find out.

d. It takes approximately 4.65 quarts of milk to make a pound of cheese. Express this amount as a mixed number in simplest form.

show your work.

d. _____

Write Fractions and Mixed Numbers as Decimals

For fractions with denominators that are *factors* of 10, 100, or 1,000, you can write equivalent fractions with these denominators.

Example

Tutor

5. Write $\frac{9}{12}$ as a decimal.

Method 1 Write an equivalent fraction.

$$\overset{\div 3}{\frac{9}{12}} = \frac{3}{4} \qquad \overset{\times 25}{\frac{3}{4}} = \frac{75}{100}$$
$$\underset{\div 3}{} \qquad \underset{\times 25}{}$$

Simplify $\frac{9}{12}$. Then multiply the numerator and denominator of $\frac{3}{4}$ by 25.

$$= 0.75$$ *Read 0.75 as seventy-five hundredths.*

Method 2 Divide the numerator by the denominator.

$$\frac{9}{12} \dashrightarrow \begin{array}{r} 0.75 \\ 12\overline{)9.00} \\ -\,84 \\ \hline 60 \\ -\,60 \\ \hline 0 \end{array}$$

To divide 9 by 12, place a decimal point after 9 and annex as many zeros as necessary to complete the division.

e. _____

f. _____

Got it? Do these problems to find out.

e. $\frac{3}{5}$ **f.** $\frac{14}{25}$ **g.** $\frac{102}{250}$

g. _____

Example

Real World

6. A caterpillar can have as many as 4,000 muscles, compared to humans, who have about 600. Write the length of the caterpillar as a decimal.

$1\frac{3}{8}$ in.

$$1\frac{3}{8} = 1 + \frac{3}{8}$$ Definition of a mixed number

$$= 1 + \frac{375}{1,000}$$ Multiply the numerator and the denominator by 125.

$$= 1 + 0.375 \text{ or } 1.375$$ Read 1.375 as *one and three hundred seventy-five thousandths.*

The length of the caterpillar is 1.375 inches.

Guided Practice

Check

Write each decimal as a fraction or mixed number in simplest form. (Examples 1–4)

Show your work.

1. 0.4 = _____

2. 0.64 = _____

3. 2.75 = _____

Write each fraction or mixed number as a decimal. (Examples 5 and 6)

4. $\frac{27}{75} =$ _____

5. $\frac{7}{2} =$ _____

6. $3\frac{1}{5} =$ _____

7. Mr. Ravenhead's car averages 23.75 miles per gallon of gasoline. Express this amount as a mixed number in simplest form. (Example 4) _____

_____

8. **STEM** The Siberian tiger can grow up to $10\frac{4}{5}$ feet long. Express this length as a decimal. (Example 6) _____

_____

9. **Building on the Essential Question** What is the relationship between fractions and decimals?

_____

_____

_____

Rate Yourself!

Are you ready to move on? Shade the section that applies.

YES   ?   NO

For more help, go online to access a Personal Tutor.

Tutor

Independent Practice

Go online for Step-by-Step Solutions

Write each decimal as a fraction in simplest form. (Examples 1–3)

1. 0.5 = _____

2. 0.7 = _____

3. 0.33 = _____

4. 0.875 = _____

 Show your work.

Write each fraction or mixed number as a decimal. (Examples 5 and 6)

5. $\frac{77}{200}$ = _____

6. $\frac{1}{20}$ = _____

7. $\frac{12}{75}$ = _____

8. $8\frac{21}{40}$ = _____

9. **STEM** Mercury orbits the Sun in $87\frac{24}{25}$ Earth days. Venus orbits the Sun in $224\frac{7}{10}$ Earth days, and Mars orbits the Sun in $686\frac{49}{50}$ days. Write each mixed number as a decimal. (Example 6)

_____

10. **STEM** Last week, a share of stock gained a total of 1.64 points. Express this gain as a mixed number in simplest form. (Example 4)

_____

11. **MP Use Math Tools** The table shows the ingredients in an Italian sandwich.

a. What fraction of a pound is each ingredient?

_____

_____

Ingredient	Amount (lb)
meat	0.35
vegetables	0.15
secret sauce	0.05
bread	0.05

b. How much more meat is in the sandwich than vegetables? Write the amount as a fraction in simplest form.

_____

_____

c. What is the total weight of the Italian sandwich? Write the amount as a fraction in simplest form. _____

_____

12. Paloma can run the 100-meter dash in $16\frac{1}{5}$ seconds. Savannah's best time is 19.8 seconds. How much faster is Paloma than Savannah in the 100-meter dash? _____

13 **STEM** The average length of a ladybug can range from 0.08 to 0.4 inch. Find two lengths that are within the given span.

Write them as fractions in simplest form. _____

🔥 H.O.T. Problems Higher Order Thinking

14. 🔴**MP** **Find the Error** Mei is writing 4.28 as a mixed number. Find her mistake and correct it.

_____
_____
_____
_____

$$4.28 = 4\frac{28}{1,000}$$

$$\text{or } 4\frac{7}{250}$$

15. 🔴**MP** **Persevere with Problems** Decide whether the following statement is *always*, *sometimes*, or *never* true. Explain your reasoning.

>*Any decimal that ends with a digit in the thousandths place can be written as a fraction with a denominator that is divisible by both 2 and 5.*

_____
_____
_____

16. 🔴**MP** **Reason Inductively** Write a fraction with a decimal value between $\frac{1}{2}$ and $\frac{3}{4}$. Write both the fraction and the equivalent decimal.

_____
_____

17. 🔴**MP** **Use Math Tools** Evelyn is making a costume for her school play. She needs to buy 2 yards of cotton fabric at a cost of $3.49 per yard, and $\frac{1}{2}$ yard of satin fabric at $5.98 per yard. She has $15 to spend on the fabric. Use mental math to determine if she will have enough money. Explain.

_____

Extra Practice

Write each decimal as a fraction or mixed number in simplest form.

18. $0.3 = \dfrac{3}{10}$

0.3 is three tenths.

Homework Help ➡

19. $0.65 = $ _____

20. $0.425 = $ _____

21. $9.35 = $ _____

Write each fraction or mixed number as a decimal.

22. $\dfrac{19}{25} = $ _____

23. $\dfrac{311}{500} = $ _____

24. $\dfrac{5}{8} = $ _____

25. $14\dfrac{3}{5} = $ _____

26. Evita lives 0.85 mile from her school. Write this distance as a fraction in simplest form.

_____

27. Rancho Middle School has an average of $23\dfrac{3}{8}$ students per teacher. Write this mixed number as a decimal.

_____

28. Alan bought 20 yards of fencing. He used 5.9 yards to surround one flower garden and 10.3 yards to surround another garden. Write the amount remaining as a fraction in simplest form.

_____

29. In a survey, 9 out of 15 students named math as their favorite class. Express this rate as a decimal.

_____

30. **MP Use Math Tools** The frequency table shows the favorite college football teams of middle school students. What fraction of the students chose the Sooners? Write the fraction as a decimal.

_____

_____

Team	Tally	Frequency
Buckeyes	|||	3
Gators	⊮|	6
Sooners	⊮	5
Tigers	||	2
Lions	||||	4

31. Lori ran the distances shown in the table. Write the total distance, in miles, as a fraction in simplest form.

[   ]

Day	Distance (mi)
Monday	0.35
Wednesday	0.2
Friday	0.25

32. Shade 0.25 of the design. Write a fraction in simplest form to represent the shaded part of the design.

[   ]

Spiral Review

Simplify each fraction. 4.NF.1

33. $\dfrac{20}{100} = \dfrac{\square}{\square}$

34. $\dfrac{35}{100} = \dfrac{\square}{\square}$

35. $\dfrac{72}{100} = \dfrac{\square}{\square}$

36. $\dfrac{48}{100} = \dfrac{\square}{\square}$

37. Jasper made 230 flyers for a dance. He handed two flyers out to each student. How many students received flyers? 4.NBT.6

_____

38. Look for a pattern and complete the table. 5.NBT.2

Multiplication Problem	Product
36 × 100	3,600
36 × 10	
36 × 1	
36 × 0.1	
36 × 0.01	

Inquiry Lab
Model Percents

 HOW can you model a percent?

Jackie is using 1-inch tiles to make the mosaic shown at the right. She needs a total of 100 tiles. What percent of the tiles are green?

What do you know? _____

_____

What do you need to find? _____

_____

Content Standards
Preparation for
6.RP.3c

MP **Mathematical Practices**
1, 3, 4

Hands-On Activity 1

A 10 × 10 grid can be used to represent *hundredths*. It can also represent percents. The word *percent* (%) means *out of one hundred*. For example, 50% means 50 out of one hundred.

Step 1 Use a 10 × 10 grid to model the percent of tiles in the mosaic that are green.

Step 2 In the mosaic, ☐ tiles out of 100 are green.

As a fraction, this is $\dfrac{\boxed{}}{100}$. When the denominator is 100, the numerator gives the numerical value of the percent.

So, ☐ % of the squares are green.

Common percents are 1%, 10%, 25%, and $33\frac{1}{3}$%. Using these percents and their multiples makes mental math easier.

Hands-On Activity 2

Model 25% with a 10 × 10 grid.

Step 1   25% means ☐ out of 100.

Step 2   Shade the squares, filling one column at a time.

Shade ☐ squares out of 100.

What fraction of the 10 × 10 grid is shaded? _____

What decimal represents the shaded part of the grid? _____

Hands-On Activity 3

Percents can also be modeled with bar diagrams. The entire bar represents 100%. The bar diagram below is divided into 10 equal sections, each representing 10%. The shaded region represents 40%.

0%				40%					100%
10%	10%	10%	10%	10%	10%	10%	10%	10%	10%

Model 60% with a bar diagram.

Step 1   The bar diagram below is divided into ☐ equal sections. To find the value of each section, divide. 100% ÷ 5 = 20%.

So, each section represents ☐ %.

Step 2   ☐ % + ☐ % + ☐ % = 60%

Shade ☐ sections of the diagram.

0%   100%

Investigate

Collaborate

Work with a partner. Identify each percent modeled.

1. _____

2. _____

3. _____

0%   50%   100%

4. _____

0%   50%   100%

Work with a partner. Model each percent.

5. 37%

6. 8%

7. 45%

0%   50%   100%

8. 5%

0%   50%   100%

Analyze and Reflect

Collaborate

Work with a partner to determine the number of shaded sections for each model. The first one is done for you.

	Percent	Number of Shaded Sections using each Model		
		10 × 10 Grid	Bar Diagram with 10 Equal Sections	Bar Diagram with 20 Equal Sections
	45	45	4.5	9
9.	15			
10.	30			
11.	55			
12.	70			
13.	85			
14.	65			

15. Write the percent shown by each model. Explain your reasoning.

a. _____

b. _____

c. _____

16. **MP Reason Inductively** How can you use a model to write a percent as a fraction with a denominator of 100? _____

_____

Create

On Your Own

17. **MP Model with Mathematics** Write a real-world problem that involves a percent. Then model the percent used in the problem. _____

_____

_____

18. **Inquiry** HOW can you model a percent? _____

_____

Percents and Fractions

 Real-World Link Tools

Sports Students were asked to choose their favorite sport to play.

1. For each sport, shade a 10 × 10 grid that represents the number of students out of 100 that chose the sport.

Basketball: 3 out of 20   **Football:** 3 out of 25

Gymnastics: 1 out of 20   **Swimming:** 9 out of 100

2. What fraction with a denominator of 100 represents the number of students who chose each sport?

Basketball: _____   Football: _____

Gymnastics: _____   Swimming: _____

 Essential Question

WHEN is it better to use a fraction, a decimal, or a percent?

Vocab **Vocabulary**

percent

Content Standards
Preparation for 6.RP.3c

 Mathematical Practices
1, 3, 4, 5

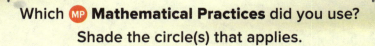

Which MP **Mathematical Practices did you use?**
Shade the circle(s) that applies.

① Persevere with Problems   ⑤ Use Math Tools

② Reason Abstractly   ⑥ Attend to Precision

③ Construct an Argument   ⑦ Make Use of Structure

④ Model with Mathematics   ⑧ Use Repeated Reasoning

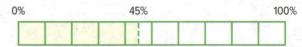

Key Concept > Percents as Fractions

Words   A **percent** is a ratio that compares a number to 100.

Example   45% ⇒ 45 out of 100 or $\frac{45}{100}$

Models

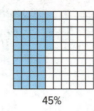

45%

0%   45%   100%

To write a percent as a fraction, first write the percent as a rate per 100. Then simplify.

Examples

1. **Write 50% as a fraction in simplest form.**

50% means *50 out of 100*.

$50\% = \dfrac{50}{100}$ Definition of percent

$= \dfrac{\overset{1}{\cancel{50}}}{\underset{2}{\cancel{100}}}$ or $\dfrac{1}{2}$ Simplify. Divide the numerator and the denominator by the GCF, 50.

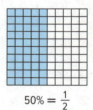

$50\% = \dfrac{1}{2}$

2. **In a recent survey, 55% of cell phone owners said they text message. What fraction of cell phone owners is this?**

$55\% = \dfrac{55}{100}$ Definition of percent

$= \dfrac{11}{20}$ Simplify.

So, $\dfrac{11}{20}$ of cell phone owners text message.

> **Got it?** Do these problems to find out.

Write each percent as a fraction in simplest form.

a. 75%   **b.** 90%   **c.** 38%

Work Zone

Check for Reasonableness

In Example 2, you can conclude that $\frac{11}{20}$ is a reasonable answer because 55% is a little more than 50%, and $\frac{11}{20}$ is a little more than $\frac{10}{20}$ or $\frac{1}{2}$.

Show your work.

a. _____

b. _____

c. _____

Example

Tutor

3. The table shows the percent of each movie type rented during a month. What fraction of the rentals were action movies?

Types of Movies	
action	35%
children's	5%
comedy	45%
drama	5%
horror	5%
romance	5%

$35\% = \dfrac{35}{100}$ *Definition of percent*

$= \dfrac{\overset{7}{35}}{\underset{20}{100}}$ *Divide the numerator and denominator by the GCF, 5.*

Action movies were rented $\dfrac{7}{20}$ of the time.

Got it? Do this problem to find out.

d. Write the fraction of rentals that were horror movies.

d. _____

Fractions as Percents

To write a fraction as a percent, find an equivalent ratio with 100 as a denominator.

Example

Tutor

4. Write the fraction $\dfrac{6}{8}$ as a percent.

$\dfrac{6}{8}$

$\dfrac{6}{8} = \dfrac{3}{4}$ *Simplify by dividing by the GCF, 2.*

$\dfrac{3}{4} = \dfrac{\blacksquare}{100}$ *Write equivalent ratios. One ratio is the fraction. The other ratio is the unknown value compared to 100.*

$\overset{\times 25}{\dfrac{3}{4} = \dfrac{75}{100}}$ *Since 4 × 25 = 100, multiply 3 by 25 to find the unknown value.*
$\times 25$

So, $\dfrac{75}{100}$ or 75% of the rectangle is shaded.

Got it? Do this problem to find out.

e. Write the fraction $\dfrac{9}{12}$ as a percent.

e. _____

Example

5. Mitch made 12 out of 40 shots during the championship game. What percent of his shots did Mitch make?

$\dfrac{12}{40} = \dfrac{3}{10}$ Simplify $\dfrac{12}{40}$ by dividing the numerator and denominator by the GCF, 4.

$\overset{\times 10}{\dfrac{3}{10}} = \dfrac{\blacksquare}{100}$ Write equivalent ratios.

$\underset{\times 10}{\dfrac{3}{10}} = \dfrac{\mathbf{30}}{100}$ Since $10 \times 10 = 100$, multiply 3 by 10 to find the unknown value.

So, $\dfrac{12}{40} = \dfrac{30}{100}$ or 30%.

Got it? Do this problem to find out.

 Show your work.

f. Alana spelled 19 out of 25 words correctly. What percent of the words did Alana spell correctly?

f. _____

Guided Practice

 Check ✓

Write each percent as a fraction in simplest form. (Examples 1–3)

1. 15% = _____

2. 80% = _____

3. 33% = _____

 Show your work.

Write each fraction as a percent. Use a model if needed. (Example 4)

4. $\dfrac{3}{10}$ = _____

5. $\dfrac{3}{20}$ = _____

6. $\dfrac{2}{5}$ = _____

7. Elsa ran 7 out of 10 days. What percent of the days did she run? (Example 5)

_____

8. **Building on the Essential Question** Why is it helpful to write a fraction as a percent?

_____

_____

_____

Independent Practice

Go online for Step-by-Step Solutions

Write each percent as a fraction in simplest form. (Examples 1–3)

1. 2% = _____

Show your work.

2. 20% = _____

3. 85% = _____

4. 4% = _____

Write each fraction as a percent. Use a model if needed. (Example 4)

5. $\frac{2}{10}$ = _____

6. $\frac{3}{4}$ = _____

7. $\frac{7}{20}$ = _____

8. $\frac{11}{25}$ = _____

9. During his workout, Elan spent 28% of the time on the treadmill. What fraction of his workout was on the treadmill? (Examples 1–3)

_____

10. A cat spends about 7 out of 10 hours sleeping. About what percent of a cat's day is spent sleeping? (Example 5)

_____

11. A survey showed that 82% of youth most often use the Internet at home. What fraction of youth surveyed most often use the Internet somewhere else?

_____

12. Cedro collects state quarters. He has 42 out of 50 available quarters. What is 42 out of 50 as a percent?

_____

13. Use the table to determine what percent of students prefer school uniforms and what percent do not prefer school uniforms. What is the relationship between these two percents?

_____

_____

Prefer School Uniforms	
No	**Yes**
⌗⌗ ⌗⌗ ⌗⌗ I	IIII

14. **Multiple Representations** The table shows the percent of Earth's atmosphere that is each element.

Element	Percent
Nitrogen	78
Oxygen	21
Other	1

   a. Bar Diagram Model 21% using a bar diagram.

0%   100%

   b. Number Write the percent of Earth's atmosphere that is nitrogen as a fraction in simplest form. _____

H.O.T. Problems Higher Order Thinking

15. MP **Reason Inductively** Write three fractions that can be written as percents between 50% and 75%. Justify your solution.

_____

16. MP **Persevere with Problems** For each model below, write the portion of the grid that is shaded as a percent and as a fraction.

  a. _____   **b.** _____   **c.** _____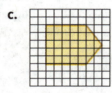

17. MP **Which One Doesn't Belong?** Identify the number that does not belong with the other three. Explain your reasoning.

$\frac{9}{20}$ $\frac{45}{100}$ 45% $\frac{8}{45}$

_____

18. MP **Persevere with Problems** Complete each blank to find an expression that is equal to 16%.

  a. _____ for every 100   **b.** _____ for every 50

  c. 1 for every _____   **d.** 0.5 for every _____

19. MP **Reason Inductively** Explain the difference between $33\frac{1}{3}$% and 33%.

_____

_____

Extra Practice

Write each percent as a fraction in simplest form.

20. $14\% = \dfrac{7}{50}$

21. $47\% = $ _____

22. $86\% = $ _____

23. $88\% = $ _____

Homework Help

Write each fraction as a percent. Use a model if needed.

24. $\dfrac{7}{10} = $ _____

25. $\dfrac{21}{25} = $ _____

26. $\dfrac{3}{5} = $ _____

27. $\dfrac{18}{25} = $ _____

28. In a recent year, 22% of E-mail users said they spend less time using E-mail because of spam. What fraction of E-mail users is this?

_____

_____

29. About $\dfrac{19}{20}$ of celery is water. What percent is this?

_____

_____

30. **MP** **Use Math Tools** Mrs. Lane took a survey of the types of pants her students were wearing. She collected the data at the right. What percent of her students were wearing shorts?

_____

_____

Type of Pants	Number of Students
Jeans	14
Shorts	9
Capris	2

31. **STEM** The circle graph shows the fraction of each type of weather during September.

a. What percent of the days were sunny? _____

b. What percent of the days were rainy? _____

c. What percent of the days were sunny or rainy? _____

d. What percent of the days were cloudy or partly cloudy?

_____

Weather During September

Rainy $\dfrac{4}{25}$

Cloudy $\dfrac{1}{10}$

Partly Cloudy $\dfrac{3}{10}$

Sunny $\dfrac{11}{25}$

32. On Friday, 65% of the sudents at Plainview Middle School bought a hot lunch and the rest of the students packed their lunch. What fraction of students packed their lunch? Select all that apply.

☐ $\frac{26}{40}$ ☐ $\frac{7}{20}$ ☐ $\frac{18}{60}$ ☐ $\frac{28}{80}$

33. The student council published the results of the survey about the new school mascot. Jesse spilled water on the paper, but he knows that 72% of the students chose a mascot other than a tiger. How many students chose a Viking as their new mascot?

Mascot	Number of Students
Eagle	11
Polar Bear	6
Tiger	14
Viking	?

Spiral Review

Multiply. 5.NBT.2

34. 0.685 × 100 = _____

35. 0.09 × 10 = _____

36. 3.255 × 100 = _____

37. Refer to the table. Which lap has the slowest time? 5.NBT.3b

Lap	Time (minutes)
1	1.59
2	1.85
3	1.64

_____

_____

38. Ruby has $10. She buys the items shown. How much will Ruby have left? 5.NBT.7

_____

_____

$1.29

$5.95

Percents and Decimals

 Real-World Link Watch ▶

School A recent survey tells the favorite subjects of students at Martin Middle School.

Math: 28%
Art: 16%

Science: 21%
English: 13%

Social Studies: 15%
Other: 7%

 Essential Question

WHEN is it better to use a fraction, a decimal, or a percent?

Content Standards
Preparation for 6.RP.3c

MP Mathematical Practices
1, 3, 4, 5, 6

1. Write a fraction with a denominator of 100 to represent the percent for each subject.

Math: $\dfrac{}{100}$ Science: $\dfrac{}{100}$

Art: $\dfrac{}{100}$ Social Studies: $\dfrac{}{100}$

English: $\dfrac{}{100}$ Other: $\dfrac{}{100}$

2. Write each fraction from Exercise 1 as a decimal.

Math: ☐   Science: ☐

Art: ☐   Social Studies: ☐

English: ☐   Other: ☐

3. **MP Make a Conjecture** Look back at Exercise 2. Compare the decimals to the percents. Explain how to write a percent as a decimal. _____

_____

Which MP Mathematical Practices did you use?
Shade the circle(s) that applies.

① Persevere with Problems   ⑤ Use Math Tools

② Reason Abstractly   ⑥ Attend to Precision

③ Construct an Argument   ⑦ Make Use of Structure

④ Model with Mathematics   ⑧ Use Repeated Reasoning

Write Percents as Decimals

Work Zone

Words To write a percent as a decimal, divide by 100 and remove the % sign. This is the same as moving the decimal point two places to the left.

Example 48% = 48%
= 0.48

Another way to write a fraction as a decimal is to write the percent as a fraction. Then write the fraction as a decimal.

Examples

Tutor

Write each percent as a decimal.

1. 56%

Method 1 Write the percent as a fraction.

$56\% = \frac{56}{100}$ Rewrite the percent as a fraction with a denominator of 100.

$= 0.56$ Write *56 hundredths* as a decimal.

Method 2 Move the decimal point.

56% = 56%   Move the decimal point two places to the left.

$= 0.56$ Remove the percent sign.

. . . .

2. 8%

$8\% = \frac{8}{100}$ Rewrite the percent as a fraction with a denominator of 100.

$= 0.08$ Write *8 hundredths* as a decimal.

. . . .

Show your work.

3. 2%

2% = 02%   Move the decimal point two places to the left.

$= 0.02$ Remove the percent sign.

a. _____

b. _____

Got it? Do these problems to find out.

c. _____

a. 32%   **b.** 6%   **c.** 93%

Write Decimals as Percents

Words   To write a decimal as a percent, multiply by 100 and add a % sign. This is the same as moving the decimal point two places to the right.

Example $0.36 = 0.36\%$
                $= 36\%$

Another way to write a decimal as a percent is to write the decimal as a fraction with a denominator of 100. Then write the fraction as a percent.

Examples

Tutor

4. Write 0.38 as a percent.

Method 1 **Write the decimal as a percent.**

$0.38 = \dfrac{38}{100}$ *Write 38 hundredths as a fraction.*

$\qquad = 38\%$ *Write the fraction as a percent.*

Method 2 **Move the decimal point.**

$0.38 = 0.38$ *Move the decimal point two places to the right.*

$\qquad = 38\%$ *Add the percent sign.*

5. Write 0.2 as a percent.

$0.2 = \dfrac{2}{10}$ *Write 2 tenths as a fraction.*

$\qquad \overset{\times 10}{\curvearrowright}$

$\dfrac{2}{10} = \dfrac{20}{100}$ *Write the equivalent fraction with a denominator of 100.*

$\qquad \underset{\times 10}{\curvearrowright}$

$\qquad = 20\%$ *Write the fraction as a percent.*

Got it? Do these problems to find out.

Write each decimal as a percent.

d. 0.47   **e.** 0.73   **f.** 0.5

STOP and Reflect

Why does it help to write a decimal as a fraction with a denominator of 100 when writing decimals as percents?

Show your work.

d. _____

e. _____

f. _____

Example

6. The United States produces more corn than any other country, producing **0.4** of the total corn crops. Write **0.4** as a percent.

$0.4 = 0.4\mathbf{0}$ Annex a zero.

    $= 0.40\%$ Multiply by 100 and add a % sign.

    $= 40\%$ Simplify.

Check $0.4 = \dfrac{40}{100}$ Write the decimal as a fraction with a denominator of 100.

      $= 40\%$ ✓   Write the fraction as a percent.

Guided Practice

 Check

Write each percent as a decimal. (Examples 1–3)

1. 27% = _____

 Show your work.

2. 15% = _____

3. 4% = _____

Write each decimal as a percent. (Examples 4 and 5)

4. 0.3 = _____

5. 0.82 = _____

6. 0.51 = _____

7. **STEM** About 0.7 of the human body is water. What percent is equivalent to 0.7? (Example 6) _____

8. **Building on the Essential Question** What is the relationship between percents and decimals?

_____

_____

_____

Rate Yourself!

How well do you understand percents and decimals? Circle the image that applies.

Clear   Somewhat Clear   Not So Clear

For more help, go online to access a Personal Tutor. Tutor

FOLDABLES Time to update your Foldable!

Independent Practice

Go online for Step-by-Step Solutions

eHelp

Write each percent as a decimal. (Examples 1–3)

1 35% = _____

show your work.

2. 2% = _____

3. 31% = _____

4. 95% = _____

Write each decimal as a percent. (Examples 4 and 5)

5. 0.22 = _____

6. 0.79 = _____

7. 0.1 = _____

8. 0.16 = _____

9. Financial Literacy A bank offers an interest rate of 4% on a savings account. Write 4% as a decimal. (Examples 1–3)

_____

10. When making a peanut butter and jelly sandwich, 96% of people put the peanut butter on first. Write 96% as a decimal. (Examples 1–3)

_____

11 In a recent year, 0.12 of Americans downloaded a podcast from the Internet. What percent is equivalent to 0.12?

(Example 6) _____

12. In a recent year, the number of homes with digital cameras grew 0.44 from the previous year. Write 0.44 as a percent. (Example 6)

_____

13. Financial Literacy The formula $I = prt$ gives the simple interest I earned on an account where an amount p is deposited at an interest rate r for a certain number of years t. Use the table to order the accounts from least to greatest interest earned after 5 years.

_____

_____

Accounts at First Savings Bank		
Account	p ($)	r (%)
A	350	4
B	500	3.5
C	280	4.25

14. **MP** **Persevere with Problems** Daphne wants to buy a coat that costs $80. The store that sells the coat has multiple locations. The sales tax in each county is shown in the table. How much more would the coat cost in Delaware county than Fairfield county?

County	Tax Rate (%)
Delaware	7.25
Fairfield	6.5
Franklin	6.75

_____

15. Dante took three tests on Friday. He got a 92% on his English test, an 88% on his math test and a 90% on his science test. Write each percent as a decimal in order from least to greatest.

_____

_____

H.O.T. Problems Higher Order Thinking

16. **MP** **Reason Inductively** Write a decimal between 0.5 and 0.75. Then write it as a fraction in simplest form and as a percent.

_____

_____

17. **MP** **Persevere with Problems** How would you write $43\frac{3}{4}\%$ as a decimal?

_____

_____

18. **MP** **Model with Mathematics** Write a percent between 25% and 50%. Then write it as a decimal and as a fraction in simplest form. _____

19. **MP** **Reason Inductively** Explain why percents are rational numbers.

_____

_____

20. **MP** **Model with Mathematics** Write a problem about a real-world situation in which you would either write a percent as a decimal or write a decimal as a percent.

_____

_____

_____

21. **MP** **Reason Abstractly** Each square below is divided into sections of equal size. In which square is 25% of the square *not* shaded? _____

Square A	Square B	Square C	Square D

Extra Practice

Write each percent as a decimal.

22. 17% = _0.17_

$17\% = \dfrac{17}{100}$

Homework Help ➡ $= 0.17$

23. 3% = _____

24. 1% = _____

25. 11% = _____

Write each decimal as a percent.

26. 0.99 = _____

27. 0.62 = _____

28. 0.6 = _____

29. 0.87 = _____

30. In one day at a store, 7% of the sales were from shoes. Write 7% as a decimal.

_____

31. In one hour on a certain street, 65% of the cars that passed were black. Write 65% as a decimal.

_____

32. In a recent year, 0.57 of those registered to vote in the United States voted in an election. Write 0.57 as a percent.

_____

33. In a recent study, 0.82 of Americans own a cell phone. What percent is equivalent to 0.82?

_____

34. **MP** **Be Precise** In the United States, sales tax is added to items that you purchase. The rate of sales tax varies by state and sometimes by county or region. Use the table to order the counties from least to greatest sales tax.

_____

_____

County Sales Tax	
County	**Sales Tax**
A	6.75%
B	0.0625
C	$\dfrac{7}{100}$

35. At baseball practice, Neil caught 16 out of 20 hits to the outfield. Select all the ways of expressing 16 out of 20.

☐ $\frac{3}{5}$ ☐ 0.8 ☐ 80% ☐ 0.08

36. The Ling family wants to purchase a television that costs $449. The store that sells the television has multiple locations in different counties. The sales tax for each county is shown. Complete the table below by arranging the counties from least to greatest sales tax rate. Write each sales tax as a percent.

County	Tax Rate
Cuyahoga	$7\frac{3}{4}$%
Geauga	$\frac{13}{200}$
Lake	0.0675

	County	Sales Tax Rate (%)
Least		
Greatest		

Spiral Review

Fill in each ◯ with <, >, or = to make a true statement. **5.NBT.3b**

37. 2.50 ◯ 2.5

38. 0.006 ◯ 0.1

39. 0.015 ◯ 0.005

40. The table shows results for the 100 meter sprint. Who had the fastest time? **5.NBT.3b**

_____

_____

Athlete	Time (s)
Bryson	12.14
Malik	11.84
Marcell	11.94
Wyatt	12.44

41. Aliah ate 0.75 sandwich. Her brother ate 1.5 sandwiches. Who ate more? **5.NBT.3b**

_____

_____

Percents Greater than 100% and Percents Less than 1%

 Real-World Link

Plants There are over 220,000 species of plants on Earth. Of those, 590 are carnivorous. Plants such as a Venus Flytrap catch their prey as food.

1. Write the fraction of species of carnivorous plants in simplest form.

$$\frac{\boxed{}}{220,000} \div \frac{\boxed{}}{\boxed{}} = \frac{\boxed{}}{22,000}$$

2. Write your answer to Exercise 1 as a decimal rounded to the nearest thousandth. Use division to find your answer.

$$\overline{\smash{)}}$$

$$\boxed{} \approx \boxed{}$$

3. Write your answer to Exercise 2 as a fraction.

_____

_____

4. **Make a Conjecture** Since 0.3 = 30% and 0.03 = 3%, what percent is equal to 0.003? Explain.

_____

_____

Essential Question

WHEN is it better to use a fraction, a decimal, or a percent?

Content Standards
Preparation for 6.RP.3c

MP Mathematical Practices
1, 3, 4, 5

Which **MP Mathematical Practices** did you use?
Shade the circle(s) that applies.

① Persevere with Problems

② Reason Abstractly

③ Construct an Argument

④ Model with Mathematics

⑤ Use Math Tools

⑥ Attend to Precision

⑦ Make Use of Structure

⑧ Use Repeated Reasoning

Percents as Decimals and Fractions

Percents greater than 100% or less than 1% can also be written as decimals or as fractions.

Examples

 Tutor

1. **Write 0.2% as a decimal and as a fraction in simplest form.**

$$0.2\% = 00.2 \qquad \text{Divide by 100 and remove \% symbol.}$$
$$= 0.002 \qquad \text{Decimal form}$$
$$= \frac{2}{1,000} \text{ or } \frac{1}{500} \qquad \text{Fraction form}$$

2. **Write 170% as a mixed number in simplest form and as a decimal.**

$$170\% = \frac{170}{100} \qquad \text{Definition of percent}$$
$$= 1\frac{70}{100} \text{ or } 1\frac{7}{10} \qquad \text{Mixed number form}$$
$$= 1.7 \qquad \text{Decimal form}$$

Got it? Do these problems to find out.

Write each percent as a decimal and as a mixed number or fraction in simplest form.

a. 0.25%　　　　**b.** 300%　　　　**c.** 530%

 Tutor

Example

Real World

3. **Jimmy's savings increased by 250%. Write 250% as a mixed number in simplest form and as a decimal.**

$$250\% = \frac{250}{100} \qquad \text{Definition of a percent}$$
$$= 2\frac{50}{100} \text{ or } 2\frac{1}{2} \qquad \text{Mixed number form}$$
$$= 2.5 \qquad \text{Decimal form}$$

So, Jimmy more than doubled his savings.

Got it? Do this problem to find out.

d. The stock price for a corporation increased by 0.11%. Write 0.11% as a decimal and as a fraction in simplest form.

Percents

A percent less than 1% equals a number less than 0.01 or $\frac{1}{100}$. A percent greater than 100% equals a number greater than 1.

Show your work.

a. _____

b. _____

c. _____

d. _____

Mixed Numbers and Decimals as Percents

To write a decimal as a percent, multiply by 100 and add a percent sign. To write a mixed number as a percent, first write the mixed number as an improper fraction.

Alternative Method

$1 = 100\%$

$\frac{1}{4} = 25\%$

So, $1\frac{1}{4} = 125\%$.

Example

4. Write $1\frac{1}{4}$ as a percent.

$1\frac{1}{4} = \frac{5}{4}$ Write $1\frac{1}{4}$ as an improper fraction.

$\frac{5}{4} = \frac{\blacksquare}{100}$ Find an equivalent fraction.

$\overset{\times 25}{\frac{5}{4} = \frac{125}{100}}$ Since $4 \times 25 = 100$, multiply 5 by 25 to find an equivalent fraction.

So, $1\frac{1}{4}$ is $\frac{125}{100}$ or 125%.

Got it? Do these problems to find out.

Write each mixed number as a percent.

e. $2\frac{9}{10}$ f. $3\frac{2}{5}$

Show your work.

e. _____

f. _____

Examples

5. Write 1.68 as a percent.

$1.68 = 1.68$ Multiply by 100.

$= 168\%$ Add % symbol.

 and Reflect

Is the decimal 6.7 equal to 67%? Explain below.

6. Write 0.0075 as a percent.

$0.0075 = 0.0075$ Multiply by 100.

$= 0.75\%$ Add % symbol.

Got it? Do these problems to find out.

g. 2.5   h. 0.004   i. 0.0016

g. _____

h. _____

i. _____

Example

7. **STEM** The cheetah is the fastest land mammal in the world. The peregrine falcon is the fastest bird in the world. Its speed is 2.1 times as fast as the cheetah. Write this number as a percent.

2.1 = 2.10   Multiply by 100.

= 210%   Add % symbol.

The peregrine falcon's speed is 210% of the cheetah's speed.

Got it? Do this problem to find out.

 Show your work.

j. **STEM** The slowest land mammal is the sloth. Its speed is about 0.0016 that of a cheetah. Write this number as a percent.

j. _____

Guided Practice

Check ✓

Write each percent as a decimal and as a mixed number or fraction in simplest form. (Examples 1–3)

 Show your work.

1. 325% = _____

2. 480% = _____

3. 0.6% = _____

Write each mixed number or decimal as a percent. (Examples 4–6)

4. $1\frac{4}{5}$ = _____

5. 0.0015 = _____

6. 2.75 = _____

7. A manufacturing company finds that 0.0019 of the light bulbs it makes are defective. Write this as a percent. (Example 7) _____

_____

8. **ⓔ** **Building on the Essential Question** How are percents greater than 100% used in real-world contexts?

_____

_____

Rate Yourself!

Are you ready to move on? Shade the section that applies.

I have a few questions. | I'm ready to move on.

I have a lot of questions.

For more help, go online to access a Personal Tutor.

Tutor

Name _____ My Homework _____

Write each percent as a decimal and as a mixed number or fraction in simplest form. (Examples 1–3)

 Show your work.

1. 350% = _____

2. 600% = _____

3 0.15% = _____

4. 0.55% = _____

Write each mixed number as a percent. (Example 4)

5. $2\frac{1}{2}$ = _____

6. $9\frac{3}{4}$ = _____

7. $4\frac{1}{5}$ = _____

8. $7\frac{3}{10}$ = _____

Write each decimal as a percent. (Examples 5 and 6)

9. 8.5 = _____

10. 2.64 = _____

11. 0.009 = _____

12. 0.0034 = _____

13 The size of a large milkshake is 1.4 times the size of a medium milkshake. Write 1.4 as a percent. (Example 7)

_____

14. **STEM** Fresh water from lakes accounts for only 0.001 of the world's water supply. Write this decimal as a percent. (Example 7)

_____

15. In a recent year, the United States Census Bureau reported that 0.3% of the population in the United States was Japanese. Write this percent as a decimal and as a fraction. Then interpret its meaning as a ratio of the United States population.

_____

16. Adrienne answered all 21 multiple-choice questions correctly on her science test. If her teacher decided to let one of the questions count as a bonus, worth the same number of points as the other problems on the test, what was Adrienne's test score? Write your answer as a decimal and as a percent.

_____

17. **Use Math Tools** Refer to the table at the right.

   a. Write the percent of magnesium found in the human body as a decimal.

   _____

   _____

   b. Which element makes up $\frac{1}{400}$ of the human body?

   _____

   _____

Elements in the Human Body

Element	Percent
Magnesium	0.05
Potassium	0.35
Sodium	0.15
Sulfur	0.25

H.O.T. Problems Higher Order Thinking

18. **MP Find the Error** Raj is writing $\frac{3}{2,000}$ as a percent. Find his mistake and correct it.

   _____

   _____

   _____

   _____

$$\frac{3}{2,000} = 0.0015$$
$$= 15\%$$

19. **MP Persevere with Problems** The speed of a giraffe is 250% of the speed of a squirrel. If a squirrel's speed is 12 miles per hour, find the speed of a giraffe.

   _____

   _____

20. **MP Model with Mathematics** Write a real-world problem involving a

   percent greater than 100%. Then solve the problem. _____

   _____

   _____

21. **MP Reason Inductively** Explain how you would show 135% on a 10-by-10

   grid. Then use the grids below to show 135%. _____

   _____

   _____

Extra Practice

Write each percent as a decimal and as a mixed number or fraction in simplest form.

22. 475% = ___4.75; $4\frac{3}{4}$___

$$475\% = \frac{475}{100}$$
$$= 4\frac{75}{100} \text{ or } 4\frac{3}{4}$$
$$= 4.75$$

23. 400% = _____

24. 0.05% = _____

25. 0.04% = _____

Write each decimal as a percent.

26. 1.07 = _____

27. 35 = _____

28. 0.003 = _____

29. 0.0077 = _____

30. A collectible action figure sold for 193% of its original price. Write this percent as a decimal and as a mixed number or fraction in simplest form.

_____

_____

31. A car's tire pressure decreased by 0.098 of its original pressure. Write 0.098 as a percent.

_____

Write each percent as a decimal.

32. $\frac{3}{4}\% = $ _____

33. $\frac{3}{25}\% = $ _____

MP Use Math Tools One complete figure represents 100%. Write a percent to represent the shaded portion of each figure below.

34. _____

35. _____

36. Shade the model to show 140%.

37. About 0.036% of the water on Earth is found in lakes and rivers. Which of the following show 0.036% written as a fraction? Select all that apply.

☐ $\frac{9}{25}$ ☐ $\frac{36}{100}$ ☐ $\frac{18}{50,000}$ ☐ $\frac{27}{75,000}$

What is 0.036% written as a fraction in simplest form? ☐

Spiral Review

Compare the fractions using <, >, or =. **4.NF.2**

38. $\frac{3}{6}$ ◯ $\frac{1}{8}$

39. $\frac{10}{17}$ ◯ $\frac{11}{12}$

40. $\frac{7}{9}$ ◯ $\frac{5}{11}$

41. Evangeline walked $\frac{3}{10}$ of a mile on Monday, $\frac{5}{10}$ of a mile on Tuesday, and $\frac{25}{100}$ of a mile on Wednesday. Plot each distance on the number line. **4.NF.6**

0   0.1   0.2   0.3   0.4   0.5   0.6   0.7   0.8   0.9   1

42. The flute players are $\frac{3}{10}$ of the band and the trumpet players are $\frac{1}{12}$ of the band. Is a greater fraction of the band flute players or trumpet players? **4.NF.2**

_____

_____

Problem-Solving Investigation
Solve a Simpler Problem

Content Standards
6.RP.3, 6.RP.3c

MP **Mathematical Practices**
1, 3, 4

Case #1 First Place Pizza

The daily lunch report indicated that 80% of the 300 students at Midtown Middle School chose pizza for lunch.

How many students bought pizza for lunch?

Understand *What are the facts?*

- The lunch report says 80% chose pizza.
- There are 300 students at the school.

Plan *What is your strategy to solve this problem?*

Solve a simpler problem by finding 10% of the total students. Then use the result to find 80% of the total students.

Solve *How can you apply the strategy?*

Complete the bar diagram. Fill in the value of each section.

|---------------- 300 students ------------------|

0% 10% 20% 30% 40% 50% 60% 70% 80% 90% 100%

There are 300 ÷ 10, or 10 groups with ☐ students in each group.

Multiply. ☐ × 8 = ☐

So, ☐ students chose pizza for lunch.

Check *Does the answer make sense?*

You know that 80% is close to 75%, which is $\frac{3}{4}$. Since $\frac{1}{4}$ of 300 is 75, $\frac{3}{4}$ of 300 is ☐. So, my answer is reasonable.

Analyze the Strategy ▶ Watch 💬 Tutor

MP **Reason Inductively** Explain when you would use the *solve a simpler problem* strategy. _____

_____

Top Tip

Heidi's dad wants to leave an 18% tip for a $24.60 restaurant bill.

```
├------------- $24.60 --------------┤
   ┌──┬──┬──┬──┬──┬──┬──┬──┬──┬──┐
   │  ┊  │  │  │  │  │  │  │  │  │
   └──┴──┴──┴──┴──┴──┴──┴──┴──┴──┘
  0% 10% 20% 30% 40% 50% 60% 70% 80% 90% 100%
```

About how much money should he leave?

1 Understand

Read the problem. What are you being asked to find?

I need to estimate _____.

Underline key words and values. What information do you know?

Heidi's dad wants to leave an _____ on a _____ bill.

Is there any information that you do *not* need to know?

I do not need to know _____.

2 Plan

Choose a problem-solving strategy.

I will use the _____ strategy.

3 Solve

Use your problem-solving strategy to solve the problem.

Solve a simpler problem by finding 20% of $25.00. Use the result to estimate

18%. The whole is _____. Make a bar diagram that is divided into _____.

Each part represents _____. The two shaded parts represent _____.

```
 10% 10% 10% 10% 10% 10% 10% 10% 10% 10%
 ┌──┬──┬──┬──┬──┬──┬──┬──┬──┬──┐
 │  │  │  │  │  │  │  │  │  │  │
 └──┴──┴──┴──┴──┴──┴──┴──┴──┴──┘
 └─────────────┬─────────────┘
              $25
```

Because the whole is $25.00,

each part is _____.

So, 18% of $24.60 is about _____.

4 Check

Use information from the problem to check your answer.

0.18 × 24.60 = _____. So, $5 is a reasonable estimate.

**Work with a small group to solve the following cases.
Show your work on a separate piece of paper.**

Case #3 Time

Three 24-hour clocks show the time to be 12 noon. One of the clocks is always correct, one loses a minute every 24 hours, and one gains a minute every 24 hours.

How many days will pass before all three clocks show the correct time again?

Case #4 Number Sense

The number 272 is a palindrome because it reads the same forward or backward.

How many numbers from 10 to 1,000 are palindromes?

Case #5 Sandwiches

In one day, a restaurant made the sandwiches shown in the table.

How many sandwiches are chicken?

Type	Number	Percent
Veggie	60	5
Cheese	204	17
Turkey	348	29
Chicken	?	35
Roast Beef	168	14

Case #6 Border

Part of a strip of border for a bulletin board is shown. All of the sections of the border are the same width.

If the first shape on the strip is a triangle and the strip is 74 inches long, what is the last shape on the strip?

Use any strategy!

Mid-Chapter Check

Vocabulary Check

1. Define *percent.* Write $\frac{25}{100}$ as a percent then write $\frac{25}{100}$ as a decimal. (Lesson 2)

_____

_____

Skills Check and Problem Solving

Write each fraction as a decimal and each decimal as a fraction in simplest form. (Lesson 1)

2. $\frac{8}{20}$ = _____

3. 0.64 = _____

4. $\frac{3}{100}$ = _____

Write each percent as a decimal and each decimal as a percent. (Lessons 3 and 4)

5. 73% = _____

6. 0.1 = _____

7. 254% = _____

8. The number of chorus students increased by a factor of 1.2 from the previous year. Write 1.2 as a percent. (Lesson 4)

_____

9. **MP Use Math Tools** The graph shows the pie sales during one week for Polly's Pies. (Lessons 2 and 3)

   a. What fraction of the pies sold was apple?

   _____

   b. Write the percent of strawberry pies sold as a decimal.

   _____

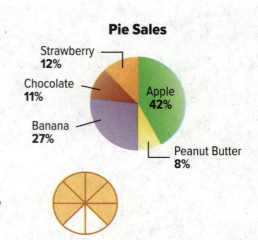

Pie Sales

Strawberry 12%
Chocolate 11%
Banana 27%
Apple 42%
Peanut Butter 8%

10. **MP Persevere with Problems** The circle at the right is divided into sections of equal size. What percent of the circle is not shaded? (Lesson 2)

_____

Compare and Order Fractions, Decimals, and Percents

Vocabulary Start-Up

The **least common denominator**, or LCD, is the least common multiple of the denominators of two or more fractions.

Complete the graphic organizer. Write the meaning of each word in the appropriate box. Provide examples.

Least	Common
_____	_____
_____	_____
_____	_____

least common denominator

Denominator	Multiple
_____	_____
_____	_____
_____	_____

Essential Question

WHEN is it better to use a fraction, a decimal, or a percent?

Vocabulary

least common denominator (LCD)

Content Standards
Preparation for 6.RP.3c

MP Mathematical Practices
1, 2, 3, 4, 5, 6

 ### Real-World Link

• Earnest is baking, but he wants to use only one measuring cup. He needs $\frac{1}{2}$ cup of sugar and $\frac{3}{4}$ cup of flour. What is the least common multiple of the denominators? ☐

• What size measuring cup should he use: $\frac{1}{2}$ cup, $\frac{1}{3}$ cup, or $\frac{1}{4}$ cup? Explain. _____

_____

Which MP Mathematical Practices did you use?
Shade the circle(s) that applies.

① Persevere with Problems
② Reason Abstractly
③ Construct an Argument
④ Model with Mathematics
⑤ Use Math Tools
⑥ Attend to Precision
⑦ Make Use of Structure
⑧ Use Repeated Reasoning

Compare and Order Fractions

To compare fractions, you can follow these steps.

1. Find the least common denominator (LCD) of the fractions. That is, find the least common multiple of the denominators.

2. Write an equivalent fraction for each fraction using the LCD.

3. Compare the numerators.

Tutor

Example

Fill in each ◯ with <, >, or = to make a true statement.

1. $\frac{5}{8}$ ◯ $\frac{7}{12}$

The LCM of the denominators, 8 and 12, is 24. So, the LCD is 24. Write an equivalent fraction with a denominator of 24 for each fraction.

$$\overset{\times 3}{\frac{5}{8}} = \frac{15}{24} \qquad \overset{\times 2}{\frac{7}{12}} = \frac{14}{24}$$
$$\underset{\times 3}{\qquad} \qquad \underset{\times 2}{\qquad}$$

$\frac{15}{24} > \frac{14}{24}$, since 15 > 14. So, $\frac{5}{8} > \frac{7}{12}$.

Got it? Do these problems to find out.

a. $\frac{2}{3}$ ◯ $\frac{4}{9}$ **b.** $\frac{5}{12}$ ◯ $\frac{7}{8}$ **c.** $\frac{1}{6}$ ◯ $\frac{5}{18}$

Least Common Multiple

$2 = \boxed{2}$
$14 = \boxed{2} \quad \times \boxed{7}$
$4 = \boxed{2} \times 2$
$7 = \qquad \boxed{7}$

The LCM is $2 \times 2 \times 7$ or 28.

Example

2. Order the fractions $\frac{1}{2}, \frac{9}{14}, \frac{3}{4}$, and $\frac{5}{7}$ from least to greatest.

Rewrite each fraction using the LCD of 28.

$$\overset{\times 14}{\frac{1}{2}} = \frac{14}{28} \quad \overset{\times 2}{\frac{9}{14}} = \frac{18}{28} \quad \overset{\times 7}{\frac{3}{4}} = \frac{21}{28} \quad \overset{\times 4}{\frac{5}{7}} = \frac{20}{28}$$
$$\underset{\times 14}{\qquad} \quad \underset{\times 2}{\qquad} \quad \underset{\times 7}{\qquad} \quad \underset{\times 4}{\qquad}$$

Since $\frac{14}{28} < \frac{18}{28} < \frac{20}{28} < \frac{21}{28}$, the order of the original fractions from least to greatest is $\frac{1}{2}, \frac{9}{14}, \frac{5}{7}, \frac{3}{4}$.

Got it? Do this problem to find out.

Show your work.

d. Order $\frac{1}{2}, \frac{5}{6}, \frac{2}{3}$, and $\frac{3}{5}$ from least to greatest.

d. _____

Compare Fractions, Decimals, and Percents

It may be easier to compare fractions, decimals, and percents when they are all written as decimals.

$\frac{1}{5} = 0.2 = 20\%$	$\frac{2}{5} = 0.4 = 40\%$	$\frac{3}{5} = 0.6 = 60\%$	$\frac{4}{5} = 0.8 = 80\%$
$\frac{1}{8} = 0.125 = 12.5\%$	$\frac{3}{8} = 0.375 = 37.5\%$	$\frac{1}{3} = 0.\overline{3} = 33.\overline{3}\%$	$\frac{2}{3} = 0.\overline{6} = 66.\overline{6}\%$

Examples

Tutor

Fill in each ◯ with <, >, or = to make a true statement.

3. $\frac{3}{4}$ ◯ 0.7

$\frac{3}{4}$ ◯ 0.7   Write the sentence.

0.75 ◯ 0.70   Write $\frac{3}{4}$ as a decimal, Annex a zero to 0.7.

0.75 > 0.70   Compare the hundredths place. 5 > 0

Since 0.75 is to the right of 0.7 on the number line, $\frac{3}{4}$ > 0.7.

4. Lucita made 85% of her free throws. Henri made $\frac{7}{8}$ of his free throws. Who has the better average? Explain.

85% ◯ $\frac{7}{8}$ Write the sentence.

0.850 ◯ 0.875   Write each number as a decimal. Annex a zero to 0.85.

0.850 < 0.875   Compare the hundredths place. 5 < 7

Since 0.850 < 0.875, Henri has the better average.

Check

Since 0.85 is to the left of 0.875, the answer is correct. ✔

Got it? Do these problems to find out.

e. $\frac{2}{3}$ ◯ 0.6

f. 0.7 ◯ $\frac{8}{11}$

g. $\frac{1}{5}$ ◯ 0.2

h. 42% ◯ 0.44

i. 7% ◯ $\frac{7}{10}$

j. 6.5 ◯ 650%

Example Tutor

5. The table shows the school carnival attendance. Which grade has the greatest part of the class attending the carnival?

Grade	Attendance
6	$\frac{5}{8}$
7	0.5
8	58.3%

Order the numbers from least to greatest. Express each number as a decimal with the same number of places.

$\frac{5}{8} = 0.625$ $0.5 = 0.500$ $58.3\% = 0.583$

Graph the numbers on a number line.

0.500   0.525   0.550   0.575   0.600   0.625

From least to greatest, the numbers are 0.5, 58.3%, and $\frac{5}{8}$.

Since $\frac{5}{8}$ represents Grade 6, Grade 6 has the greatest part of the class attending the school carnival.

Got it? Do this problem to find out.

Show your work.

k. _____

k. Hiroshi found that $\frac{3}{5}$ of his class prefers vanilla ice cream, 26% prefers chocolate, and 0.14 prefers strawberry. Which kind of ice cream do students prefer the least?

Guided Practice

 Check ✓

1. Order the fractions $\frac{4}{5}$, $\frac{1}{2}$, $\frac{9}{10}$, and $\frac{3}{4}$ from least to greatest. (Examples 1 and 2)

_____

Show your work.

2. Cora spends $\frac{2}{3}$ of her free time blogging on the Internet. Leah spends 60% of her free time blogging on the Internet. Who spends more of her free time blogging?

(Examples 3 and 4) _____

3. The table shows the wins for some middle school football teams. Which team has the greatest fraction of wins? (Example 5)

Team	Wins
Eagles	95%
Wolves	$\frac{9}{10}$
Mustangs	0.89

_____

4. **Building on the Essential Question** How do you compare fractions, decimals, and percents?

_____

_____

_____

 Rate Yourself!

Are you ready to move on? Shade the section that applies.

YES   ?   NO

For more help, go online to access a Personal Tutor.

 Tutor

Name _____   My Homework _____

Independent Practice

Go online for Step-by-Step Solutions

Fill in each ◯ **with <, >, or = to make a true statement.** (Examples 1 and 3)

Show your work.

1. $\frac{1}{3}$ ◯ $\frac{3}{5}$

2. $\frac{7}{12}$ ◯ $\frac{1}{2}$

3 $\frac{1}{4}$ ◯ 0.4

4. 0.7 ◯ $\frac{7}{9}$

Order the fractions from least to greatest. (Example 2)

5. $\frac{1}{2}, \frac{2}{3}, \frac{1}{4}, \frac{5}{6}$

_____

6. $\frac{2}{3}, \frac{2}{9}, \frac{5}{6}, \frac{11}{18}$

_____

7. Darius spends 35% of his time doing math homework. Alex spends $\frac{2}{5}$ of his time doing math homework. Who spends more homework time on math? Explain. (Example 4)

_____

_____

8. Three snack bars contain $\frac{1}{5}$, 0.22, and 19% of their Calories from fat. Which snack bar contains the least amount of Calories from fat? (Example 5)

_____

_____

9. **MP** **Model with Mathematics** Use the graphic novel frame below for Exercises a–b.

a. Write each score as a decimal. _____

b. Compare the three scores. _____

10. **MP** **Be Precise** Complete the graphic organizer. Write the original numbers to complete the statement.

Number	Steps to Write the Number as a Decimal with Three Places	Decimal
$\frac{3}{8}$	Divide the _____ by the _____.	0.375
0.3	The number is a decimal. Annex _____ zeros.	0.300
38.7%	Move the _____ point _____ places to the left. Remove the _____ symbol.	0.387

So, _____ < _____ < _____.

11. Order the portion of responses listed in the table from least to greatest.

_____

Number of Times Eating Fast Food per Week	0	1–2	3–4	5+
Portion of Responses	17%	$\frac{11}{20}$	0.2	8%

H.O.T. Problems Higher Order Thinking

12. **MP** **Reason Abstractly** Specify three fractions with different denominators that have an LCD of 24. Then arrange the fractions in order from least to greatest.

_____

13. **MP** **Persevere with Problems** Order $\frac{3}{8}, \frac{3}{7}$, and $\frac{3}{9}$ from least to greatest without writing equivalent fractions with a common denominator. Explain your strategy. _____

_____

14. **MP** **Persevere with Problems** Are the fractions $\frac{3}{9}, \frac{3}{10}, \frac{3}{11}$, and $\frac{3}{12}$ arranged in order from least to greatest or from greatest to least? Explain.

_____

_____

15. **MP** **Construct an Argument** Is 0.4 less than, greater than, or equal to 44%? Explain your reasoning. _____

_____

Extra Practice

Fill in each ◯ with <, >, or = to make a true statement.

16. $\frac{7}{8}$ ⟩ $\frac{5}{6}$

$\frac{7}{8} = \frac{21}{24}; \frac{5}{6} = \frac{20}{24}$

$\frac{21}{24} > \frac{20}{24}$ so $\frac{7}{8} > \frac{5}{6}$

Homework Help

17. $\frac{14}{18}$ ◯ $\frac{7}{9}$

18. 0.75 ◯ $\frac{1}{2}$

19. $\frac{1}{3}$ ◯ 0.33

Order the fractions from least to greatest.

20. $\frac{1}{6}, \frac{2}{5}, \frac{3}{5}, \frac{3}{7}$

21. $\frac{5}{8}, \frac{3}{4}, \frac{1}{2}, \frac{9}{16}$

_____

22. Shop Rite has jeans on sale for $\frac{3}{10}$ off. Save More has jeans on sale for 33% off. Which store has a better sale on jeans? Explain.

_____

_____

23. A city's population rose 3% one year, 0.08 the next year, and by $\frac{2}{50}$ the next year. Order these increases from least to greatest.

_____

_____

Order each set of values from least to greatest.

24. 0.4, $\frac{5}{8}$, 38%

25. $\frac{1}{2}$, 0.55, $\frac{5}{7}$

_____

26. **MP** **Use Math Tools** The table shows the favorite subjects of students in a recent survey.

a. Did more students choose art or math? Explain.

_____

_____

b. Which subject did most students choose? Explain.

_____

_____

c. Order the subjects from least to greatest.

_____

_____

Favorite Subject	
Subject	**Portion of Students**
Art	$\frac{4}{25}$
English	13%
Math	0.28
Other	7%
Science	$\frac{21}{100}$
Social Studies	0.15

27. A plumber needs to drill a hole that is just slightly larger than $\frac{3}{16}$ inch in diameter. Which measure is the smallest but still larger than $\frac{3}{16}$ inch?

☐ $\frac{5}{32}$ inch   ☐ $\frac{5}{16}$ inch   ☐ $\frac{13}{64}$ inch   ☐ $\frac{9}{32}$ inch

28. Fairview Elementary started a recyling program. The display shows the portion of the each item that is recycled at the school. Sort the items from least to greatest amount.

PLASTIC 0.16

PAPER 50%

ALUMINUM 0.24

GLASS $\frac{1}{10}$

	Recycled Item
Least	
Greatest	

Spiral Review

Round each decimal to the nearest hundredth. 5.NBT.4

29. $0.623 \approx$ _____

30. $4.288 \approx$ _____

31. $5.105 \approx$ _____

32. In a survey, $\frac{9}{25}$ of students ride the bus to school and $\frac{19}{50}$ walk to school. What fraction of students ride the bus or walk to school? **4.NF.3d**

_____

33. The student council bought 7 bags of apples for their fall party. How much did they pay for the apples? **5.NBT.7**

_____

_____

$3.45

Estimate with Percents

 Real-World Link

Movies Josefina surveyed 298 students and found that 52% like scary movies. Estimate the number of students that like scary movies.

 Essential Question

WHEN is it better to use a fraction, a decimal, or a percent?

Content Standards
6.RP.3, 6.RP.3c

MP **Mathematical Practices**
1, 3, 4, 5

1. Write the common percents from 0% to 100% at the top of the bar diagram.

2. What common percent is 52% close to? ☐
   Shade the bar diagram above to show your answer.

3. Round 298 to the nearest hundred. 298 ≈ ☐
   Write your answer in the box below 100%.

4. Use the bar diagram to estimate 52% of 298. Explain.

   _____

   _____

   _____

   _____

5. Use the bar diagram below to estimate 73% of 400. _____

Help!

Which MP Mathematical Practices did you use?
Shade the circle(s) that applies.

① Persevere with Problems   ⑤ Use Math Tools

② Reason Abstractly   ⑥ Attend to Precision

③ Construct an Argument   ⑦ Make Use of Structure

④ Model with Mathematics   ⑧ Use Repeated Reasoning

Estimate the Percent of a Number

Estimating with percents will provide a reasonable solution to many real-world problems. Choose compatible numbers when estimating the percent of a number.

Examples

1. **Estimate 47% of 692.**

47% is close to 50% or $\frac{1}{2}$. Round 692 to 700.

$\frac{1}{2}$ of 700 is 350. $\frac{1}{2}$ or *half* means to divide by 2.

0%	50%	100%
0	350	700

So, 47% of 692 is about 350.

2. **Estimate 60% of 27.**

60% is $\frac{3}{5}$.

Round 27 to 25 since it is divisible by 5.

$\frac{1}{5}$ of 25 is 5. $\frac{1}{5}$, or *one fifth,* means divide by 5.

0%	20%	40%	60%	80%	100%
0	5	10	15	20	25

Show your work.

So, $\frac{3}{5}$ of 25 is 3 × 5 or 15.

So, 60% of 27 is about 15.

Got it? Do these problems to find out.

Estimate each percent.

   a. 48% of 76   **b.** 18% of 42   **c.** 73% of 41

a. _____

b. _____

c. _____

Example

Tutor

3. **STEM** Polar bears can eat as much as 10% of their body weight in less than one hour. If an adult male polar bear weighs 715 pounds, about how much food can he eat in one hour?

To determine how much food a polar bear can eat in one hour, you need to estimate 10% of 715.

Method 1 Find equivalent ratios.

$10\% = \frac{1}{10}$ and $715 \approx 700$

$\frac{1}{10} = \frac{\blacksquare}{700}$ Write the equivalent ratios.

$$\overset{\times 70}{\frac{1}{10} = \frac{\blacksquare}{700}}_{\times 70}$$ Since $10 \times 70 = 700$, multiply 1 by 70.

The unknown value is 70.

Method 2 Use mental math.

$10\% = \frac{1}{10}$ and $715 \approx 700$

$\frac{1}{10}$ of 700 is 70.

So, a polar bear can eat about 70 pounds of food in one hour.

Got it? Do this problem to find out.

Show your work.

d. Kayleigh decided to donate 30% of her savings. If she has $238 in her savings account, about how much will she donate?

d. _____

Estimate Using the Rate per 100

You can also estimate with percents using a rate per 100.

Examples

Tutor

4. **Estimate 17% of 198.**

17% = 17 out of 100   Write the percent as a rate per 100.

198 ≈ 200   Round to the nearest hundred.

Since 200 is 100 + 100, add 17 + 17 to estimate 17% of 198.

34 is about 17% of 198.

5. An airline records the snack orders of passengers. Last year 9% of all passengers ordered ginger ale to drink. There are 408 passengers on the flight to Houston, Texas. About how many passengers does the airline expect to order ginger ale on this flight?

Estimate 9% of 408.

9% = 9 out of 100   *Write the percent as a rate per 100.*

408 ≈ 400   *Round to the nearest hundred.*

Since 400 is 100 × 4, multiply 9 × 4 to estimate 9% of 408.

36 is about 9% of 408. So, about 36 passengers will order ginger ale.

Show your work.

Got it? Do these problems to find out.

e. _____

Estimate using a rate per 100.

   e. 27% of 307   **f.** 76% of 192

f. _____

   g. Last year 24% of the zoo visitors were under the age of 3. Last week, the zoo had 996 visitors. About how many of the zoo visitors were under the age of 3?

g. _____

Guided Practice

Check ✓

Estimate each percent. (Examples 1 and 2)

Show your work.

1. 19% of $53 ≈ _____

2. 21% of 96 ≈ _____

3. 59% of 16 ≈ _____

4. A purse that originally cost $29.99 is on sale for 50% off. About how much is the sale price of the purse? (Example 3)

_____

5. Mr. Marcucci received a bonus of $496 from his employer. He has to pay 33% of his bonus to taxes. How much will Mr. Marcucci pay in taxes? (Examples 4 and 5)

_____

6. ⓔ **Building on the Essential Question** When is an estimate more useful than an exact answer?

_____

_____

Rate Yourself!

How confident are you about estimating with percents? Shade the ring on the target.

I'm on target.

I need help.

For more help, go online to access a Personal Tutor.

Tutor

Independent Practice

Go online for Step-by-Step Solutions

Estimate each percent. (Examples 1 and 2)

1. 47% of $118 ≈

_____

 Show your work.

2. 19% of 72 ≈

_____

3. 42% of 16 ≈

_____

4. 67% of 296 ≈

_____

Estimate using a rate per 100. (Example 4)

5. 24% of 289 ≈ _____

6. 67% of 208 ≈ _____

7. **STEM** Penguins spend almost 75% of their lives in the sea. An Emperor Penguin in the wild has a life span of about 18 years. About how many years does this penguin spend in the sea? (Example 3)

_____

8. In Nathan's baseball card collection, 58% of the cards are players from the National League. He has 702 baseball cards. About how many baseball cards are players from the National League? Use a rate per 100 to estimate. (Example 5)

_____

9. **MP** **Model with Mathematics** Refer to the graphic novel frame below for Exercises a–b.

a. Suppose Angel is shooting baskets and makes 40% of the 15 shots. Does he win a prize? Explain your reasoning.

_____

b. About what percent of the baskets need to be made in order to win a prize? _____

10. About 42% of Alaska's population lives in the city of Anchorage. If Alaska has a total population of 648,818, about how many people live in Anchorage?

11 During the basketball season, Tyrone made 37 baskets out of 71 attempts. About what percent of his shots did he miss?

MP **Use Math Tools** **Estimate the percent that is shaded in each figure.**

12.

13.

14.

H.O.T. Problems   Higher Order Thinking

15. **MP** **Reason Inductively**   Rachel wants to buy a shirt regularly priced at $32. It is on sale for 40% off. Rachel estimates that she will save $\frac{2}{5}$ of $30 or $12. Will the actual amount be more or less than $12? Explain.

16. **MP** **Persevere with Problems**   Order 10% of 20, 20% of 20, and $\frac{1}{5}$% of 20 from least to greatest.

17. **MP** **Construct an Argument**   A classmate is trying to estimate 42% of $122. Explain how your classmate should solve the problem.

18. **MP** **Model with Mathematics**   Melissa's homeroom has raised 63% of its goal for the school fundraiser. Matt's homeroom has raised 48%. Create a situation in which Matt's homeroom raised more money than Melissa's homeroom.

Extra Practice

Estimate each percent.

19. 53% of 59 ≈

$\frac{1}{2}$ of 60 is 30.

Homework Help →

53% is close to 50% or $\frac{1}{2}$. Round 59 to 60.

20. 35% of 147 ≈

21. 26% of 125 ≈

22. 79% of 82 ≈

Estimate using a rate per 100.

23. 19% of 288 ≈

24. 74% of 315 ≈

25. 61% of 407 ≈

26. 89% of 195 ≈

27. Trevon spent 8 hours and 15 minutes at an amusement park yesterday. He spent 75% of the time at the park on rides. About how much time did he spend on rides?

28. A group of friends went on a hiking trip. They planned to hike a total of 38 miles. They want to complete 25% of the hike by the end of the first day. About how many miles should they hike the first day?

29. Briana has just finished her sixth grade scrapbook. In her scrapbook, 47% of the pages include her twin sister, Bethany. The scrapbook has 896 photos. About how many photos include Bethany? Use a rate per 100 to estimate.

30. The community garden has 596 vegetables. In the garden, 64% of the vegetables are green vegetables. About how many vegetables in the garden are green? Use a rate per 100 to estimate.

MP **Use Math Tools** Estimate the percent that is shaded in each figure.

31.

32.

33. In a survey of teens, 21% said their friends like to read and talk about books. About how many teens out of 1,095 would say their friends read and talk about books?

34. Everett wants to buy the items shown in the table. He has a coupon that will take 20% off his total purchase and he has $50 to spend. What items can he purchase to spend the most of the $50, after applying the coupon? Explain.

Item	Cost ($)
jeans	25
khaki pants	20
package of socks	10
polo shirt	15
3 T-shirts	15

Spiral Review

Write each fraction as a decimal. 4.NF.6

35. $\frac{22}{100} =$ _____

36. $\frac{7}{100} =$ _____

37. $\frac{67}{100} =$ _____

38. $\frac{15}{100} =$ _____

39. $\frac{12}{100} =$ _____

40. $\frac{6}{100} =$ _____

41. At a clothing store, T-shirts are on sale for $9.97 each. What is the cost of 3 T-shirts? **5.NBT.7** _____

42. The Sylvester family planted a garden with the dimensions shown. What is the area of the garden? **5.NBT.7**

7.5 ft

3 ft

_____

Inquiry Lab
Percent of a Number

Content Standards
6.RP.3, 6.RP.3c

MP **Mathematical Practices**
1, 3, 4

 HOW can you model the percent of a number?

There were 180 people in a movie theater. Twenty percent of them received the student discount and 10% received the senior citizen discount. The rest did not receive a discount. How many people did not receive a discount?

What do you know? _____

_____

What do you need to find? _____

_____

Hands-On Activity

Model the situation using two bar diagrams.

Step 1   Use a bar diagram to represent 100%. Then use another bar diagram of equal length to represent 180 people.

	student discount		senior discount								
percent	10%	10%	10%	10%	10%	10%	10%	10%	10%	10%	100%
people	18	18	18	18	18	18	18	18	18	18	180

Step 2   Divide each bar into 10 equal parts. Think: 180 ÷ 10 = ☐

So, each part of 180 represents ☐ people.

Step 3   Determine how many people did not receive a discount. Shade 2 sections of each bar diagram to represent the student discount. Shade 1 section of each bar diagram to represent the senior discount.

There are ☐ unshaded sections in each bar diagram.

☐ × ☐ = ☐

So, ☐ people at the movie did not receive a discount.

MP Model with Mathematics Work with a partner. Find the part of each whole using two bar diagrams.

1. 50% of 80 children = _____

		100%
		children

2. 25% of $32 = _____

Analyze and Reflect

3. **MP Reason Inductively** Explain how to use two bar diagrams to find 45% of $60.

_____

_____

_____

Create

4. **MP Model with Mathematics** Write and solve a real-world problem about finding the percent of a number.

_____

_____

_____

5. **Inquiry** HOW can you model the percent of a number?

_____

Percent of a Number

 Real-World Link

Snacks In a survey, 200 students chose their favorite snacks. Use the table to find the number of students who chose each snack.

Snack	Percent	Fraction	Equivalent Fraction	Number of Responses
Fruit	23%	$\frac{23}{100}$	$\frac{46}{200}$	46 out of 200
Cheese	15%	$\frac{}{100}$	$\frac{}{200}$	☐ out of 200
Veggies	17%	$\frac{}{100}$	$\frac{}{200}$	☐ out of 200
Cookies	15%	$\frac{}{100}$	$\frac{}{200}$	☐ out of 200
Chips	18%	$\frac{}{100}$	$\frac{}{200}$	☐ out of 200
No Snack	12%	$\frac{}{100}$	$\frac{}{200}$	☐ out of 200

Check Add the number of responses in the last column.

46 + ☐ + ☐ + ☐ + ☐ + ☐ = 200 ✓

1. How does finding the percent as a rate per 100 help you find the number of responses out of 200?

_____

_____

_____

 Essential Question

WHEN is it better to use a fraction, a decimal, or a percent?

Content Standards
6.RP.3, 6.RP.3c

MP **Mathematical Practices**
1, 3, 4, 5

Which **MP** **Mathematical Practices** did you use?
Shade the circle(s) that applies.

① Persevere with Problems

② Reason Abstractly

③ Construct an Argument

④ Model with Mathematics

⑤ Use Math Tools

⑥ Attend to Precision

⑦ Make Use of Structure

⑧ Use Repeated Reasoning

Find the Percent of a Number

You can use fractions and decimals to find the percent of a number. To find the percent of a number, write the percent as a fraction with a denominator of 100. Then multiply the fraction by the number.

Example

1. **Refer to the circle graph. Suppose there are 300 students at York Middle School. Find the number of students that have cheese as a snack.**

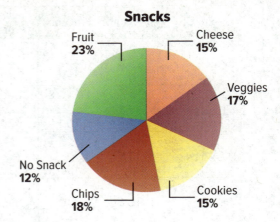

Snacks

Fruit 23%
Cheese 15%
Veggies 17%
No Snack 12%
Chips 18%
Cookies 15%

Method 1   **Write the percent as a fraction.**

$15\% = \dfrac{15}{100}$ or $\dfrac{3}{20}$ *Write the percent as a rate per 100.*

$\dfrac{3}{20}$ of $300 = \dfrac{3}{20} \times 300$ *Multiply.*

$\qquad\qquad = 45$

Method 2   **Write the percent as a decimal.**

$15\% = 0.15$

0.15 of $300 = 0.15 \times 300$

$\qquad\qquad = 45$

So, 45 students have cheese as a snack.

Check   Use a bar diagram.

| 30 | 15 | | | | | | | | | | 300 |

$30 + \dfrac{1}{2} \times 30 = 30 + 15$ or 45 ✔

Got it?   Do this problem to find out.

a. Find the number of students at York Middle School that have chips as a snack.

Percents Greater Than 100% and Less Than 1%

You may choose whether to write a percent as a fraction or as a decimal based on the problem.

Examples

Tutor

2. **Find 145% of 320.**

$145\% = \dfrac{145}{100}$ or $\dfrac{29}{20}$ — Write 145% as a rate per 100. Then simplify.

145% of $320 = \dfrac{29}{20} \times 320$ — Write the multiplication problem.

$\qquad = \dfrac{29}{20} \times \dfrac{\overset{16}{\cancel{320}}}{1}$ — Divide the numerator and denominator by 20.

$\qquad\quad \underset{1}{}$

$\qquad = 29 \times 16$ — Simplify.

$\qquad = 464$ — Multiply.

So, 145% of 320 is 464.

- -

3. **Find 220% of 65.**

$220\% = \dfrac{220}{100}$ or $\dfrac{\boxed{}}{\boxed{}}$ — Write 220% as a fraction in simplest form.

220% of $65 = \dfrac{\boxed{}}{\boxed{}} \times \boxed{}$ — Write the multiplication problem.

$\qquad = \underline{\hspace{2cm}}$ — Divide by the GCF.

$\qquad = \underline{\hspace{1cm}} \times \underline{\hspace{1cm}}$ — Simplify.

$\qquad = \underline{\hspace{1cm}}$ — Multiply.

So, 220% of 65 is $\underline{\hspace{1.5cm}}$.

- -

4. **Find 0.25% of 58.**

$0.25\% = 0.0025$ — Write 0.25% as a decimal.

0.25% of $58 = 0.0025 \times 58$ — Write the multiplication problem.

$\qquad\quad = 0.145$ — Multiply.

So, 0.25% of 58 is 0.145.

Got it? Do these problems to find out.

Find the percent of each number.

b. 128% of 550   **c.** 0.3% of 200   **d.** 0.85% of 600

Show your work.

b. _____

c. _____

d. _____

Tutor

Example

5. In a recent state Special Olympics meet, Franklin County sent a team of 70 players. Twenty percent of the team competed in soccer. How many athletes competed in soccer?

20% = 0.20   Write 20% as a decimal.

20% of 70 = 0.2 × 70   Write the multiplication problem.

= 14   Multiply.

So, 14 team members were soccer players.

Got it? Do this problem to find out.

 Show your work.

e. In the same meet, 15% of the team from Delaware County competed in tennis. If there were 20 members on the team, how many competed in tennis?

e. _____

Guided Practice

Check ✓

Find the percent of each number. (Examples 1–4)

 Show your work.

1. 32% of 60 = _____

2. 0.55% of 220 = _____

3. 275% of 4 = _____

4. Troy wants to buy a jersey of his favorite MLS team. The jersey is 30% off the original price. If the original price of the jersey is $35, what is the

amount Troy will save? (Example 5) _____

_____

5. **Building on the Essential Question** How do you find a percent of a number?

_____

_____

_____

Rate Yourself!

Are you ready to move on? Shade the section that applies.

YES   ?   NO

For more help, go online to access a Personal Tutor.

Tutor

FOLDABLES Time to update your Foldable!

Name _____ My Homework _____

The cafeteria at Midtown Middle School surveyed 575 students about their favorite food. Find the number of students that responded for each of the following. (Example 1)

1. chicken: 8% =

2. salad: 20% =

3. burgers: 16% =

4. fruit: 24% =

 Show your work.

Find the percent of each number. (Examples 2–4)

5. 0.9% of 1,000 =

6. 0.46% of 80 =

7. 350% of 96 =

8. 222% of 55 =

9. The original price of a pair of shoes is $42. The sale price is 20% off the original price. What is the amount off the original price?

(Example 5) _____

_____

10. Torri had $20 to buy a birthday present for her dad. She decided to buy a DVD for $18. The sales tax is 7%. Does she have enough money? Explain your reasoning.

_____

_____

11. Twenty-four students in Jamal's class are wearing tennis shoes. There are thirty students in his class. Jamal says that 70% of his class is wearing tennis shoes. Is Jamal correct? Explain your reasoning.

_____

_____

12. **MP** **Use Math Tools** Marisol keeps track of her weekly quiz grades as shown in the table.

  a. Complete the table.

  b. In which class did Marisol have the higher score?

  _____

Test	Number Correct	Score	Total
Math	68		85
Science		90%	70

  c. Suppose Marisol scored a 96% on an English test. There were 50 questions on the test. How many did Marisol answer correctly?

  _____

13. **MP** **Use Math Tools** Use the graphic organizer to compare and contrast percents and fractions. Use the phrases *less than*, *equal to*, and *greater than* to complete each statement. Write an example in the space provided.

Percent	Shared Concept	Fraction
A whole is represented by a percent that is _____ 100%. Example: _____	whole	A whole is represented by a fraction with a numerator that is _____ the denominator. Example: _____
Part of a whole is represented by a percent that is _____ 100%. Example: _____	part of a whole	Part of a whole is represented by a fraction with a numerator that is _____ the denominator. Example: _____
An amount that is greater than one is represented by a percent that is _____ 100%. Example: _____	more than one	An amount that is greater than one is represented by a fraction with a numerator that is _____ the denominator. Example: _____

🔥 H.O.T. Problems Higher Order Thinking

14. **MP** **Model with Mathematics** Write and solve a real-world problem in which the part of a whole results in a number greater than the whole itself.

_____

_____

_____

15. **MP** **Justify Conclusions** Is 16% of 40 the same as 40% of 16? Explain your

reasoning. _____

_____

16. **MP** **Persevere with Problems** Find 15% of 15% of 15% of 500. How

does this compare to finding 45% of 500? _____

_____

17. **MP** **Persevere with Problems** A number n is 25% of some number a and 35% of a number b. Is $a > b$, $a < b$, or is it impossible to determine the relationship?

Explain. _____

_____

Extra Practice

Find the percent of each number.

18. 6% of 95 =

5.7

Homework Help 6% = 0.06

0.06 × 95 = 5.7

19. 15% of 110 =

20. 75% of 260 =

21. 28% of 575 =

22. 0.6% of 36 =

23. 108% of 148 =

24. 102% of 750 =

25. 0.03% of 1,500 =

26. Brenna completes 65% of her first serves. If she attempted 80 first serves last match, how many did she complete?

27. Jack is mixing a cleaning solution that is 12% bleach. After mixing the solution, Jack has 150 ounces of cleaning solution. How many ounces of bleach did Jack use?

28. What is 38% of 250?

29. 76% of 524 is what number?

30. What is 26% of 360?

31. 55% of 387 is what number?

32. **MP** **Use Math Tools** Mr. Blackwell tracks sales of ski equipment each week for a month. Complete the table to determine which week had the highest percent of ski equipment sales.

Week Number	Ski Equipment Sales ($)	Percent of Total Sales	Total Sales ($)
1		50	400
2	175		250
3		65	300
4	110		275

33. What expression can be used to find the total discount? Select all that apply.

☐ 0.3×42 ☐ $\frac{3}{10} \times 42$ ☐ 0.03×42 ☐ 3×4.2

34. There are 450 vehicles in a car lot. Select the correct value for the number of vehicles of each type that are in the lot.

Type of Vehicle	Percent of All Cars	Number of Cars
hybrid	28	
sport utility	20	
sedan	38	

90	171
95	180
126	200
135	

Spiral Review

Multiply. 5.NBT.7

35. $1.63 \times 20 =$ _____

36. $7.5 \times 12 =$ _____

37. $0.6 \times 15 =$ _____

38. Ella has 4 trading cards. Connor has 8 trading cards. How many times more cards does Connor have than Ella? 4.NBT.6 _____

39. The art club had the members vote on three places to take a field trip. The results are in the table. If all of the members voted, what part of the club voted for the Carnegie Museum of Art? 5.NBT.7

Trip	Part of Club
Carnegie Museum of Art	
Fallingwater	0.20
Westemoreland Museum of American Art	0.48

Solve Percent Problems

Vocabulary Start-Up

A **proportion** is an equation that shows that two ratios are equivalent. In a **percent proportion**, one ratio compares a part to the whole. The other ratio is the equivalent percent written as a fraction with a denominator of 100.

How do you compare part and whole?

fraction	ratio	percent
$\dfrac{2}{5}$ $\dfrac{\text{part}}{\text{whole}}$	Using the information in the first ratio, fill in the others.	$\dfrac{2}{5} = \dfrac{\square}{100}$
What do you call the part? _____ the whole? _____	$\dfrac{2}{5}$ $\square$ to $\square$: $\square$	$\square$ % of 5 = 2

Real-World Link

Basketball Kara is on her school basketball team. She has completed 9 out of 12 free throw shots successfully. Write the ratio as a percent and as a fraction in simplest form.

_____

 Which **MP** **Mathematical Practices** did you use?
Shade the circle(s) that applies.

① Persevere with Problems
② Reason Abstractly
③ Construct an Argument
④ Model with Mathematics
⑤ Use Math Tools
⑥ Attend to Precision
⑦ Make Use of Structure
⑧ Use Repeated Reasoning

 Essential Question

WHEN is it better to use a fraction, a decimal, or a percent?

Vocabulary

proportion
percent proportion

Content Standards
6.RP.3, 6.RP.3c

MP Mathematical Practices
1, 2, 3, 4, 7

Use Number Lines to Find the Whole

If you know the part and the percent, you can find the whole, or the total. You have used bar diagrams to solve percent problems. Double number lines are another way to illustrate percents.

Examples

Tutor

1. **10 is 25% of what number?**

Use double number lines to model 25% and 10.

To model 25%, divide the number line into four parts.

Write 10 at the 25% mark. Add 10 at each mark to find the whole.

The number 40 is at the 100% mark.

So, 10 is 25% of 40.

- -

2. **Country music makes up 75% of Landon's music library. If he has downloaded 90 country music songs, how many songs does Landon have in his music library?**

Use double number lines to model 75% and 90.

To model 75%, divide the number line into four parts.

$90 \div 3 = 30$. Add 30 at each mark to find the whole.

The number 120 is at the 100% mark.

So, Landon has 120 songs in his music library.

Check Look back at the number lines. The number 90 should line up with 75%. ✔

Show your work.

Got it? Do these problems to find out.

a. 30 is 50% of what number?   **b.** 60 is 20% of what number?

c. Peyton spent 60% of her money to buy a new television. If the television cost $300, how much money did she have?

a. _____

b. _____

c. _____

Use the Percent Proportion

The diagram uses a percent proportion to show that 75% of 32 is 24.

$$\text{part} \longrightarrow \frac{24}{32} = \left.\frac{75}{100}\right\} \text{percent}$$
$$\text{whole} \longrightarrow$$

Examples

 Tutor

3. **15 is 30% of what number?**

> | Words | 15 is 30% of what number? |
> |---|---|
> | | |
> | **Proportion** | $\dfrac{\text{part}}{\text{whole}} \longrightarrow \dfrac{15}{\blacksquare} = \left.\dfrac{30}{100}\right\} \text{percent}$ |

$\dfrac{15}{\blacksquare} = \dfrac{30}{100}$ Write the proportion.

$\dfrac{15}{50} = \dfrac{30}{100}$ Since 15 is one half of 30, divide 100 by 2.

So, 15 is 30% of 50.

· ·

4. **225 is 75% of what number?**

$\dfrac{\boxed{}}{\blacksquare} = \dfrac{\boxed{}}{100}$ Write the proportion.

$\dfrac{\boxed{}}{\boxed{}} = \dfrac{\boxed{}}{100}$ Since $75 \times \boxed{} = 225$, multiply 100 by $\boxed{}$.

So, 225 is 75% of _____.

Got it? Do these problems to find out.

d. 75 is 15% of what number?

e. 9 is 36% of what number?

f. 7 is 70% of what number?

g. 7 is 35% of what number?

 Show your work.

d. _____

e. _____

f. _____

g. _____

Example

5. Before 1982, pennies were 95% zinc and 5% copper. If 100 pennies minted in 1980 have an approximate mass of 15 grams of copper, what is the total mass of 100 pennies?

The percent is 5 and the part is 15. You need to find the whole.

$$\frac{15}{\blacksquare} = \frac{5}{100}$$ Write the proportion.

$$\frac{15}{300} = \frac{5}{100}$$ ×3   Since 5 × 3 = 15, multiply 100 by 3.

The total mass of 100 pennies is 300 grams.

Check ✓

Use double number lines to find the whole. (Example 1)

1. 40 is 20% of what number? _____

Show your work.

2. 90 is 25% of what number? _____

Write a percent proportion and solve each problem. (Examples 3 and 4)

3. 120 is 30% of what number?

_____

4. 60 is 15% of what number?

_____

5. In the first year of ownership, a new car can lose 20% of its value. If a car lost $4,200 of value in the first year, how much did the car originally cost? (Examples 2 and 5)

_____

_____

6. **Building on the Essential Question** How can you use proportions to solve percent problems?

_____

_____

Rate Yourself!

How well do you understand percent problems? Circle the image that applies.

Clear   Somewhat Clear   Not So Clear

For more help, go online to access a Personal Tutor.

Tutor

FOLDABLES Time to update your Foldable!

Independent Practice

Go online for Step-by-Step Solutions

Use double number lines to find the missing number. (Example 1)

1. 63 is 90% of what number? _____

2. 80 is 25% of what number? _____

Write a percent proportion and solve each problem. (Examples 3 and 4)

3. 22 is 44% of what number?

_____

4. 450 is 75% of what number?

_____

5. A store is having a sale where winter clothes are 60% of the original price. A sweater is on sale for $30. What was the original price of the sweater? (Examples 2 and 5)

_____

_____

6. Kai calculates that he spends 15% of a school day in science class. If he spends 75 minutes in science class, how many minutes does Kai spend in school?

(Examples 2 and 5) _____

_____

For Exercises 7–9, use the table.

7. If you have 3 cups of pineapple juice, how many total

cups of punch can you make? _____

_____

8. How many cups of sorbet are used in 8 cups of punch?

_____

_____

Punch Recipe	
Ginger Ale	40%
Orange Juice	25%
Pineapple Juice	20%
Sorbet	15%

9. Elise does not like sorbet, so she omits that ingredient and adds 5 percent of each of the other ingredients. How many cups of punch will she have if she uses

6 cups of orange juice? _____

_____

10. **MP Identify Structure** Complete the following graphic organizers. Identify the missing information.

a.
$\frac{3}{4}$	part	3
	whole	

b.
47%	part	
	whole	100%

c.
12% of 225	part	12%
	whole	

d.
120 out of 400	part	
	whole	

e. How does identifying the part and the whole help you to write the percent proportion? _____

_____

🔥 H.O.T. Problems Higher Order Thinking

11. **MP Reason Abstractly** Write a percent proportion where the part and the whole are known. Solve the problem to find the percent. _____

_____

12. **MP Persevere with Problems** Using what you know about percents, explain why a commercial that says "80% of dentists use this toothpaste" might be misleading. _____

_____

_____

_____

13. **MP Reason Inductively** The purity of gold is listed in karats. Refer to the table. If a necklace is 75% gold, what karat is it? Explain your reasoning. _____

_____

Karats	Pure Gold (%)
24	100
12	50

14. **MP Construct an Argument** Omar scored an 82% on his first test of the quarter. Will a score of 38 out of 50 on the second test help or hurt his grade? Explain your reasoning. _____

_____

15. **MP Persevere with Problems** At a zoo, an Asian elephant is about 3 tons and eats about 300 pounds of food a day. What percentage of its body weight does the elephant eat each day? _____

Extra Practice

Use double number lines to find the missing number.

16. 140 is 70% of what number? _200_

17. 240 is 60% of what number? _____

Write a percent proportion and solve each problem.

18. 95 is 95% of what number?

_____

19. 270 is 90% of what number?

_____

20. **MP Justify Conclusions** Action movies make up 85% of Devon's movie collection. If he has 20 movies, how many action movies are in Devon's collection? Explain your reasoning to a classmate.

_____

21. The drama club held a car wash on Saturday and Sunday. They washed a total of 60 cars. If they washed 40% of the cars on Sunday, how many cars did they wash on Sunday?

_____

_____

22. A tiger can eat food that weighs up to 15% of its body weight. If a tiger can eat 75 pounds of food, how much does a tiger weigh?

_____

_____

23. According to the school survey, 12% of the students at Rockwood Junior High School speak Spanish. There are 36 students at the school who speak Spanish. How many students were surveyed?

_____

_____

24. Miley's Music has a sale on music CDs. All music CDs are discounted 15%. Mariana's receipt indicates that she saved $3 on her CD purchase. What is the full price of her music CD before the discount?

_____

_____

25. The interior paint color, Melon Madness, is 30% yellow. Raul used 72 ounces of yellow paint to mix the last batch. How many ounces of Melon Madness did he make in the last batch?

_____

_____

26. A 4-H club held a canned food drive. On Wednesday, they collected 63 cans, which was 21% of the total cans collected during the food drive. Complete the table to show the number of cans collected on Thursday and Friday.

Day	Percent of Total Collected	Number of Cans
Wednesday	21	63
Thursday	46	
Friday	33	

27. Refer to the survey results shown. Suppose 150 students were surveyed. How many students chose math as their favorite subject?

Favorite Subject	
English	23%
Science	30%
Social Studies	15%
Math	■
Music	12%

Spiral Review

Find the equivalent fraction. 4.NF.1

28. $\dfrac{84}{120} = \dfrac{\boxed{}}{10}$

29. $\dfrac{60}{98} = \dfrac{30}{\boxed{}}$

30. $\dfrac{40}{64} = \dfrac{5}{\boxed{}}$

31. $\dfrac{32}{41} = \dfrac{96}{\boxed{}}$

32. $\dfrac{13}{15} = \dfrac{\boxed{}}{60}$

33. $\dfrac{24}{32} = \dfrac{12}{\boxed{}}$

34. A store has a sale for $\dfrac{3}{10}$ off gloves. Write $\dfrac{3}{10}$ as a decimal. 4.NF.5

35. Xavier runs 0.75 mile each day. How far has he run at the end of 6 days?
5.NBT.7

21ST CENTURY CAREER
in Movies

Special Effects Animator

Are you fascinated by how realistic the special effects in movies are today? If you have creative talent and are good with computers, a career in special effects animation might be a great fit for you. Special effects animators use their artistic ability and expertise in computer-generated imagery (CGI) to simulate real-life objects like water and fire. They are also able to create fantastic images like flying superheroes, exploding asteroids, and monsters taking over cities.

Is This the Career for You?

Are you interested in a career as a special effects animator? Take some of the following courses in high school.
- ◆ Digital Animation
- ◆ Calculus
- ◆ Geometry
- ◆ Physics
- ◆ Art/Sculpture

Turn the page to find out how math relates to a career in Movies.

MP The Effects are Amazing!

Special effects animators must specify when objects fade or change color. Table 1 shows when an object starts fading out. Table 2 shows the percent of an object's total lifetime that it has the initial color, cross-fading of colors, and the final color. Use the tables to solve each problem.

1. Express the part of total lifetime for each object in Table 1 as a fraction in simplest form. _____

2. At what percent of the light beam's total lifetime does it begin to fade out? _____

3. In Table 2, express the percents for the cross-fading of both objects as decimals.

_____

4. Which best describes the part of the robot's lifetime in which it has the initial color: $\frac{3}{100}$, $\frac{3}{10}$, or $1\frac{3}{10}$? _____

5. What fraction of the tornado's lifetime does it have the initial color? _____

6. What fraction of the robot's lifetime does it have the final color? _____

Table 1 Fading Out an Object	
Object	**Part of Total Lifetime**
Explosion	0.72
Fog	0.24
Light beam	0.65

Table 2 Changing Color of an Object			
Object	**Percent of Total Lifetime**		
	Initial Color	**Cross-Fading**	**Final Color**
Robot	30%	15%	55%
Tornado	12%	77%	11%

MP Career Project

It's time to update your career profile! Choose one of your favorite movies. Use the Internet to research how the movie's special effects were created. Write a brief description of the processes used by the special effects animators.

_____
_____
_____
_____

List several jobs that are created by the movie industry.

- _____
- _____
- _____
- _____
- _____

Chapter Review

Vocabulary Check

Unscramble each of the clue words. After unscrambling all of the terms, use the numbered letters to find the phrase.

INROALAT NUEMBR

☐☐☐☐☐☐☐☐ ☐☐☐☐☐☐
10　　　　6　9　　　12　　　1

TEEPNCR

☐☐☐☐☐☐☐
　　　　4

LESTA MOCNOM EOMITNORNAD

☐☐☐☐☐ ☐☐☐☐☐☐ ☐☐☐☐☐☐☐☐☐☐☐
　　8　　　　　　　　　　　　　　　　　　　　2

PIONORTOPR

☐☐☐☐☐☐☐☐☐☐
　　　　　5　　7

CTREENP ROIPPTORNO

☐☐☐☐☐☐☐ ☐☐☐☐☐☐☐☐☐☐
11　　3　　　　　　　　　　　　13

☐☐☐☐☐☐☐☐ ☐☐☐ ☐☐
F   F
1 2 3 4 5 6 7 8　9 10 11　12 13

Complete each sentence using one of the unscrambled words above.

1. A _____ is a ratio that compares a number to 100.

2. A _____ is an equation that shows that two ratios are equivalent.

3. In a _____, one ratio compares a part to a whole.

4. A number that can be written as a fraction is a _____.

5. The _____ is the least common multiple of the denominators of two or more fractions.

Use Your FOLDABLES

Use your Foldable to help review the chapter.

Tape here

Fractions, Decimals, and Percents

Examples

Examples

Examples

Got it?

The problems below may or may not contain an error. If the problem is correct, write a "✓" by the answer. If the problem is not correct, write an "X" over the answer and correct the problem.

1. $\frac{4}{5} = 0.4$

The first one is done for you.

$\frac{4}{5} \rightarrow 5\overline{)4.0}$
$\quad\quad \underline{-40}$
$\quad\quad\quad\; 0$

0.8

2. $0.55 = \frac{11}{20}$

3. $120\% = \frac{3}{25}$

Power Up! Performance Task

Shooting Stars

Aliya, Benito, Cassie, and Dylan participated in a free throw contest. Each participant had to shoot at least ten shots within a certain time period.

The participants filled out their individual performance. The results are shown in the table.

Name	Shots Made	Shots Attempted	Fraction	Decimal	Percent
Aliya	14	20			
Benito		36	$\frac{1}{2}$		
Cassie	18	25			72%
Dylan	9		$\frac{3}{5}$	0.6	

Write your answers on another piece of paper. Show all of your work to receive full credit.

Part A

The paper got wet after the participants filled in the table and some of the numbers cannot be read. Fill in the missing information. Express the fractions in simplest form. Who had the greatest percent of shots made?

Part B

Use a bar diagram to model the scores of each participant. Suppose Marco recorded his scores late and made 15 out of 27 shots. Draw a bar diagram to represent his score. Then order the players from least to greatest score.

Part C

The coach decided to extend the contest another day. Aliya is the only participant to shoot more free throws. How many free throws must she make in a row to have the highest percentage of the four participants? Explain how you know.

Reflect

Answering the Essential Question

Use what you learned about fractions, decimals, and percents to complete the graphic organizer.

Essential Question

WHEN is it better to use a fraction, a decimal, or a percent?

Converting Fractions, Decimals, and Percents

Fraction $\frac{1}{8}$	→	Decimal:
		Percent:
Decimal 0.8	→	Percent:
		Fraction:
Percent 55%	→	Fraction:
		Decimal:

Answer the Essential Question. WHEN is it better to use a fraction, a decimal, or a percent?

UNIT PROJECT

Watch ▶ **People Everywhere** Federal, state, and local governments use population data that are gathered from the U.S. Census to help them plan what services communities need. In this project you will:

- **Collaborate** with your classmates to collect data and to compare populations within our state, county, or city.
- **Share** the results of your research in a creative way.
- ⓔ **Reflect** on how you use mathematics to describe change and model real-world situations.

By the end of this project, you will have a better understanding of the population of people that live around you!

Collaborate

⏻ **Go Online** Work with your group to research and complete each activity. You will use your results in the Share section on the following page.

1. Use the U.S. Census Web site to find the total population for your state. Then compare the following:
   - the ratio of males to females
   - the ratio of people in your age group to those in any other age group

2. Research the population density of your state. Compare the results with all of the surrounding states. Plot your results on a number line. How does your state compare to others?

3. Research the growth rate of the population of your state over the past ten U.S. Census reports. Display your results in a table and a line graph. Describe any patterns in the line graph.

4. Explain in a journal entry or blog how you could use population data to predict the population of your state in 2020.

5. One way federal funds are distributed to states and counties is based on their population. Research the population of your county, as well as the surrounding counties. If $1,000,000 is to be divided among the counties, about what fraction of the money do you think your county would receive? Why?

Share

With your group, decide on a way to present what you have learned from each of the activities about the population of your state according to the U.S. Census. Some suggestions are listed below, but you can also think of other creative ways to present your information. Remember to show how you used math to complete each of the activities in this project!

- Create a presentation using the data you collected. Your presentation should include a spreadsheet, graph, and one other visual display.
- Write a persuasive letter to a local government official. In the letter, explain what you have learned in this project. Then lobby to have a new service provided for your community. This could be a new park, school, hospital, or whatever you think your community needs.

Check out the note on the right to connect this project with other subjects.

 with Science

Environmental Literacy Select two states, other than your own, and research the types of landforms found in these states. Some questions to consider are:

- What different types of landforms are found in these states?

- How do the landforms found in these states compare with those found in your state?

Reflect

On Your Own

6. **Answer the Essential Question** How can you use mathematics to describe change and model real-world situations?

   a. How did you use what you learned about ratios and rates to describe change and model the real-world situations in this project?

   _____

   _____

   _____

   b. How did you use what you learned about fractions, decimals, and percents to describe change and model the real-world situations in this project?

   _____

   _____

   _____

UNIT 2
The Number System

Essential Question

HOW can mathematical ideas be represented?

Chapter 3
Compute with Multi-Digit Numbers

The standard algorithm used to multiply and divide whole numbers can be applied to operations with decimals. In this chapter, you will multiply and divide multi-digit decimals.

Chapter 4
Multiply and Divide Fractions

Models and equations can be used to represent real-world situations involving operations with fractions. In this chapter, you will multiply and divide fractions by whole numbers and by fractions.

Chapter 5
Integers and the Coordinate Plane

Integers, terminating decimals, and repeating decimals are rational numbers. In this chapter, you will compare and order rational numbers and graph points in four quadrants of the coordinate plane.

Unit Project Preview

 Watch

Get Out the Map! Maps have been around for thousands of years. Over time maps have become more detailed and accurate. Today, using a map is as simple as turning on a GPS device and typing in an address.

Let's make a map of your neighborhood. Choose three or four locations in your neighborhood that you visit, such as a friend's house or the library. On the coordinate plane below, plot points to represent these locations in relation to your home.

At the end of Chapter 5, you'll complete a project that will teach you about using a map to travel somewhere new. So, bring your sense of adventure and get ready to explore someplace new!

My Neighborhood

Your House

Chapter 3
Compute with Multi-Digit Numbers

Essential Question

HOW can estimating be helpful?

Content Standards
6.NS.2, 6.NS.3

 Mathematical Practices
1, 2, 3, 4, 5, 6

Math in the Real World

Skyscrapers A certain skyscraper in Chicago has 1,200,000 square feet of space. On average, there are 29,268 square feet of space on each floor. Estimate to find the number of floors in the building.

floors

29,268 square feet

 FOLDABLES
Study Organizer

1 Cut out the Foldable on page FL7 of this book.

 2 Place your Foldable on page 250.

 3 Use the Foldable throughout this chapter to help you learn about computing with multi-digit numbers.

Vocab abc **Vocabulary**

compatible numbers

Review Vocabulary

Graphic Organizer One way to remember vocabulary terms is to connect them to an everyday meaning or an example. Use this information to complete the graphic organizer.

Everyday meaning:

Math meaning:

Product

Example:

Non-example:

What Do You Already Know?

List three things you already know about multi-digit numbers in the first section.
Then list three things you would like to learn about multi-digit numbers in the second section.

Compute with Multi-Digit Numbers

What I know	What I want to find out

When Will You Use This?

Here are a few examples of how multi-digit numbers are used in the real world.

Activity 1 Use the Internet to find the cost of something that you really want. Describe a plan that you could use to save the money needed to buy it.

_____

_____

_____

_____

David & Raj in *Money Challenge*

This carwash job is great!

Great idea!

CAR WASH

Activity 2 Go online at **connectED.mcgraw-hill.com** to read the graphic novel *Money Challenge*. How much does the video game system Raj wants to buy cost? _____

Are You Ready?

Try the Quick Check below.
Or, take the Online Readiness Quiz.

Quick Review

4.NBT.5, 5.NBT.6

Example 1

Find 13 × 15.

```
    13
  × 15
  ─────
    65     Multiply the ones.
 + 130     Multiply the tens.
  ─────
   195     Add.
```

Example 2

Find 323 ÷ 17.

```
        19
    ┌───────
 17 ) 323       Divide the tens.
    − 17
    ───────
      153       Divide the ones.
    − 153
    ───────
        0
```

Quick Check

Multiply Find each product.

1. 15 × 20 = _____

2. 19 × 51 = _____

3. 49 × 22 = _____

Divide Find each quotient.

4. 112 ÷ 8 = _____

5. 539 ÷ 11 = _____

6. 779 ÷ 19 = _____

7. A musician sold 64 million albums in 16 months. She sold the same amount in each month. How many albums did she sell in each month?

_____

How Did You Do?

**Which problems did you answer correctly in the Quick Check?
Shade those exercise numbers below.**

① ② ③ ④ ⑤ ⑥ ⑦

Add and Subtract Decimals

Real-World Link

Swimming One event in competitive swimming is the women's 100-meter butterfly. The table shows the times of different swimmers at a recent Olympics.

Women's 100-Meter Butterfly

	Swimmer's	Time (s)
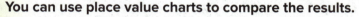	Lisbeth Trickett	56.73
	Christine Magnuson	?
	Gabriella Silva	58.10

Essential Question

HOW can estimating be helpful?

Content Standards
6.NS.3

MP **Mathematical Practices**
1, 2, 3, 4, 5, 6

You can use place value charts to compare the results.

1. It took Christine Magnuson 0.37 second longer to finish than Lisbeth Trickett. What was Magnuson's time, in seconds?

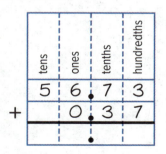

tens	ones	tenths	hundredths
5	6 .	7	3
+	0 .	3	7

2. At a high school meet, a swimmer swam the women's 100-meter butterfly in 72.34 seconds. How many seconds faster did Gabriella Silva swim her race?

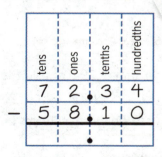

tens	ones	tenths	hundredths
7	2 .	3	4
− 5	8 .	1	0

Almost there!

Which **MP** **Mathematical Practices** did you use?
Shade the circle(s) that applies.

① Persevere with Problems

② Reason Abstractly

③ Construct an Argument

④ Model with Mathematics

⑤ Use Math Tools

⑥ Attend to Precision

⑦ Make Use of Structure

⑧ Use Repeated Reasoning

Add Decimals

To add decimals, line up the decimal points. Then, add digits in the same place-value position.

Examples

1. Find the sum of 23.1 and 5.8.

Estimate $23.1 + 5.8 \approx 23 + 6$ or 29

$$
\begin{array}{r}
23.1 \\
+\ 5.8 \\
\hline
28.9
\end{array}
$$

Line up the decimal points.

Add as with whole numbers.

Check for Reasonableness $28.9 \approx 29$ ✓

The sum of 23.1 and 5.8 is 28.9.

- -

2. Find the sum of 29.6 and 14.7.

Estimate $29.6 + 14.7 \approx$ ⬚ $+$ ⬚ or ⬚

Line up the decimal points.

Add as with whole numbers.

Check for Reasonableness   ⬚ $\approx$ ⬚

The sum of 29.6 and 14.7 is ⬚.

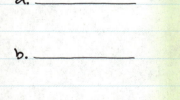 Show your work.

Got it? Do these problems to find out.

Find each sum.

   a. 54.7 + 21.4   **b.** 14.2 + 23.5   **c.** 17.3 + 33.5

a. _____

b. _____

c. _____

Subtract Decimals

To subtract decimals, line up the decimal points. Then, subtract digits in the same place-value position. You may need to *annex*, or place zeros at the end of a decimal, in order to subtract.

Examples

Tutor

3. **Find the difference of 5.774 and 2.371.**

Estimate	$5.774 - 2.371 \approx 6 - 2$ or 4

$$\begin{array}{r} 5.774 \\ -\ 2.371 \\ \hline 3.403 \end{array}$$

Line up the decimal points.

Subtract as with whole numbers.

Check for Reasonableness $3.403 \approx 4$ ✔

So, $5.774 - 2.371 = 3.403$.

....

4. **Find 6 − 4.78.**

Estimate $6 - 4.78 \approx 6 - 5$ or 1

$$\begin{array}{r} 6.00 \\ -\ 4.78 \\ \hline 1.22 \end{array}$$

Annex zeros so that both numbers have the same number of decimal places.

Check for Reasonableness $1.22 \approx 1$ ✔

So, $6 - 4.78 = 1.22$.

....

5. **Find 23 − 4.216.**

Annex the zeros so that both numbers have the same number of decimal places.

Subtract as with whole numbers.

So, $23 - 4.216 =$ _____.

Got it? Do these problems to find out.

Find each difference.

d. $9.543 - 3.671$ **e.** $\$50.62 - \$39.81$ **f.** $14 - 9.09$

Show your work.

d. _____

e. _____

f. _____

Real World

 Tutor

Example

6. Reagan is creating a video. The first video clip was 22.36 minutes long. The second video clip was 17.03 minutes long. What is the total length of the video?

Estimate $22.36 + 17.03 \approx 22 + 17$ or 39

$$
\begin{array}{r}
22.36 \\
+\ 17.03 \\
\hline
39.39
\end{array}
$$

Line up the decimal points.

Add as with whole numbers.

Check for reasonableness $39.39 \approx 39$ ✓

So, the video is 39.39 minutes long.

> *Show your work.*

Got it? Do this problem to find out.

g. Jonathan is traveling for work. This morning his GPS indicated that the total distance to his destination is 589.4 miles. Before lunch he drove 208.62 miles. How much farther does Jonathan need to travel?

g. _____

Guided Practice

 Check ✓

Find each sum or difference. (Examples 1–5)

1. $14.7 + 87.9 =$ _____

> *Show your work.*

2. $66.5 - 24.1 =$ _____

3. $52.1 - 31.47 =$ _____

4. Grayson is making a snack mix for his family camping trip. He added 14.52 ounces of peanuts to 27.35 ounces of granola. How many ounces of snack mix does he have? (Example 6)

_____

_____

5. **Building on the Essential Question** How is estimation helpful when adding and subtracting decimals?

_____

_____

_____

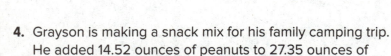

Rate Yourself!

☐ I understand how to add and subtract decimals.

▶▶ Great! You're ready to move on!

☐ I still have questions about adding and subtracting decimals.

▌▌ No Problem! Go online to access a Personal Tutor. Tutor

Independent Practice

Go online for Step-by-Step Solutions

Find each sum. (Examples 1 and 2)

1. 7.2 + 9.5 = _____

2. 1.34 + 2 = _____

3 54.5 + 48.51 = _____

 Show your work.

Find each difference. (Examples 3–5)

4. 5.6 − 3.5 = _____

5. 97 − 16.98 = _____

6. 58.67 − 28.72 = _____

7 The table shows the top three finishers in barrel racing. What is the time difference between Nicolas and Sancho? (Example 6)

_____

Barrel Racing Results	
Rider	**Time (s)**
Nicolas	14.67
Becki	15.98
Sancho	16.40

8. In two months, Mica spent a total of $305.43 on groceries. She spent $213.20 in the first month. How much did she spend in the second month? (Example 6)

_____

9. Financial Literacy A hat costs $10.95 and a T-shirt costs $14.20. How much change will you receive if you pay for both items with a $50 bill?

_____

 Show your work.

10. **MP** **Use Math Tools** The 4-by-100 meter relay is a track and field event involving four runners on each team.

a. What is the combined time of Carter and Frater?

_____

b. How much faster did Powell run his leg of the race than Bolt?

_____

c. What is the combined time of all the runners?

_____

4 by 100 Meter Times	
Runner	**Time (s)**
Carter	13.4
Frater	12
Bolt	12
Powell	11.7

11. **MP** **Find the Error** Luis is finding 8.9 − 3.72. Find his mistake and correct it.

$$\begin{array}{r} 8.9 \\ -\ 3.72 \\ \hline 5.22 \end{array}$$

12. **MP** **Reason Abstractly** Write two different pairs of decimals whose sums are 14.1. One pair should involve regrouping.

13. **MP** **Reason Inductively** Explain how you know that the sum of 12.6, 3.1, and 5.4 is greater than 20.

14. **MP** **Persevere with Problems** Josie found that 3.28 + 3.28 + 3.28 = 9.84. What is the missing factor in the related multiplication problem

3.28 × ☐ = 9.84? Explain.

15. **MP** **Reason Abstractly** Without subtracting 8.5 − 4.64, determine what digit will be in the hundredths place. Explain.

16. **MP** **Reason Inductively** Kiera is buying the items shown at the right for her kitchen. Will $35 be enough to purchase all three items? Explain your reasoning.

Item	Cost ($)
mixing bowl	14.95
spatula	8.49
measuring cups	10.75

Extra Practice

Find each sum.

17. $4.9 + 3.0 =$ ___7.9___

Homework Help

$$\begin{array}{r} 4.9 \\ +3.0 \\ \hline 7.9 \end{array}$$

18. $0.796 + 13 =$ _____

19. $15.63 + 24.36 =$ _____

Find each difference.

20. $19.86 - 4.94 =$ _____

21. $82 - 67.18 =$ _____

22. $14.39 - 12.16 =$ _____

23. Financial Literacy The current balance of Tami's checking account is $237.80. Find the new balance after Tami writes a check for $29.95.

_____

Show your work.

24. The annual rainfall for Kayston Falls was 50.38 inches in 2012. In 2013, the annual rainfall was 55.76 inches. What is the difference in rainfall between the two years?

_____

25. STEM The melting point of sodium is 97.8 degrees Celsius. The melting point of potassium is 63.65 degrees Celsius. How much higher is the melting point of sodium?

_____

_____

26. MP Be Precise Collin needs three wooden boards to repair his porch. The lengths he needs are 2.2 meters, 2.82 meters, and 4.25 meters. He purchases a board that is 10 meters long and cuts the three sections. How much of the board that Collin purchased will be left?

_____

_____

°C

97.8 °C

63.65 °C

27. Dominic is traveling through the state of Oregon. He drove 66.4 miles from Salem to Eugene and then continued to Medford. The total distance of his trip was 233.25 miles. What is the distance from Eugene to Medford?

☐

28. The table shows the top four finishers at a swimming event.

Boy's 50 Yard Freestyle	
Swimmer	Time (s)
Andrew	22.63
Kendrick	22.20
Ty	22.58
Will	22.31

Determine if each statement is true or false.

a. Will swam 0.32 seconds faster than Andrew.   ☐ True   ☐ False

b. Ty swam 0.27 seconds faster than Will.   ☐ True   ☐ False

c. Andrew swam 0.5 seconds slower than Ty.   ☐ True   ☐ False

d. Will swam 0.11 seconds slower than Kendrick.   ☐ True   ☐ False

Spiral Review

Round each decimal to the nearest whole number. 5.NBT.4

29. 4.75 ≈ _____

30. 34.1 ≈ _____

31. 22.48 ≈ _____

32. Plot the number 2.78 on the number line below. **4.NF.7**

2.7   2.8

33. The table shows the distances Juliana biked several days this week. Which day of the week did she bike the greatest distance?
5.NBT.3b

_____

_____

Day	Distance Biked (mi)
Monday	9.34
Wednesday	9.47
Thursday	9.74
Sunday	9.32

Estimate Products

 Real-World Link

Skateboarding The record for the greatest distance traveled on skateboard in 24 hours was set in a recent year by James Peters. He traveled about 7.6 miles per hour.

1. Plot 7.6 on the number line.

7   7.5   8

2. What whole number is 7.6 closest to? ▢

3. Estimate how many miles James Peters traveled in 24 hours.

▢ × ▢ = _____

4. Is your estimate higher or lower than the actual distance he traveled? Explain.

_____

_____

5. A new record was set later by Ted McDonald. He traveled about 10.1 miles per hour for 24 hours. About how much farther did Ted McDonald travel?

▢ × 24 = _____

▢ − ▢ = _____

So, Ted McDonald traveled about _____ miles farther.

Which MP Mathematical Practices did you use?
Shade the circle(s) that applies.

① Persevere with Problems   ⑤ Use Math Tools

② Reason Abstractly   ⑥ Attend to Precision

③ Construct an Argument   ⑦ Make Use of Structure

④ Model with Mathematics   ⑧ Use Repeated Reasoning

Essential Question

HOW can estimating be helpful?

Content Standards
Preparation for 6.NS.3

MP Mathematical Practices
1, 2, 3, 4, 5

Estimate Products Using Rounding

To estimate products of decimals, round each number. First underline the digit to be rounded. Then look at the digit to the right of the place being rounded.

- If the digit is 4 or less, the underlined digit remains the same.
- If the digit is 5 or greater, add 1 to the underlined digit.
- After rounding, all place values to the right of the underlined digit have a value of zero.

After the numbers are rounded, multiply.

Examples

Tutor

1. **Estimate 8.7 × 2.8.**

Round to the nearest whole number to make it easier to compute mentally.

$$
\begin{array}{ccc}
8.7 & \rightarrow & 9 \\
\times\ 2.8 & \rightarrow & \times\ 3 \\
\hline
& & 27
\end{array}
$$
Round 8.7 to 9.
Round 2.8 to 3.

The product is about 27.

2. **Estimate 42.6 × 37.2.**

Round to the greatest place value to make it easier to compute mentally.

42.6 ≈ ☐

37.2 ≈ ☐

☐☐   Multiply.

☐☐

× ──────

☐☐☐☐

The product is about ☐.

Show your work.

> ## Got it? Do these problems to find out.

Estimate each product.

a. 9.6 × 1.8   **b.** 4.2 × 3.1   **c.** 68.4 × 21.3

Rounding Decimals

When rounding decimals, such as 99.96 to the tenths, the 9 must round up. So, 99.96 rounded to the nearest tenth is 100.0.

a. _____

b. _____

c. _____

Examples

Tutor

3. **A greyhound can travel 39.3 miles per hour. At this speed, about how far could a greyhound travel in 6.5 hours?**

39.3	→	40	Round 39.3 to 40.
× 6.5	→	× 7	Round 6.5 to 7.
		280	

The greyhound could travel about 280 miles in 6.5 hours.

4. **Suppose one U.S. dollar is equal to 5.8 Egyptian pounds. About how many Egyptian pounds would you receive for $48.50?**

Round to the greatest place value to make it easier to compute mentally.

5.8 ≈ ☐

48.50 ≈ ☐

☐☐

×   ☐

☐☐☐   Multiply.

So, $48.50 is equal to about ☐ Egyptian pounds.

STOP and Reflect

You could estimate 18.76 × 5 by multiplying 19 × 5 or by multiplying 20 × 5. Which answer will be more exact? Explain below.

Got it? **Do these problems to find out.**

d. **STEM** Earth is rotating around the Sun about 18.6 miles per second. About how many miles does it travel in 4.8 seconds?

e. The average walking speed of a person is 4.8 kilometers per hour. Estimate the number of kilometers you could walk in 3 hours.

f. **STEM** A King Palm can grow about 2.1 feet a year. Estimate the height of the King Palm, in yards, after 15 years.

Show your work.

d. _____

e. _____

f. _____

Example

Tutor

5. **Patrice has $20 to buy 5 binders. Binders cost $4.29 each. Does she have enough money? Explain your reasoning.**

Estimate.

$5 \times \$4 = \$20$ Estimate 4.29 as 4.

$5 \times \$5 = \$25$ Estimate 4.29 as 5.

The actual cost is between $20 and $25. So, Patrice does not have enough money to buy the binders.

Guided Practice

Check ✓

Estimate each product. (Examples 1 and 2)

1. $5.8 \times 4 \approx$ _____

2. $13.92 \times 2.7 \approx$ _____

3. $94.89 \times 3.11 \approx$ _____

Show your work.

4. **Financial Literacy** A grocery store sells American cheese for $3.89 per pound. About how much would 1.89 pounds of the cheese cost?
(Examples 3 and 4)

_____

5. Greg has 52 megabytes left on his MP3 player. He wants to download 7 songs that each use 7.9 megabytes of memory. He estimates that he will need 56 megabytes of memory. Is his estimate reasonable? Explain your reasoning. (Example 5)

_____

_____

6. **❓ Building on the Essential Question** How do you determine which place value to use when estimating products?

_____

_____

_____

Rate Yourself!

How confident are you about estimating products? Check the box that applies.

For more help, go online to access a Personal Tutor.

Independent Practice

Go online for Step-by-Step Solutions

Estimate each product. (Examples 1 and 2)

1. $9.7 \times 3.3 \approx$ _____

2. $3.4 \times 5.6 \approx$ _____

3 $17.5 \times 8.4 \approx$ _____

Show your work.

4. $44.8 \times 5.1 \approx$ _____

5. $28.21 \times 8.02 \approx$ _____

6. $71.92 \times 2.01 \approx$ _____

7. On average, the U.S. produces 36.5 million tons of fruit each year. About how much fruit does it produce in 2.25 years? (Examples 3 and 4)

_____

_____

8. Lisha is making headbands using ribbon. She would like to make 12 headbands. Each one requires 15.5 inches of ribbon. She estimates that she will need to buy 160 inches of ribbon. Is her estimate reasonable? Explain your reasoning. (Example 5)

_____

_____

9 **Financial Literacy** Hannah's hourly wage at the ice cream shop is $5.85. The table shows the number of hours she worked. She estimates her earnings to be $120. Without calculating her actual earnings, determine if her estimate is more or less than her actual earnings. Explain your reasoning.

_____

_____

Day	Hours Worked
Monday	3.5
Tuesday	4.25
Wednesday	3.75
Thursday	2.5
Friday	4.75

10. **STEM** A car releases 19.6 pounds of carbon dioxide for every 1 gallon of gasoline burned. Estimate the number of pounds of carbon dioxide released if 14.5 gallons are burned.

_____

_____

11. **MP Model with Mathematics** Refer to the graphic novel frame below for Exercises a–b.

CAR WASH Paycheck

Hourly wage	Hours worked
$ 5.50	$

I've already saved $68 for the $200 video game system.

a. How much more does Raj need until he has enough to buy the video game system? _____

b. Raj estimates that if he works for 20 hours, he will have enough to buy the video game system. Is he correct? Explain. _____

 H.O.T. Problems Higher Order Thinking

12. **MP Reason Abstractly** Name three decimals with a product that is about 40.

_____

13. **MP Persevere with Problems** A scooter can travel between 22 and 28 miles on each gallon of gasoline. If one gallon of gasoline costs between $3.75 and $3.95 per gallon, about how much will it cost to travel 75 miles? _____

14. **MP Justify Conclusions** Suppose your friend multiplied 1.2 and 2.6 and got 31.2 as the product. Is your friend's answer reasonable? Justify your response.

_____

_____

15. **MP Reason Inductively** Green peppers are on sale for $2.89 per pound. Mrs. Moseley bought 1.75 pounds of peppers. Is it more reasonable to say that she spent between $5 and $6 or between $6 and $7 on the peppers? Explain your reasoning.

_____

_____

_____

Extra Practice

Estimate each product.

16. 26.3 → 26
     × 9.7 → × 10
     26 × 10 = 260

Homework Help

17. 33.6
     × 82.1
     _____

18. 99.1
     × 11.2
     _____

19. **STEM** A single year on Saturn is equal to 29.4 years on Earth. About how many Earth-years are equal to 3.2 years on Saturn?

_____

_____

20. Miguel received a $50 gift card to a bookstore. He would like to buy 3 books that cost $15.75 each including tax. He estimates that he cannot buy all three books because each book costs about $20, and all three books would cost $60. Is his estimate reasonable? Explain your reasoning.

_____

_____

Use estimation to determine whether each answer is reasonable. If the answer is reasonable, write yes. If not, write no and provide a reasonable estimate.

21. $22.8 \times 4.7 = 107.16$ _____

22. $2.1 \times 4.9 \times 7.2 = 105.84$ _____

23. $7.8 \times 1.1 \times 4.2 = 50$ _____

24. $43.8 \times 2.8 \times 3.1 = 371.8$ _____

25. **MP** **Use Math Tools** The table shows some nutritional facts about orange juice. Estimate each value for 1 quart of orange juice. (*Hint:* 4 cups is equal to 1 quart.)

_____

_____

Orange Juice (1 cup)	
Calories	112
Vitamin C	96.9 mg
Carbohydrates	26.8 g
Calcium	22.4 mg

26. Medina's school lunch menu is shown. Estimate the cost for the following lunch combinations.

Combination	Total Cost
2 slices of pizza, salad, fruit punch	
fish and fries, pudding, milk	
1 slice of pizza, salad, pudding, fruit punch	

Friday			
Pizza	$1.75	Fruit Punch	$0.75
Fish and Fries	$2.25	Milk	$0.80
Salad	$1.15	Pudding	$0.85

Which combination is the least expensive?

27. Mario and Andrew's hourly charge for mowing lawns is shown.

Mario	Andrew
$8.25/h	$5.60/h

Mario and Andrew each worked 25 hours. About how much more money did Mario earn?

Spiral Review

Multiply. 5.NBT.5

28.
$$65 \times 18$$

29.
$$15 \times 23$$

30.
$$198 \times 75$$

31. Marissa spent $15.63 at the bookstore. She paid with a $20 bill. How much change will she receive? 5.NBT.7 _____

_____

32. Cristian is placing photos onto scrapbook paper for his photo album. The scrapbook paper is 12 inches long and 12 inches wide. What is the area of the paper?

(*Hint:* area = length × width) 4.MD.3 _____

12 in.

12 in.

Multiply Decimals by Whole Numbers

Plants Bamboo is one of the fastest growing plants. It can grow about 4.9 feet in height per day. It is a favorite food of panda bears. You can use repeated addition to find the total height a bamboo plant can grow over a number of days. Complete the table below. The first one is done for you.

 Essential Question

HOW can estimating be helpful?

Content Standards
6.NS.3

MP Mathematical Practices
1, 3, 4, 5, 6

Number of Days	Repeated Addition					Multiplication
2	4.9 + 4.9 = 9.8					2 × 4.9 = 9.8
1. 3	+	+	=			× 4.9 =
2. 4	+	+	+	=		× 4.9 =
3. 5	+	+	+	+	=	× 4.9 =

4. Use the pattern in the table to predict 6 × 4.9.

   Check by using repeated addition. _____

   _____

5. **MP Make a Conjecture** Look back at Exercises 1–4. Compare the number of decimal places in each factor to the number of decimal places in the product. How do you determine the placement of the decimal point in a product?

   _____

   _____

   _____

   _____

Which **MP Mathematical Practices** did you use?
Shade the circle(s) that applies.

① Persevere with Problems
② Reason Abstractly
③ Construct an Argument
④ Model with Mathematics

⑤ Use Math Tools
⑥ Attend to Precision
⑦ Make Use of Structure
⑧ Use Repeated Reasoning

Multiply Decimals

Using repeated addition can help you place the decimal point. The whole number represents the number of times the decimal is used as an addend. So, place the decimal point in the product the same number of places from the right as the decimal factor.

Examples

1. **Find 4 × 0.83.**

Estimate 4 × 1 = 4

0.83   ← two decimal places
× 4
─────
3.32   Place the decimal point two places from the right.

4 groups of 0.83

| 0.83 | 0.83 | 0.83 | 0.83 |

Check for Reasonableness 3.32 ≈ 4 ✓

2. **Find 3 × 14.2.**

Estimate 3 × 14 = 42

14.2   ← one decimal place
× 3
─────
42.6   Place the decimal point one place from the right.

3 groups of 14.2

| 14.2 | 14.2 | 14.2 |

Check for Reasonableness 42.6 ≈ 42 ✓

3. **Find 4 × 0.95.**

Estimate 4 × ☐ = ☐

Use the bar diagram to find the product.

☐ groups of ☐

☐ ☐
0 . 9 5
× 　　4
──────
☐ ☐ ☐

0.95 has ☐ decimal places.

Place the decimal point ☐ places from the right.

Check for Reasonableness ☐ ≈ ☐ ✓

Got it? Do these problems to find out.

a. 5 × 0.25   b. 8 × 4.47   c. 9 × 2.63

a. _____

b. _____

c. _____

Annex Zeros in the Product

If there are not enough decimal places in the product, you need to annex zeros to the left. To annex a zero means to place a zero at the beginning or end of a decimal.

Examples

4. **Find 2 × 0.018.**

0.018 ← three decimal places
× 2
0.036

↑ Annex a zero on the left of 36 to make three decimal places.

Check by Adding

	0	.	0	1	8
+	0	.	0	1	8

✓

5. **Find 4 × 0.012.**

0 . 0 1 2   0.012 has ☐ decimal places.
×   4
☐ ☐ ☐ ☐   Annex a ☐ to make ☐ decimal places.

Check by Adding

	0	.	0	1	2
	0	.	0	1	2
	0	.	0	1	2
+	0	.	0	1	2

✓

Got it? Do these problems to find out.

d. 3 × 0.02   **e.** 0.12 × 8   **f.** 11 × 0.045

Show your work.

d. _____

e. _____

f. _____

Example

Tutor

6. A batch of trail mix calls for 1.2 pounds of dry cereal. Nigela is making 5 batches of trail mix. She already has 2.2 pounds of cereal. How many more pounds of dry cereal does she need?

Step 1 Multiply.

$$
\begin{array}{r}
1.2 \quad \leftarrow \text{ one decimal place}\\
\times\ 5\\
\hline
6.0 \quad \leftarrow \text{ one decimal place}
\end{array}
$$

Step 2 Subtract.

$$
\begin{array}{r}
6.0\\
-\ 2.2\\
\hline
3.8
\end{array}
$$

So, Nigela will need 3.8 more pounds of dry cereal.

Guided Practice

 Check

Multiply. (Examples 1–5)

1. $2.7 \times 6 =$ _____

 Show your work.

2. $0.52 \times 3 =$ _____

3. $5 \times 0.09 =$ _____

4. $4 \times 0.027 =$ _____

5. $0.071 \times 8 =$ _____

6. $0.065 \times 18 =$ _____

7. A bee hummingbird has a mass of 1.8 grams. How many grams are 6 hummingbirds and a 4-gram nest? (Example 6)

_____

8. Justin buys 12 pencils for $0.56 each. He pays with a $10 bill. How much change will he receive? (Example 6)

_____

9. **Building on the Essential Question** How can estimating products help you to place the decimal

correctly? _____

_____

Rate Yourself!

Are you ready to move on? Shade the section that applies.

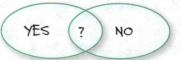

YES   ?   NO

For more help, go online to access a Personal Tutor.

Tutor

Independent Practice

Go online for Step-by-Step Solutions

Multiply. (Examples 1–5)

1. 1.2 × 7 = _____

2. 0.7 × 9 = _____

3. 2 × 1.3 = _____

4. 0.8 × 9 = _____

5. 3 × 0.02 = _____

6. 0.0036 × 19 = _____

7. The table shows the number of gallons of gasoline the Beckleys purchased on their road trip. What was the total cost for gas for the trip? (Example 6)

_____

Fuel	
Number of Gallons	**Cost per Gallon ($)**
12	4.89
17	4.72
15	5.09

8. Sharon buys 14 folders for $0.75 each. How much change

will she receive if she pays with $15? (Example 6) _____

9. **STEM** The hottest temperature recorded in the world, in degrees Fahrenheit, can be found by multiplying 13.46 by 10. Find this temperature.

Justify your procedure. _____

_____

_____

10. **MP Justify Conclusions** Asher recently bought the poster shown at the right. What is its area? Explain your reasoning to a classmate.

(*Hint:* Use area = length × width.) _____

_____

_____

4 ft

3.2 ft

11. **MP Use Math Tools** The thickness of each type of coin is shown in the table. How much thicker is a stack of a dollar's worth of nickels than a dollar's worth of quarters? Explain your answer.

Coin	Thickness (mm)
penny	1.55
nickel	1.95
dime	1.35
quarter	1.75

_____

_____

 H.O.T. Problems Higher Order Thinking

12. **MP Model with Mathematics** Write a real-world problem involving multiplication by a decimal factor. Then solve the problem.

_____

_____

13. **MP Persevere with Problems** Discuss two different ways to find the value of the expression 5.4 × 1.17 × 100 that do not require you to first multiply 5.4 × 1.17.

_____

_____

_____

_____

_____

14. **MP Reason Inductively** Use the product of 123 × 47 to find the product of 123 × 0.47. Explain the difference in the two products.

_____

_____

15. **MP Construct an Argument** Your friend thinks that 1.5 × 8 = 1.20 because you do not count the zero when placing the decimal point. Is your friend correct? Justify your reasoning.

_____

_____

16. **MP Construct an Argument** Explain how you can mentally determine whether 125 × 0.9 is less than, greater than, or equal to 125.

_____

_____

_____

Extra Practice

Multiply.

17. $1.7 \times 5 =$ _8.5_

Homework Help

$$
\begin{array}{r}
3 \\
1.7 \\
\times \quad 5 \\
\hline
8.5
\end{array}
$$

18. $0.9 \times 4 =$ _____

19. $2.4 \times 8 =$ _____

20. $3 \times 0.5 =$ _____

21. $7 \times 0.012 =$ _____

22. $0.0198 \times 75 =$ _____

23. The mass of a certain monarch butterfly is 0.56 gram. What is the mass of 4 monarch butterflies?

_____

Show your work.

24. The height of Mount Everest, in meters, can be found by multiplying 8.85 by 1,000. Find the height of Mount Everest. Explain your answer.

_____

_____

25. A sheet of printer paper is 8.5 inches by 11 inches. What is the area of the paper? (*Hint:* area = length × width)

_____

26. Sofia bought 12 pens for $0.59 each. She paid with a $10 bill. How much change will she receive?

_____

27. **MP** **Be Precise** One kilometer is about 0.62 mile. It is 12 kilometers from Noah's house to the ice skating rink. About how many miles is it from Noah's house to the ice skating rink?

_____

28. The school store is selling the items shown in the table. Miguel has $12. Which of the following can he buy? Select all that apply.

☐ two pennants, two bumper stickers, and four magnets

☐ three pennants and two bumper stickers

☐ two pennants, three bumper stickers, and one magnet

☐ four pennants, a bumper sticker, and one magnet

Item	Price
Pennant	$2.49
Bumper Stucker	$1.79
Magnet	$0.89

29. Mrs. Bowman designed the rug shown. If it costs $22 per square foot to make, how much will Mrs. Bowman pay for the rug?

5 ft

6.25 ft

Spiral Review

30. Use number patterns and powers of ten to complete the table. **5.NBT.2**

Factor		Factor		Product
2.9	×	10	=	
3.44	×		=	344
	×	100	=	870
10.25	×		=	102.5
156.23	×	10	=	

Round each decimal to the nearest whole number. **5.NBT.4**

31. 5.7 ≈ _____

32. 0.05 ≈ _____

33. 13.49 ≈ _____

34. Several students from Southbend Middle School are visiting the Smithsonian American Art Museum. Mrs. Mabika divided the students into 5 equal groups. There are 3 boys and 4 girls in each group. Fill in the missing numbers to find the total number of students. **5.OA.2**

☐ × (☐ + 4) = ☐ students

Multiply Decimals by Decimals

Planets The table shows the weight of a 1-pound object on each planet.

Planet	Weight (Pounds)
Mercury	0.3
Venus	0.9
Earth	1
Mars	0.3
Jupiter	2.3
Saturn	1
Uranus	0.8
Neptune	1.1

 Essential Question

HOW can estimating be helpful?

Content Standards
6.NS.3

 Mathematical Practices
1, 2, 3, 4, 5, 6

1. A 0.5-pound object weighs one half as much as a 1-pound object. If a cheeseburger weighs a half pound on Earth, what will it weigh on Jupiter? _____

2. What would a barbell that weighs 5 pounds on Earth weigh on Jupiter? _____

3. Use the results from Exercises 1 and 2 to find 0.05 × 2.3. Explain your answer. _____

_____

_____

Which **Mathematical Practices** did you use?
Shade the circle(s) that applies.

① Persevere with Problems
② Reason Abstractly
③ Construct an Argument
④ Model with Mathematics

⑤ Use Math Tools
⑥ Attend to Precision
⑦ Make Use of Structure
⑧ Use Repeated Reasoning

Multiply Decimals

When multiplying a decimal by a decimal, multiply as with whole numbers. To place the decimal point, find the sum of the number of decimal places in each factor. The product has the same number of decimal places.

Tutor

Examples

1. **Find 3.6 × 0.05.**

   Estimate 3.6 × 0.05 ⟶ 4 × 0 or 0

   $$\begin{array}{r} 3.6 \quad \leftarrow \text{one decimal place} \\ \times\ 0.05 \quad \leftarrow \text{two decimal places} \\ \hline 0.180 \quad \leftarrow \text{three decimal places} \end{array}$$

   The product is **0.180 or 0.18.**   Once you place the decimal point, you can drop the zero at the right.

2. **Find 0.112 × 7.2.**

   Estimate 0.112 × 7.2 ≈ ☐ × ☐ or ☐

   0.112 has ☐ decimal places.

   7.2 has ☐ decimal place.

   So the product has ☐ + ☐, or ☐ decimal places.

   $$\begin{array}{r} 0.\quad 1\quad 1\quad 2 \\ \times \qquad\qquad 7.\quad 2 \\ \hline \Box\ \Box\ \Box \\ +\quad \Box\ \Box\ \Box \\ \hline \Box\ \Box\ \Box\ \Box\ \Box \end{array}$$

   Show your work.

   The product is _____.

   Check for reasonableness _____ ≈ _____ ✔

Got it? Do these problems to find out.

a. 5.7 × 2.8   b. 4.12 × 0.05   c. 0.014 × 3.7

STOP and Reflect

How is the product of 4.2 × 6.7 similar to the product of 42 × 67? How are they different? Explain below.

a. _____

b. _____

c. _____

Annex a Zero

If there are not enough decimal places in the product, you need to annex zeros to the left.

Examples

Tutor

3. **Find 1.4 × 0.067.**

```
    0.067    ← three decimal places
  ×   1.4    ← one decimal place
    268
  + 67
  0.0938     ← Annex a zero to make four decimal places.
```

4. **Find 0.45 × 0.053.**

The product will have ☐ decimal places. Annex zeros, if needed.

```
          0.  4  5
  ×    0.  0  5  3
        ☐  ☐  ☐
  +  ☐  ☐  ☐
  ☐  ☐  ☐  ☐  ☐  ☐
```

Check Multiply related whole numbers.

```
        4  5
  ×     5  3
     ☐  ☐  ☐
  + ☐  ☐  ☐
  ☐  ☐  ☐  ☐
```

Move the decimal to the left 5 places. What is the number? _____

Is the answer the same? ☐

Got it? Do these problems to find out.

d. 0.04 × 0.32   **e.** 0.26 × 0.205   **f.** 1.33 × 0.06

Show your work.

d. _____

e. _____

f. _____

beep!

Example

5. A certain car can travel 28.45 miles with one gallon of gasoline. The gasoline tank can hold 11.5 gallons. How many miles can this car travel on a full tank of gas? Justify your answer.

Estimate 28.45 × 11.5 ⟶ 30 × 12 or 360

```
    28.45     ← two decimal places
  × 11.5      ← one decimal place
   14225
    2845
 + 2845
  327.175     ← The product has three decimal places.
```

The car could travel 327.175 miles. Since 327.175 is close to 360, the answer is reasonable.

Guided Practice

Check ✓

Multiply. (Examples 1–4)

Show your work.

1. 0.6 × 0.5 = _____

2. 27.43 × 1.089 = _____

3. 0.98 × 7.3 = _____

4. 2.7 × 1.35 = _____

5. 0.03 × 0.09 = _____

6. 0.04 × 2.12 = _____

7. A mile is equal to approximately 1.609 kilometers. How many kilometers is 2.5 miles? Justify your answer. (Example 5) _____

_____

8. **ⓔ** **Building on the Essential Question** Why is estimating not as helpful when multiplying very small numbers such as

0.007 and 0.053? _____

_____

_____

Rate Yourself!

Are you ready to move on? Shade the section that applies.

I have a few questions.

I'm ready to move on.

I have a lot of questions.

For more help, go online to access a Personal Tutor.

Tutor

Independent Practice

Go online for Step-by-Step Solutions

Multiply. (Examples 1–4)

1. $0.7 \times 0.4 =$ _____

2. $0.4 \times 3.7 =$ _____

3. $0.52 \times 2.1 =$ _____

4. $6.2 \times 0.03 =$ _____

5. $14.7 \times 11.361 =$ _____

6. $0.28 \times 0.08 =$ _____

7. **STEM** A giraffe can run up to 46.93 feet per second. How far could a giraffe run in 1.8 seconds? Justify your answer. (Example 5)

_____

_____

8. A nutrition label indicates that one serving of apple crisp oatmeal has 2.5 grams of fat. How many grams of fat are there in 3.75 servings? Justify your answer. (Example 5)

_____

_____

9. **Financial Literacy** Pears cost $0.92 per pound and apples cost $1.10 per pound. Mr. Bonilla bought 3.75 pounds of pears and 2.1 pounds of apples. How much did he pay for the pears and apples? Explain your answer.

_____

_____

Multiply.

10. $25.04 \times 3.005 =$ _____

11. $1.03 \times 1.005 =$ _____

12. $5.12 \times 4.001 =$ _____

13. **STEM** The table shows the approximate distance around Earth.

Location	Approximate Distance (mi)
around Earth at the equator	24,889.78
around Earth through the poles	24,805.94

a. About how many more miles would a satellite travel if it circled the equator 2.5 times than if it circled around the poles 2.5 times?

_____

b. The distance around Jupiter at the equator is about 17.6 times greater than the distance around Earth at the equator. About how many more miles would a satellite travel if it circled Jupiter's equator than if it circled Earth's equator? Round to the nearest tenth.

_____

 H.O.T. Problems Higher Order Thinking

14. **MP** **Reason Abstractly** Write a multiplication problem in which the

product is between 0.05 and 0.75. _____

15. **MP** **Justify Conclusions** Place the decimal point in the answer to make it correct. Explain your reasoning. $3.9853 \times 8.032856 = 32013341\ldots$

_____

_____

16. **MP** **Construct an Argument** Determine whether the following statement is *always*, *sometimes*, or *never* true. Give examples to justify your answer.

The product of two decimals less than 1
is less than either of the factors.

_____

_____

17. **MP** **Reason Inductively** Is the product of 0.4×1.8 greater than or less

than 0.4? Explain your reasoning. _____

_____

18. **MP** **Persevere with Problems** Evaluate the expression $0.3(3 - 0.5)$. _____

19. **MP** **Model with Mathematics** Write a world problem in which you multiply two decimals. The product should be between 0 and 1.

_____

_____

Extra Practice

Multiply.

20. $1.5 \times 2.7 =$ ___4.05___

Homework
Help →

$$\begin{array}{r} 1.5 \\ \times\ \ 2.7 \\ \hline 105 \\ +30 \\ \hline 4.05 \end{array}$$

21. $3.1 \times 0.8 =$ _____

22. $2.4 \times 3.48 =$ _____

23. $5.04 \times 3.2 =$ _____

24. $27.4 \times 33.68 =$ _____

25. $0.451 \times 0.05 =$ _____

26. Katelyn has a vegetable garden that measures 16.75 feet in length and 5.8 feet in width. Find the area of the garden. Justify your answer.

_____

_____

27. **MP** **Use Math Tools** Find examples of decimals in a newspaper or magazine, on television, or on the Internet. Write a real-world problem in which multiplies decimals.

_____

_____

28. **MP** **Be Precise** Find the area of the figure at the right. Justify your procedure.

_____

Show
your
work. →

6.1 in.

3 in.

6.9 in.

3.1 in.

29. Junnie walked for 2.5 hours at a speed of 3.2 miles per hour. Maurice walked for 1.8 hours at a speed of 4.1 miles per hour. (*Hint:* Distance equals speed times time.)

a. Who walked farther? _____

_____

b. How much farther did that person walk? _____

_____

30. Renaldo can rollerblade 9.7 miles per hour. Eliza can rollerblade 8.2 miles per hour. How much farther will Renaldo rollerblade in 0.75 hour?

31. Look at the figures below. The area of the larger rectangle is 4 times the area of the smaller rectangle. Fill in the missing measurement.

16.2 cm

12.5 cm

cm

25 cm

Spiral Review

Divide. 5.NBT.6

32. $60 \div 12 =$ _____

33. $96 \div 8 =$ _____

34. $750 \div 15 =$ _____

35. Logan has 20 action figures. He is shipping them to a friend. He can fit 3 action figures in a box. How many boxes will he need? 5.NBT.6

_____

36. Three friends are dividing the cost of a kite equally. The kite costs $15.75. How much will each person pay? 5.NBT.7

_____

$15.75

Inquiry Lab
Multiply by Powers of Ten

 HOW can number patterns be used to multiply by powers of 10?

Content Standards
6.NS.3

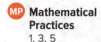 **Mathematical Practices**
1, 3, 5

Each planet in our solar system orbits around the Sun at a different distance from the Sun. Mercury orbits at an average distance of 28.6 million miles. One million is 1,000,000. What is 28.6 × 1,000,000?

What do you know? _____

What do you need to know? _____

Hands-On Activity

Numbers like 10, 100, and 1,000 are called *powers of 10* because they can be obtained by raising 10 to a power.

Step 1   Use the table to find a pattern.

Decimal		Power of 10		Product
28.6	×	0.1	=	2.86
28.6	×	1	=	28.6
28.6	×	10	=	286
28.6	×	1,000	=	28,600

Move the decimal point the _____ number of places as the number of zeros in the power of 10.

Move the decimal point to the _____ when multiplying by a power of 10 that is less than 1.

Move the decimal point to the _____ when multiplying by a power of 10 that is greater than 1.

Step 2   Determine how many zeros are in 1,000,000 and move the decimal point in 28.6 the appropriate number of places.

There are [ ] zeros in 1,000,000.

$$28.6 \text{ million} = 28.6 \times 1{,}000{,}000$$
$$= 28.600000$$
$$= 28{,}600{,}000$$

Move the decimal point [ ] places to the right.

MP **Use Math Tools** Work with a partner to complete the tables.

Decimal		Power of 10		Product
12.4	×	0.1	=	1.24
1. 12.4	×	0.01	=	
2. 12.4	×	0.001	=	
3. 12.4	×	0.0001	=	

Decimal		Power of 10		Product
1.24	×	1	=	1.24
4. 1.24	×	10	=	
5. 1.24	×	100	=	
6. 1.24	×	1,000	=	
7. 1.24	×	10,000	=	
8. 1.24	×	100,000	=	
9. 1.24	×	1,000,000	=	
10. 1.24	×	10,000,000	=	

11. **MP** **Use Math Tools** Suppose you plan to purchase 10 items that each cost $4.95. Explain how to use mental math to find the cost of the 10 items.

_____

12. **MP** **Reason Inductively** The product of 13.6 and a power of 10 is 13,600. What is the power of 10? Explain. _____

_____

Create

On Your Own

13. **MP** **Reason Abstractly** Write a rule you can use to find the product of a number and power of 10 without using paper and pencil or a calculator. _____

_____

14. **inquiry** HOW can number patterns be used to multiply by powers of 10?

_____

_____

MP Problem-Solving Investigation
Look for a Pattern

Content Standards
6.NS.3

MP Mathematical Practices
1, 3, 4, 8

Case #1 Dance Party

The Student Government is organizing a spring dance. They plan to decorate with helium-filled balloons. The cost of the balloons is shown in the table.

What is the cost of 6 bags of balloons?

Number of Bags	Total Cost ($)
1	4.75
2	9.50
3	14.25
4	19.00

1 **Understand** *What are the facts?*

The table shows the cost of the balloons. Six bags of balloons are needed.

2 **Plan** *What is your strategy to solve this problem?*

Look for a pattern in the table. Each bag costs $4.75.

3 **Solve** *How can you apply the strategy?*

Complete the table to find the cost of 6 bags of balloons.

Number of Bags	Total Cost ($)
1	4.75
2	9.50
3	14.25
4	19.00
5	
6	

+ 4.75
+ 4.75
+ 4.75
+ 4.75
+ 4.75

So, six bags of balloons cost $ _____ .

4 **Check** *Does the answer make sense?*

Use multiplication to check your answer. $4.75 \times 6 =$ _____ ✓

Analyze the Strategy

Tutor

MP **Reason Inductively** How would the results change if the store offered a discount of $0.50 for each bag of balloons? _____

_____

Case #2 Virtual DJ

The Student Government is hiring a DJ for the spring dance. They expect the dance to last for 5 hours. The cost to hire DJ Trax is shown in the table.

How much will it cost to hire DJ Trax for the dance?

Cost to Hire DJ Trax	
Number of Hours	Total Cost ($)
1	125.50
2	251.00
3	376.50

Understand

Read the problem. What are you being asked to find?

I need to _____.

Underline key words and values in the problem. What information do you know?

The cost to hire DJ Trax is $ _____ for 1 hour, $ _____ for 2 hours, and $ _____ for 3 hours.

Is there any information that you do *not* need to know?

I do not need to know _____.

Plan

Choose a problem-solving strategy.

I will use the _____ strategy.

Solve

Use your problem-solving strategy to solve the problem.

Describe the pattern. Then complete the table.

The cost to hire DJ Trax _____ by $ _____ for each hour.

Cost to Hire DJ Trax				
Number of Hours	1	2	3	4
Total Cost ($)	125.50	251.00	376.50	

So, it will cost $ _____ to hire DJ Trax for 5 hours.

Check

Use information from the problem to check your answer.

$ _____ × ☐ = $ _____

Collaborate

Work with a small group to solve the following cases. Show your work on a separate piece of paper.

Case #3 Gaming

The table shows the cost of a subscription to the Action Gamers Channel.

Action Gamers Channel Prices	
Number of Months	Total Cost ($)
2	15.90
3	23.85
4	31.80

What is the cost of a 1 year subscription?

_____

Case #4 Number Theory

The diagram to the right is known as Pascal's Triangle.

If the pattern continues, what will the numbers in the next row be from left to right?

_____

Case #5 Number Sense

Describe the pattern below. Then find the next three numbers.

3.5, 13, 41.5, 127, ☐ , ☐ , ☐

_____

Use any strategy!

Case #6 Games

Claudio wants to buy a new game system. A certain store sells the system for $235.99 and the games for $45.99 each. He bought one system and 3 games.

If Claudio used a $400 gift card, how much will be left on the card?

_____

Mid-Chapter Check

Vocabulary Check

1. Define *product*. Give an example of two whole number factors with a product of 9.

   _____

   _____

Skills Check and Problem Solving

Find each sum or difference. (Lesson 1)

2. $42.7 + 52.12 = $ _____

3. $4.7 - 3.28 = $ _____

4. $8.37 - 0.015 = $ _____

Multiply. (Lessons 3 and 4)

5. $2.3 \times 5 = $ _____

6. $3.4 \times 5.2 = $ _____

7. $1.2 \times 0.015 = $ _____

8. The table shows a list of walking trails in the United States. Latisha walked the KATY Trail 6 days last week. How many miles did she walk in a week. (Lesson 3)

   _____

9. **MP Be Precise** The length of a pool table is 7.1 feet and the width is 3.6 feet. Find the area of the surface of the pool table by multiplying the length by the width. (Lesson 4)

   _____

Location	Length of Trail (mi)
Florida Trail (FL)	4.8
Long Path (NJ)	3.3
Ohio & Erie Canal Trail (OH)	4.3
KATY Trail (MO)	5.7
Point Reyes National Seashore (CA)	5.0

10. **MP Persevere with Problems** While shopping for school supplies, Ashton buys two 3-ring binders, three packages of pencils, and two erasers. If he pays using a $20 bill, how much change will he receive? (Lesson 2) _____

School Supplies	
Item	**Cost ($)**
eraser	0.75
package of pencils	1.50
3-ring binder	2.75

Divide Multi-Digit Numbers

Vocabulary Start-Up

When one number is divided by another, the result is called a *quotient*. The *dividend* is the number that is divided and the *divisor* is the number used to divide another number.

Label the division problem with the correct vocabulary term: quotient, dividend, and divisor.

$$80\overline{)240} \quad \text{with } 3$$

 Essential Question

HOW can estimating be helpful?

Content Standards
6.NS.2

MP Mathematical Practices
1, 2, 3, 4, 5, 6

 ## Real-World Link

Circulation When you are at rest it takes about 60 seconds for a single blood cell to travel around your body and back to your heart.

1. In 120 seconds, about how many times does a single blood cell travel around your body and back to your heart? Write the dividend, divisor, and quotient in the diagram below.

quotient

divisor → dividend

2. Camila's target heart rate should be about 200 beats per minute. Estimate the number of times Camila's heart will beat in one second if her heart is working at this rate. Explain.

_____

_____

Which MP Mathematical Practices did you use?
Shade the circle(s) that applies.

① Persevere with Problems
② Reason Abstractly
③ Construct an Argument
④ Model with Mathematics
⑤ Use Math Tools
⑥ Attend to Precision
⑦ Make Use of Structure
⑧ Use Repeated Reasoning

Divide Three-Digit Dividends

In this lesson, you will divide multi-digit numbers. Use estimation to help you place the first digit in the quotient.

Examples

1. **Find 351 ÷ 9.**

Estimate $360 \div 9 = 40$. So, the first digit is in the tens place.

Write $351 \div 9$ as $9\overline{)351}$.

$$
\begin{array}{r}
39 \\
9\overline{)351} \\
-27 \\
\hline
81 \\
-81 \\
\hline
0
\end{array}
$$

Divide each place-value position from left to right.

Since $81 - 81 = 0$, there is no remainder.

So, $351 \div 9$ is 39.

Check Compare 39 to the estimate. $39 \approx 40$ ✔

- -

2. **Find $31\overline{)878}$.**

Estimate $900 \div 30 = 30$. So, the first digit is in the tens place.

$$
\begin{array}{r}
28 \text{ R10} \\
31\overline{)878} \\
-62 \\
\hline
258 \\
-248 \\
\hline
10
\end{array}
$$

Divide each place-value position from left to right.

Since $258 - 248 = 10$ and $10 < 31$, 10 is the remainder.

So, $31\overline{)878}$ is 28 R10.

Check 28 R10 $\approx$ 30 ✔

 Show your work.

Got it? Do these problems to find out.

Find each quotient.

  a. $768 \div 8$ **b.** $16\overline{)318}$

a. _____

b. _____

Divide Four-Digit Dividends

The steps for dividing three-digit dividends and four-digit dividends are the same.

Examples

3. **Find 6,493 ÷ 75.**

Estimate $6,400 \div 80 = 80$

$$
\begin{array}{r}
86 \text{ R43} \\
75\overline{)6,493} \\
-600 \\
\hline
493 \\
-450 \\
\hline
43
\end{array}
$$

Divide each place-value position from left to right.

Check for Reasonableness 86 R43 ≈ 80 ✔

The quotient of 6,493 ÷ 75 is 86 R43.

Check Your Answer

You can check division with a remainder. Multiply the quotient by the divisor. Then add the remainder.

$$
\begin{array}{r}
86 \\
\times 75 \\
\hline
6,450
\end{array}
\qquad
\begin{array}{r}
6,450 \\
+ \ \ 43 \\
\hline
6,493
\end{array}
$$

4. **The average person has 1,460 dreams a year. What is the average number of dreams a person has each night?**

Find 1,460 ÷ 365.

Estimate $1,600 \div 400 = 4$

$$
\begin{array}{r}
4 \\
365\overline{)1,460} \\
-1,460 \\
\hline
0
\end{array}
$$

Check for Reasonableness 4 = 4 ✔

The average number of dreams a person has each night is 4.

Got it? Do these problems to find out.

c. Find $56\overline{)4,321}$.

d. Find $91\overline{)8,465}$.

e. To promote its opening weekend, a water park gave the local middle school 1,050 free tickets. The middle school has 350 students. Each student will receive the same number of tickets. How many tickets will each student receive?

Show your work.

c. _____

d. _____

e. _____

Example

5. The total number of seats in a college stadium is **54,912.** There are **44** sections and each section has an equal number of seats. How many seats are in each section?

Divide 54,912 by 44.

```
        1,248
   44)54,912      Divide each place-value position from left to right.
      −44
      109
      −88
       211
      −176
        352
       −352
          0
```

There are 1,248 seats in each section.

Guided Practice

Find each quotient. (Examples 1–4)

1. 8)736

2. 11)620

3. 19)7,814

4. 37)3,511

5. Zach bought two new jet skis for $15,480. He will make 36 equal payments. How much will each payment be?

(Example 5) _____

6. **Building on the Essential Question** How is estimation helpful when dividing multi-digit numbers?

_____
_____
_____
_____

Independent Practice

Go online for Step-by-Step Solutions

Find each quotient. (Examples 1–3)

1. 174 ÷ 6 =

_____

 Show your work.

2. 453 ÷ 8 =

_____

3. 645 ÷ 43 =

_____

4. 299 ÷ 21 =

_____

5. 62)8,090

6. 31)2,480

7. 34)5,780

8. 16)3,482

9. A tour bus travels 2,160 miles in 36 hours. What is the average distance the bus travels in one hour? (Example 4)

 Show your work.

_____

10. A charity sold 475 tickets to a dinner auction. If the charity raised $16,625 in ticket sales, what was the cost of one ticket? (Example 5)

_____

11. A city phone book has 86 pages filled with residents' names. There are a total of 15,050 names in the book. Each page has an equal number of names on it. How many names are on each page? (Example 5)

_____

12. **MP** **Use Math Tools** The table shows the number of servings for different size cakes at Mimi's Bakery. Suppose a high school graduation expected 2,889 guests. How many X-large sheet cakes should the school order? Explain how you solved.

_____

_____

_____

Mimi's Bakery	
Sheet Cake Size	**Number of Servings**
Small	30
Medium	60
Large	90
X-Large	120

13. **MP** **Be Precise** How many 8-ounce cups can be filled from 9 gallons of juice? (*Hint:* There are 128 ounces in one gallon.)

_____

14. **MP** **Be Precise** Water stations will be placed every 600 meters of a fifteen kilometer race. How many water stations will be needed? (*Hint:* There are 1,000 meters in one kilometer.)

_____

H.O.T. Problems Higher Order Thinking

15. **MP** **Model with Mathematics** Write and solve a real-word problem that involves a two-digit divisor and a four-digit dividend. _____

_____

_____

16. **MP** **Persevere with Problems** If the divisor is 40, what is the least three-digit dividend that would give a remainder of 4? _____

_____

17. **MP** **Justify Conclusions** Can the remainder in a division problem ever equal the divisor? Why or why not? _____

_____

18. **MP** **Reason Abstractly** Use the digits 2, 4, and 8 one time each in the following problem.

$$\boxed{},\boxed{}00 \div \boxed{}0 = 30$$

_____

_____

19. **MP** **Reason Inductively** The table shows the mileage Karla drove on her trip. She drove for a total of 24 hours. Explain how to find her average speed. Then find the average speed.

_____

_____

Karla's Trip	
Day	Miles
1	486
2	316
3	638

Extra Practice

Find each quotient.

20. 182 ÷ 7 =

26

```
      26
  7 ) 182
    - 14
      42
    - 42
       0
```

21. 345 ÷ 6 =

_____

22. 792 ÷ 33 =

_____

23. 811 ÷ 79 =

_____

24. 44) 2,876

25. 26) 4,340

26. 33) 9,537

27. 19) 4,257

28. A city library has 9,440 nonfiction books. The librarian wants to divide the books evenly among 80 shelves. How many books will be on each shelf?

_____

29. A concession stand manager ordered 20,280 souvenir cups. He wants to divide the cups evenly among the 24 concession stands. How many cups will each concession stand receive?

_____

30. **MP Use Math Tools** The table shows the average weight of animals. How many more tons does a blue whale weigh than an African elephant? Explain how you solved. (*Hint:* There are 2,000 pounds in one ton.)

_____

_____

_____

Animal Weights	
Animal	**Weight (pounds)**
African Elephant	15,000
Blue Whale	238,000
Great White Shark	5,000
Lowland Gorilla	500

31. A toy factory assembles 19,824 toy castles over a 12-hour period of time. The same number of castles is assembled every hour.

How many toy castles were assembled each hour? [   ]

Each employee can assemble 28 castles per hour. How many employees worked each hour?

[   ]

32. The table shows the total amount spent on each item in a dance team's order. There are 15 girls on the team. Complete the table to determine the individual cost of each item.

Item	Total Cost ($)	Cost per Item
boots	960	
dress	1,350	
ribbons	105	
tights	180	
warm up	675	

How much will one team member spend on all of the items?

[   ]

Spiral Review

Divide mentally. 4.NBT.6

33. 300 ÷ 5 = _____

34. 4,800 ÷ 8 = _____

35. 4,200 ÷ 6 = _____

36. The maple tree in Logan's backyard is 58.6 feet tall. Plot 58.6 on the number line below. Then round 58.6 to the nearest whole number. 4.NF.7

Estimate Quotients

Vocabulary Start-Up

To determine what a compatible number is, first you must determine what compatible means. Fill in the table below.

Definition:

Example:

compatible

What would make numbers compatible?

Non-Example:

Essential Question

HOW can estimating be helpful?

Vocabulary

compatible numbers

Content Standards
6.NS.2

MP Mathematical Practices
1, 3, 4, 5

 ## Real-World Link

Remote Control Latasha and her two sisters want to buy their little brother a remote control helicopter. The helicopter costs $28.90. They decided to split the cost equally.

1. What number that is a multiple of 3 is close to $28.90? Explain.

2. Use your answer from Exercise 1 to determine about how much each person will pay. Explain.

Which **MP Mathematical Practices** did you use? Shade the circle(s) that applies.

① Persevere with Problems

② Reason Abstractly

③ Construct an Argument

④ Model with Mathematics

⑤ Use Math Tools

⑥ Attend to Precision

⑦ Make Use of Structure

⑧ Use Repeated Reasoning

Estimate by Rounding Dividends

To estimate quotients of decimals, use rounding and compatible numbers. **Compatible numbers** are numbers that are easy to divide mentally.

Examples

1. **Estimate 11.75 ÷ 3.**

Round the dividend, 11.75, to a whole number.

The divisor is 3. So, round 11.75 to a whole number that is a multiple of 3.

$$3\overline{)11.75} \longrightarrow 3\overline{)12}^{\,4}$$ Using multiples of 3, 12 is closest to 11.75

So, 11.75 ÷ 3 is about 4.

2. **The Jenkins family bought five tickets to a charity auction. The receipt shows the total cost of the tickets. Estimate the cost of each ticket. Justify your answer.**

$$5\overline{)61.25} \longrightarrow 5\overline{)60}^{\,12}$$ Round 61.25 to 60.

Each ticket costs about $12.
Since 5 × 12 = 60 and 60 ≈ 61.25, the answer is reasonable.

Show your work.

Got it? Do these problems to find out.

Estimate each quotient.

a. 49.3 ÷ 7

b. $25\overline{)98.1}$

c. Suppose the Jenkins family decided to purchase 6 tickets for a total price of $64.50 using a discount. Estimate the cost of each ticket. Justify your answer.

a. _____

b. _____

c. _____

Estimate by Rounding Divisors

You can also estimate quotients of decimals by rounding the divisors. Choose compatible numbers that are easy to divide mentally.

Examples

Tutor

3. **Estimate 32 ÷ 3.9.**

Round the divisor, 3.9, to a whole number.

The dividend is 32. So, round 3.9 to a whole number that is a factor of 32.

$$3.9\overline{)32} \longrightarrow 4\overline{)32}^{\,8}$$ Round 3.9 to 4 since 32 and 4 are compatible numbers.

So, 32 ÷ 3.9 is about 8.

Check by Multiplying 3.9 × 8 = 31.2

31.2 ≈ 32 ✔

4. **Estimate 56 ÷ 6.8.**

Round the divisor, ☐ , to a whole number.

The dividend is ☐ .

So, round 6.8 to a whole number that is a _____ of 56.

Round 6.8 to ☐ .

$$6.8\overline{)56} \longrightarrow \boxed{}\overline{)56}^{\,\boxed{}}$$

So, 56 ÷ 6.8 is about ☐ .

Check by Multiplying 6.8 × ☐ = ☐

☐ ≈ 56 ✔

Got it? Do these problems to find out.

Estimate each quotient.

d. 54 ÷ 9.16

e. $10.75\overline{)99}$

STOP and Reflect

How does the division fact 63 ÷ 9 = 7 help you to estimate the quotient of 63 ÷ 8.4? Answer below.

Show your work.

d. _____

e. _____

Example

5. **STEM** A Pacific Leatherback turtle can have a mass of up to 704.4 kilograms. An Olive Ridley turtle can have a mass of up to 49.9 kilograms. About how many times heavier is the Pacific Leatherback turtle? Explain why your answer is reasonable.

$$49.9\overline{)704.4} \longrightarrow 50\overline{)700}^{\,14}$$ Round 49.9 to 50 and 704.4 to 700.

The Pacific Leatherback is about 14 times heavier than the Olive Ridley turtle.

Check for Reasonableness Since 50 × 14 = 700, and 700 ≈ 704.4, your answer is reasonable. ✔

 Show your work.

Got it? Do this problem to find out.

f. There are approximately 250.9 million cars in the United States. Spain has approximately 25.1 million cars. About how many times more cars does the U.S. have than Spain? Explain why your answer is reasonable.

f. _____

Guided Practice

Estimate each quotient. (Examples 1, 3, and 4)

 Show your work.

1. 25 ÷ 4.7 ≈ _____

2. 40.79 ÷ 7 ≈ _____

3. $38.1\overline{)984.76}$ ≈ _____

4. **STEM** The average yearly precipitation for Gulfport, Mississippi, is 65.3 inches. About how much precipitation does the area receive each month? Explain why your answer is reasonable. (Example 2)

_____

5. Mauricio bought 6.75 yards of fabric for a total of $47.50. About how much was the cost per yard? Explain why your answer is reasonable. (Example 5)

_____

6. **Building on the Essential Question** When is it helpful to estimate quotients? _____

_____

Independent Practice

Go online for Step-by-Step Solutions

Estimate each quotient. (Examples 1, 3, and 4)

1. $32.4 \div 3 \approx$ _____

2. $76.2 \div 18.4 \approx$ _____

3. $11.4\overline{)35.7} \approx$ _____

 Show your work.

4. **Financial Literacy** Emily spent a total of $38.04 on four CDs. If each CD cost the same amount, what is a reasonable amount for the cost of each CD? Explain why your answer is reasonable. (Example 2)

_____

_____

5. A recipe for a smoothie calls for 0.75 pound of strawberries. If Kerry has 3.15 pounds of strawberries, how many batches of the recipe can she make? (Example 5)

_____

_____

6. **Financial Literacy** For each handmade greeting card Jacqui sells, she makes a profit of $0.35. In one week, she made a profit of $42. She sells the cards for $0.75 each.

a. About how many greeting cards did Jacqui sell that week?

_____

b. About how much did she earn before paying expenses?

_____

7. **MP Justify Conclusions** The average cow produces about 53 pounds of milk per day. If one gallon of milk weighs about 8.5 pounds, estimate the number of gallons of milk a cow produces each day. Explain why your estimate is reasonable.

_____

_____

8. When full, a 22-gallon gas tank holds 129.8 pounds of gasoline. Estimate the weight of one gallon of gasoline. If it costs $91.30 to fill the gas tank, estimate the cost per gallon.

_____

_____

9. MP **Use Math Tools** Use estimation and mental math to find the four

missing quantities from the receipt. _____

_____

Precious Pets — Receipt

Qty	Description	Unit Price	Total
▢	Hamster cage	$35.99	$35.99
▢	Exercise wheel	$5.29	$10.58
▢	Softwood bedding	$6.29	$25.16
▢	Hamster food	$4.59	$36.72
		Total	$108.45

🔥 H.O.T. Problems Higher Order Thinking

10. MP **Model with Mathematics** Write a real-world division problem involving decimals in which you would use compatible numbers to estimate the

quotient. _____

_____

_____

11. MP **Persevere with Problems** Determine where to place the decimal point in the dividend and divisor so that the quotient is between 23 and 25.

$$16023 \div 654$$

_____

12. MP **Reason Inductively** Is an estimate for the quotient of a division problem involving decimals *always, sometimes,* or *never* less than the actual quotient

of the numbers? Explain your reasoning. _____

_____

_____

_____

13. MP **Justify Conclusions** Explain how you know which compatible numbers to use when estimating a decimal quotient. Support your answer with an

example. _____

_____

_____

_____

Extra Practice

Estimate each quotient.

14. 54 ÷ 9.4 ≈ _____

Homework
Help

$$9.4\overline{)54} \rightarrow 9\overline{)54}^{\,6}$$

15. 45.8 ÷ 23.6 ≈ _____

16. $23.3\overline{)119}$ ≈ _____

17. The average annual snowfall in King Salmon, Alaska, is 45.9 inches. The snow season lasts about 7 months of the year. About how much snow does the area receive on average each month? Explain why your answer is reasonable.

_____

_____

18. Scientists at the zoo recently studied an anaconda that weighs 8,643.2 ounces. The average weight of the common rat is 11.8 ounces. About how many times heavier is the anaconda than the common rat? Explain why your answer is reasonable.

_____

_____

19. **MP** **Justify Conclusions** Aurelia would like to save $474.72 in a year to purchase a new video camera. She estimates she needs to save $40 per month. Explain why her estimate is reasonable.

_____

_____

20. A piggy bank containing only quarters has a mass of 850 grams when empty and 7,822 grams when filled. If a quarter weighs 5.6 grams, estimate the amount of money inside the piggy bank.

_____

_____

21. Melanie is making homemade stickers. She uses the recipe shown to create the glue for the stickers.

a. She has 545 milliliters of vinegar. Which is a more reasonable estimate for the number of batches she can make, 5 or 7? Explain your answer.

_____

_____

Recipe: Sticker Glue

- 4 packages gelatin
- 112 milliliters vinegar
- 14 milliliters lemon extract

b. About how many times as many milliliters of vinegar are needed than

lemon extract? _____

22. For a craft activity at a day care, each child will need 1.75 yards of ribbon and 2.8 yards of fabric. There are 25 yards of ribbon and 30 yards of fabric available. Estimate the number of children that can participate in the activity. Explain your reasoning.

23. An advertisement showing the costs of different bikes was in the local newspaper.

The cost of a 26" bike is equal to about how many bike locks?

☐ about 7 locks
☐ about 8 locks
☐ about 9 locks
☐ about 10 locks

Bike Country	
26" Bike	$135.99
Folding Bike Rack	$43.95
Seat Covers	$6.59
Bike Lock	$12.89
Helmet	$29.49

Spiral Review

Find each quotient. 5.NBT.7

24. 8.4 ÷ 10 = _____

25. 100)‾14.7‾ = _____

26. 94.5 ÷ 100 = _____

27. Describe the number pattern below. Then find the next three numbers.

7,345.6; 734.56; 73.456; **5.NBT.2** _____ ; _____ ; _____

_____

28. The movie theater sold 825 tickets to fill 3 theaters. Each theater has an equal number of seats. How many seats are in each theater?

4.NBT.6 _____

Divide Decimals by Whole Numbers

connectED.mcgraw-hill.com

Real-World Link

Movies Charlotte, Aaron, Maddie, and Catie went to the movies and ordered snacks from the menu shown.

Cinema 15		
Popcorn	small	$2.45
	large	$5.60
Candy	small	$2.25
	large	$3.20
Drink	small	$2.75
	medium	$3.35
	large	$3.95

1. How much did they pay for four small popcorns?

    _____ × 4 = _____

2. What is the total cost for two small packages and one large package of candy?

    _____

3. How much do four medium drinks cost?

    _____ × 4 = _____

4. What is the total cost for Exercises 1–3?

    Popcorn $ ☐ . ☐

    Candy $ ☐ . ☐

    Drinks   + $ ☐ . ☐
    _____
             $ ☐ . ☐

5. Estimate how much each person should pay if they split the total cost evenly.

Essential Question

HOW can estimating be helpful?

Content Standards
6.NS.3

MP Mathematical Practices
1, 3, 4, 5, 6

Which MP Mathematical Practices did you use?
Shade the circle(s) that applies.

① Persevere with Problems
② Reason Abstractly
③ Construct an Argument
④ Model with Mathematics
⑤ Use Math Tools
⑥ Attend to Precision
⑦ Make Use of Structure
⑧ Use Repeated Reasoning

Divide a Decimal by a 1-Digit Number

When dividing a decimal by a whole number, divide as with whole numbers. Then place the decimal in the quotient directly above its place in the dividend.

Example

1. **Find 6.8 ÷ 2.**   **Estimate** $6 \div 2 = 3$

$$\begin{array}{r} 3.4 \\ 2\overline{)6.8} \\ -6 \\ \hline 08 \\ -8 \\ \hline 0 \end{array}$$

6 ones divided by 2 is 3 ones.

8 tenths divided by 2 is 4 tenths.

$6.8 \div 2 = 3.4$ Compared to the estimate, the quotient is reasonable.

Show your work.

a. _____

b. _____

c. _____

Got it? **Do these problems to find out.**

a. $7.5 \div 3$ **b.** $3.5 \div 7$ **c.** $9.8 \div 2$

Divide a Decimal by a 2-Digit Number

The decimal point in the quotient is placed directly above its place in the dividend. In real-world situations where the division does not result in a remainder of zero, round the quotient to a specified place.

Example

2. **Find 7.7 ÷ 14.**   **Estimate** $10 \div 10 = 1$

$$\begin{array}{r} 0.55 \\ 14\overline{)7.70} \\ -70 \\ \hline 70 \\ -70 \\ \hline 0 \end{array}$$

← Place the decimal point.

Annex a zero and continue dividing.

$7.7 \div 14 = 0.55$ Compared to the estimate, the quotient is reasonable.

Checking Answers

To check that the answer is correct, multiply the quotient by the divisor.

In Example 2,

$0.55 \times 14 = 7.7$. ✓

Got it? Do these problems to find out.

d. 9.48 ÷ 15   **e.** 3.49 ÷ 4   **f.** 55.08 ÷ 17

Show your work.

d. _____

e. _____

f. _____

 Example

Tutor

3. Lin is mailing a care package to his brother. The table gives the cost for mailing packages. If Lin's care package weighs 3 pounds, how much is the cost per pound?

Weight (Pounds)	Cost ($)
1	4.80
2	5.63
3	6.74
4	7.87

To find the cost per pound, divide $6.74 by 3.

```
    2.246   ←  Place the decimal point after dividing to thousandths.
3)6.7400
 − 6
   07
  − 6
    14
   − 12
     20        Annex a zero and continue dividing.
    − 18
      2        The remainder will never be zero.
```

Round 2.246 to 2.25 because hundredths are the smallest denomination used in money. It costs about $2.25 per pound to mail the package.

Check Use a bar diagram and multiplication to check your work.

```
|------------- 6.75 -------------|
| 2.25 | 2.25 | 2.25 |
```

2.25 × 3 = 6.75

6.75 ≈ 6.74 ✓

Dividing Money

When dividing money, it is sometimes necessary to divide to the thousandths place and then round to the hundredths.

Got it? Do this problem to find out.

g. Find the cost per pound of a two-pound and four-pound package.

g. _____

Example

4. Ryan and his sister are sharing the cost of a video game. The video game costs $28.60. If Ryan saved $20 to buy the game, how much does he have left after paying his share?

Step 1	Determine how much Ryan will pay.

Place the decimal point.

$$
\begin{array}{r}
14.30 \\
2\overline{)28.60} \\
-2 \\
\hline
08 \\
-8 \\
\hline
06 \\
-6 \\
\hline
0
\end{array}
$$

Ryan's share is $14.30.

Step 2	Determine how much Ryan will have left.

$$
\begin{array}{r}
\$20.00 \\
-\ \$14.30 \\
\hline
\$5.70
\end{array}
$$

So, Ryan has $5.70 left.

 Show your work.

Got it? Do this problem to find out.

h. _____

h. Kristen and her two friends are sharing the cost of a funnel cake. The funnel cake costs $5.49. If Kristen has $2.00, how much will she have left after she pays her share?

Guided Practice

 Check ✓

Divide. Round to the nearest tenth if necessary. (Examples 1 and 2)

 Show your work.

1. $3.6 \div 4 =$ _____

2. $12.32 \div 22 =$ _____

3. $69.904 \div 34 =$ _____

4. Light travels 5.88 trillion miles in one year. How far will light travel in one month? (Examples 1 and 3) _____

5. Four dozen bagels costs $30.00. How much change will you receive if you pay for a dozen bagels with a ten-dollar bill? (Examples 2 and 4) _____

6. **Building on the Essential Question** How can estimating quotients help you to place the decimal correctly? _____

_____

Rate Yourself!

How confident are you about dividing decimals by whole numbers? Check the box that applies.

For more help, go online to access a Personal Tutor.

FOLDABLES Time to update your Foldable!

Independent Practice

Go online for Step-by-Step Solutions

Divide. Round to the nearest tenth if necessary. (Examples 1 and 2)

1. 39.39 ÷ 3 = _____

2. 7.24 ÷ 7 = _____

3. 118.5 ÷ 5 = _____

4. 11.4 ÷ 19 = _____

5. 55.2 ÷ 46 = _____

6. 336.752 ÷ 31 = _____

7. The Gonzalez family is taking a cruise that costs $3,082.24 for a family of four. How much does it cost per person? (Example 3)

_____

8. Find the average height of the buildings shown in the table. (*Hint:* To find the average, add the values and divide by the number of values.) (Example 4)

_____

World's Tallest Buildings (thousands of feet)				
1.667	1.483	1.483	1.451	1.381

9. **MP** **Be Precise** Mr. Jamison will stain the deck in his backyard. The deck has an area of 752.4 square feet. If the deck is 33 feet long, how wide is it? Justify your procedure. _____

_____

10. **MP** **Be Precise** The Verrazano-Narrows Bridge in New York City is 4.26 thousand feet long and is the seventh longest suspension bridge in the world. There are 3 feet in a yard. How long is the bridge in yards? Justify your procedure. _____

_____

11. The Student Council is raising money by selling bottled water at a band competition. The table shows the prices for different brands. Which brand costs the least per bottle? Explain your reasoning. _____

_____

_____

Cost of Bottled Water (20 oz bottles)		
Brand A	6-pack	$3.45
Brand B	12-pack	$5.25
Brand C	24-pack	$10.99

🔥 H.O.T. Problems Higher Order Thinking

12. **MP Persevere with Problems** Find each of the following quotients to detect a pattern. Then explain how you can use this pattern to mentally divide 0.0096 by 3.

$844 \div 2$ $0.844 \div 2$ $84.4 \div 2$ $0.0844 \div 2$ $8.44 \div 2$ $0.00844 \div 2$

_____

_____

_____

13. **MP Find the Error** Amanda is finding $11.2 \div 14$. Find her mistake and correct it. _____

_____

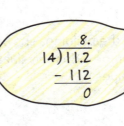

_____

14. **MP Reason Inductively** Is the quotient $2.7 \div 3$ greater than or less than 1? Explain.

_____

_____

15. **MP Use Math Tools** Explain how you can use estimation to place the decimal point in the quotient $42.56 \div 22$.

_____

_____

16. **MP Model with Mathematics** Write a real-world problem that involves dividing a decimal by a whole number.

_____

_____

_____

Extra Practice

Divide. Round to the nearest tenth if necessary.

17. $36.8 \div 2 = $ _18.4_

Homework Help →

$$
\begin{array}{r}
18.4 \\
2\overline{)36.8} \\
-2 \\
\hline
16 \\
-16 \\
\hline
0\,8 \\
-8 \\
\hline
0
\end{array}
$$

18. $124.2 \div 9 = $ _____

19. $6.271 \div 4 = $ _____

20. $10.22 \div 14 = $ _____

21. $59.84 \div 32 = $ _____

22. $751.2 \div 25 = $ _____

23. The Franklin Middle School jazz band plans to have a car wash to raise $468.75 for a new sound system. In the past, they washed an average of 125 cars at each car wash. What should they charge per car so they can reach their goal?

_____

_____

24. Marcel is weeding a rectangular vegetable garden. The garden has an area of 599.5 square feet. If the garden is 22 feet wide, how long is the garden? Justify your procedure.

_____

_____

25. **MP** **Use Math Tools** The table shows the prices for different party toy packages from the Tomtown Toy Company. Which item costs the least per toy? Explain your reasoning. _____

_____

_____

Cost of Party Toy Packages		
Dominoes	6-pack	$3.98
Peg Games	12-pack	$9.99
Mini Footballs	24-pack	$17.98

26. The table shows the number of subscribers to several Internet providers. Determine if each statement is true or false.

Internet Provider	Subscribers (millions)
Company A	2.45
Company B	3.12
Company C	2.83

a. Company B earned $119 million last month. Each subscriber paid about $40.　　☐ True　　☐ False

b. Company A earned $126 million last month. Each subscriber paid about $60.　　☐ True　　☐ False

c. Company C earned $108 million last month. Each subscriber paid about $30.　　☐ True　　☐ False

27. Marvin completed 8 rounds of a trivia game. The scores he earned each round are shown in the table.

Round	1	2	3	4	5	6	7	8
Score	10.2	11.4	10.5	12.6	12.2	11.6	13.1	12.8

What was his average score? ▢

Spiral Review

Add or subtract mentally. Use compensation. **5.NBT.7**

28. $0.47 + 0.36 =$ _____

29. $26.5 - 9.3 =$ _____

30. $29.4 + 1.7 =$ _____

31. Use $<$, $>$, or $=$ to compare 34.3 and 34.32. **5.NBT.3b**

32. A king cobra has a mass of 8.845 kilograms. Round the mass to the nearest tenth kilogram. **5.NBT.4**

33. The same king cobra is 4.237 meters long. Round the length to the nearest meter. **5.NBT.4**

Divide Decimals by Decimals

 Real-World Link

Art An art studio has 36 gallons of acrylic paint. They separate it into 9 containers. How many gallons are in each container?

Division Problem	Quotient
36 ÷ 9	

Use the division problem to find patterns and complete the tables below. Then use these patterns to describe the dividends, divisors, and quotients in each set.

 Essential Question

HOW can estimating be helpful?

Content Standards
6.NS.3

MP Mathematical Practices
1, 3, 4, 5

1.

Division Problem	Quotient
36 ÷ 0.9	40
36 ÷ 0.09	400
36 ÷ 0.009	4,000
36 ÷ 0.0009	

2.

Division Problem	Quotient
3.6 ÷ 9	0.4
0.36 ÷ 9	0.04
0.036 ÷ 9	0.004
0.0036 ÷ 9	

3.

Division Problem	Quotient
3.6 ÷ 0.9	4
0.36 ÷ 0.09	4
0.036 ÷ 0.009	4
0.0036 ÷	4

Which MP Mathematical Practices did you use?
Shade the circle(s) that applies.

① Persevere with Problems

② Reason Abstractly

③ Construct an Argument

④ Model with Mathematics

⑤ Use Math Tools

⑥ Attend to Precision

⑦ Make Use of Structure

⑧ Use Repeated Reasoning

Divide by Decimals

When dividing by decimals, change the divisor into a whole number. To do this, multiply both the divisor and the dividend by the same power of 10. Then divide as with whole numbers.

Examples

Tutor

1. Find **1.71 ÷ 0.9.**   **Estimate** 2 ÷ 1 = 2

Multiply by 10 to make a whole number.

$$0.9\overline{)1.71} \longrightarrow 9\overline{)17.1}$$

Multiply by the same number, 10.

1.9
9)17.1
− 9
81
− 81
0

Place the decimal point.
Divide as with whole numbers.

1.71 divided by 0.9 is 1.9.   Compared to the estimate, the quotient is reasonable.

Check 1.9 × 0.9 = 1.71 ✓

2. Find **2.64 ÷ 0.6.**   **Estimate** ☐ ÷ ☐ = ☐

$$0.6\overline{)2.64}$$

Multiply 0.6 by ☐ to make a whole number.

Multiply the dividend, ☐, by the same power of 10.

☐
0.6)2.64   Place the decimal point in the quotient.
☐   Divide as with whole numbers.
___
☐
___
☐

Show your work.

2.64 divided by 0.6 is ☐.

Compared to the estimate, is the quotient reasonable? _____

Got it? Do these problems to find out.

a. 54.4 ÷ 1.7   **b.** 8.424 ÷ 0.36   **c.** 0.0063 ÷ 0.007

a. _____

b. _____

c. _____

Zeros in the Quotient and Dividend

Line up the numbers by place value as you divide. Annex zeros in the quotient in order to keep digits with the correct place value. Annex zeros in the dividend to continue dividing after the decimal point.

Examples

3. Find 52 ÷ 0.4.

$$0.4 \overline{)52.0}$$

Multiply each by 10.

```
   130.
4)520.
 − 4
   12
 − 12
   00
```

Place the decimal point.

Write a zero in the ones place of the quotient because 0 ÷ 4 = 0.

So, 52 ÷ 0.4 = 130.

4. Find 0.009 ÷ 0.18.

$$0.18 \overline{)0.009}$$

Multiply each by 100.

```
      0.05
18)0.90
 − 0
   09
 − 00
   90
 − 90
    0
```

Place the decimal point.

9 tenths divided by 18 is 0, so write a 0 in the tenths place.

Annex a 0 in the dividend and continue to divide.

So, 0.009 ÷ 0.18 is 0.05.

5. Find 11.2 ÷ 0.07.

Multiply 0.07 and 11.2 by ⬚.

$$0.07 \overline{)11.20}$$

Place the decimal point in the quotient.
Divide as with whole numbers.

Checking Answers

You can always check your answer to a division problem by multiplying the quotient by the divisor.

Show your work.

d. _____

e. _____

f. _____

Got it? Do these problems to find out.

d. 5.6 ÷ 0.014   **e.** 6.24 ÷ 200   **f.** 0.4 ÷ 25

Example

 Tutor

6. How many times as many Internet users are there in Japan than in Spain? Round to the nearest tenth.

Find 127.4 ÷ 40.4.

$$
\begin{array}{r}
3.15 \\
40.4\overline{)127.4} \rightarrow 404\overline{)1274.00} \\
-\ 1212 \\
\hline
620 \\
-\ 404 \\
\hline
2160 \\
-\ 2020 \\
\hline
140
\end{array}
$$

Internet Users in 2008 (millions)	
China	1,321.9
United States	301.1
Japan	127.4
France	63.7
Spain	40.4
Canada	33.4

To the nearest tenth, 127.4 ÷ 40.4 = 3.2. So, there are about 3.2 times as many Internet users in Japan than in Spain.

Guided Practice

 Check ✓

Divide. (Examples 1–5)

1. 3.69 ÷ 0.3 = _____

Show your work.

2. 0.0338 ÷ 1.3 = _____

3. 2.943 ÷ 2.7 = _____

4. Alicia bought 5.75 yards of fleece fabric to make blankets for a charity. She needs 1.85 yards of fabric for each blanket. How many blankets can Alicia make with the fabric she bought? (Example 6)

_____

5. **Q** **Building on the Essential Question** When is it helpful to round the quotient to the nearest hundredth?

_____
_____
_____
_____

Rate Yourself!

Are you ready to move on? Shade the section that applies.

YES   ?   NO

For more help, go online to access a Personal Tutor. Tutor

FOLDABLES Time to update your Foldable!

Independent Practice

Go online for Step-by-Step Solutions

Divide. (Examples 1–5)

 Show your work.

1. $1.44 \div 0.4 =$ _____

2. $16.24 \div 0.14 =$ _____

3. $0.6 \div 0.0024 =$ _____

4. $96.6 \div 0.42 =$ _____

5. $13.5 \div 0.03 =$ _____

6. $0.12 \div 0.15 =$ _____

7. **MP** **Use Math Tools** The average person's *stride length,* the distance covered by one step, is approximately 2.5 feet long. How many steps would the average person take to travel 50 feet? (Example 6)

_____

8. **STEM** Alaska has a coastline of about 6.64 thousand miles. Florida has about 1.35 thousand miles of coastline. How many times more coastline does Alaska have than Florida? Round to the nearest tenth if necessary. Justify your procedure.

_____

9. **MP** **Model with Mathematics** Refer to the graphic novel frame below for Exercises a–b.

CAR WASH Paycheck

Hourly wage | Hours worked
$5.50 |

I've already saved $68 for the $200 video game system.

a. How many hours does Raj need to work to earn the remainder of the

money he needs to buy the video game system? _____

b. Suppose Raj receives a raise for his hard work and now earns $6.25 per hour. How many hours would he need to work to earn $132?

_____

10. A necklace is being made with beads that are 1.25 centimeters in diameter each. The necklace is 30 centimeters long. How many beads are needed? _____

11. 🏠 **MP** **Use Math Tools** Use the table that shows popular sports car colors in North America.

Popular Sports Car Colors	
Color	**Portion of Responses**
Silver	0.2
Blue	0.16
Black	0.14
Red	0.09
Other	0.41

    a. How many times more respondents chose silver than red? Round to the nearest tenth if necessary.

    _____

    b. How many times more respondents chose either silver or black than red? Round to the nearest tenth if necessary.

    _____

🔥 H.O.T. Problems Higher Order Thinking

12. **MP** **Persevere with Problems** Find two positive decimals a and b that make the following statement true. Then find two positive decimals a and b that make the statement false.

$$\text{If } a < 1 \text{ and } b < 1, \text{ then } a \div b < 1.$$

_____

_____

_____

13. **MP** **Which One Doesn't Belong?** Identify the problem that does not have the same quotient as the other three. Explain your reasoning.

| $49 \div 7$ | $4.9 \div 7$ | $0.49 \div 0.7$ | $0.049 \div 0.07$ |

_____

_____

14. **MP** **Reason Inductively** Rodrigo researched the number of pets owned in the U.S. and displayed his results in the bar graph shown. In his research, he found that $7,737.6 million was spent annually on veterinary visits for dogs and $5,913 million was spent annually on visits for cats. How much more was spent on average by each person for dogs than for cats?

_____

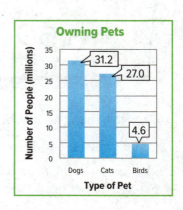

Owning Pets

Extra Practice

Divide.

15. $0.68 \div 3.4 =$ _0.2_

Homework Help →

$$
\begin{array}{r}
0.2 \\
3.4\overline{\smash{)}0.68} \\
-68 \\
\hline
0
\end{array}
$$

16. $2.07 \div 0.9 =$ _____

17. $0.16728 \div 3.4 =$ _____

18. $1.08 \div 2.7 =$ _____

19. $8.4 \div 0.02 =$ _____

20. $0.242 \div 0.4 =$ _____

21. A submarine sandwich 1.5 feet long is cut into 0.25-foot pieces. How many pieces will there be?

Show your work. →

_____

22. **MP Use Math Tools** Find examples of decimals in a newspaper or magazine. Write a real-world problem in which you would divide decimals.

_____

_____

23. The table shows the five most populated countries in the world. How many times as many people live in China than in the United States? Round to the nearest tenth if necessary.

_____

Most Populated Countries	
Country	**Approximate Population (billions)**
China	1.325
India	1.13
United States	0.304
Indonesia	0.235
Brazil	0.19

24. **MP Justify Conclusions** Lake Superior, along the U.S.-Canadian border, has a maximum depth of 1.333 thousand feet. There are 5,280 feet in one mile. How deep is Lake Superior in miles? Round to the nearest hundredth if necessary. Explain your answer.

_____

_____

25. About 24.8 million people live in Texas. About 0.6 million people live in Vermont. How many times as many people live in Texas than Vermont? Round to the nearest tenth if necessary. [ ]

26. The table shows the time it took several members of a cycling club to bike different distances. Sort the cyclists from least to greatest unit rate.

Member	Distance (mi)	Time (min)
Abby	5.8	23.2
Devon	7.2	40
Max	6.25	31.25
Selena	6.75	28.125

	Cyclist	Rate (mi/min)
Least		
Greatest		

Spiral Review

Fill in each ◯ with <, >, or = to make a true sentence. **4.NF.2**

27. $\frac{2}{4}$ ◯ $\frac{5}{8}$

28. $\frac{6}{12}$ ◯ $\frac{5}{10}$

29. $\frac{1}{2}$ ◯ $\frac{6}{14}$

30. Plot the fraction $\frac{7}{12}$ on the number line. Is $\frac{7}{12}$ closer to 0, $\frac{1}{2}$, or 1?

4.NF.2 _____

31. Lily spent $\frac{1}{6}$ of her free time practicing soccer and $\frac{5}{12}$ of her free time playing a video game. What fraction of her free time did she spend on these two activities? **5.NF.2**

_____

21ST CENTURY CAREER
in Design

Sports Equipment Designer

Do you have a passion for sports and a strong interest in science? Are you a creative thinker who always has new ideas or better ways of doing things? If so, then you should consider a career designing sports equipment. Sports equipment designers combine creativity and engineering principles to create equipment that is cutting edge and helps improve athletic performance. They design everything from baseball bats and footballs to lacrosse protective gear and racing wheelchairs.

Is This the Career for You?

Are you interested in a career as a sports equipment designer? Take some of the following courses in high school.

◆ Algebra
◆ Biology
◆ Calculus
◆ Computer Science
◆ Physics

Find out how math relates to a career in Design.

🔴MP Gaining a Competitive Edge

When a punter kicks a football, the ball has both horizontal motion and vertical motion. The table shows these values when a football is kicked at 25 meters per second.

Use the information in the table to solve each problem. Assume that each football is kicked at 25 meters per second. Round to the nearest tenth if necessary.

1. The *hang time*, or time that a football is in the air, of a football that is kicked at a 27° angle is given by 0.204 × 11.3. What is the approximate hang time? _____

2. How much greater is the hang time of a football that is kicked at a 62° angle than one that is kicked at a 45° angle? Use the expressions 0.204 × 22.1 and 0.204 × 17.7.

   _____

3. The final distance from the punter to a football kicked at a 27° angle is approximately 22.3 × 11.3 × 0.2. What is the distance from the punter to the football?

   _____

4. Find the distance of a football that is kicked at an angle of 62° if the distance is found by using the expression

   11.7 × 22.1 × 0.2. _____

5. The hang time of a football is about 3 seconds. Find 3 ÷ 0.204 to determine the vertical motion of the football. _____

6. A football reaches its maximum height in $y ÷ 9.8$ seconds. A football is kicked at a 62° angle. At the same time, another football is kicked at a 27° angle. Which reaches its maximum height first? Explain.

   _____

   _____

Punting A Football		
Angle of Kick	Horizontal Motion (m/s)	Vertical Motion (m/s)
	x	y
27°	22.3	11.3
45°	17.7	17.7
62°	11.7	22.1

🔴MP Career Project

It's time to update your career portfolio! Choose a piece of sports equipment and describe how it has changed over the past 20 years. List the reasons for the changes.

_____

_____

_____

_____

_____

Suppose you are an employer hiring a sports equipment designer. What questions would you ask a potential employee?

• _____

• _____

• _____

Chapter Review

Vocabulary Check

Write the correct term for each clue in the crossword puzzle.

Across

1. easy to divide mentally

5. a number that has a digit in the tenths place, hundredths place, or beyond

6. to find an approximate value for a number

Down

2. the answer to a multiplication problem

3. a number with more than one digit

4. the number by which the dividend is being divided

5. a number that is being divided

Use Your  FOLDABLES

Use your Foldable to help review the chapter.

Tape here

Divide with Decimals

Examples

Examples

Got it?

Complete the cross number puzzle by solving the problems.

Across

1. 34.5×14
3. $569.6 \div 3.2$
4. 18.5×40
5. $50.4 \div 2.4$
6. $562.39 + 304.61$
9. 42.5×116
10. 339.2×2.5
12. $1{,}584 \div 4.5$
13. $1{,}218 \div 6$

Down

1. $24.3 + 15.7$
2. 21.2×17.5
3. 33.75×3.2
5. $146.53 + 92.47$
7. $2{,}628 \div 36.5$
8. 24×4.5
9. $263.4 + 199.6$
11. 35.2×25

Power Up! Performance Task

Calorie Counter

Juan is calculating how many Calories he consumes in a day. Through Internet research, he determines that a person his age should consume 68,820 Calories per month.

Write your answers on another piece of paper. Show all of your work to receive full credit.

Part A
Based on a 31-day month, how many Calories should Juan consume in a day?

Part B
Juan wants to eat fewer than 800 Calories at lunch. The table shows the menu at the school cafeteria with the Calorie totals listed for each item. Find two different meal options for Juan to choose for lunch. Then find the cost for each meal.

Food Item	Calories	Cost ($)
Carrots	41	0.80
Chocolate Cake (slice)	513	2.10
Cookie	253	1.45
Corn Dog	212	1.50
French Fries	103	1.10
Green Beans	39	0.80
Hamburger	449	2.50
Pepperoni Pizza (slice)	334	1.75
Salisbury Steak	342	2.30

Part C
Juan has a budget of $20.00 for lunch each week. Choose one of the meals from Part B for Juan's lunch on Monday. How much does he have left for the rest of the week? How much could he spend each remaining day, if he spends the same amount each day?

Part D
On game day, the entire basketball team eats together. There are 12 players on the team. Each player eats two slices of pizza, carrots, and a cookie. Find the total cost of the team's lunch.

Reflect

Answering the Essential Question

Use what you learned about computing with multi-digit numbers to complete the graphic organizer.

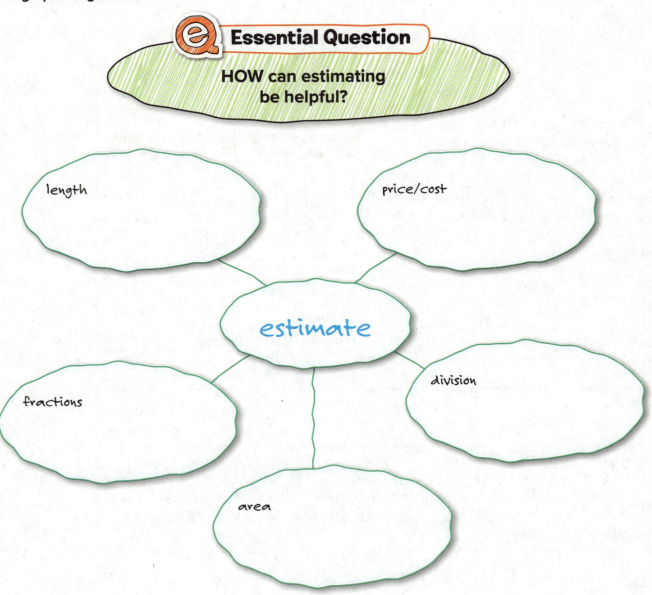

Essential Question

HOW can estimating be helpful?

length

price/cost

estimate

fractions

division

area

Answer the Essential Question. HOW can estimating be helpful?

_____

_____

Chapter 4
Multiply and Divide Fractions

 Essential Question

WHAT does it mean to multiply and divide fractions?

Content Standards
6.NS.1, 6.RP.3, 6.RP.3d

MP **Mathematical Practices**
1, 2, 3, 4, 5, 6, 7, 8

 Math in the Real World

Construction Builders measure using fractions and convert measurements between different units.

A $12\frac{2}{3}$-foot board needs to be divided into 19-inch pieces. On the board below, make the divisions to show the number of pieces.

$12\frac{2}{3}$ ft

FOLDABLES
Study Organizer

1 Cut out the Foldable on page FL9 of this book.

2 Place your Foldable on page 336.

3 Use the Foldable throughout this chapter to help you learn about multiplying and dividing fractions.

Vocabulary

Commutative Property	**reciprocals**
dimensional analysis	**unit ratio**

Study Skill: Writing Math

Explain Your Answer When you explain your answer, you give reasons why your answer is correct.

Sal wants to buy 5 packages of the limited edition cards. Is $20 enough for 5 packages, or does Sal need to bring $25 to the store to buy them? Explain your answer.

Football Card Prices	
Package	**Price ($)**
All-Star	3.75
Limited Edition	4.59
Deluxe	5.99

Step 1 Estimate.

$5 \times \$4 =$ _____ Round down.

$5 \times \$5 =$ _____ Round up.

Step 2 Answer the question.

Sal should bring _____ to the store.

Step 3 Explain why. Write your explanation in complete sentences.

Using the estimate, Sal knows that the actual cost is between _____

and _____. So _____ is not enough.

Practice explaining your answer.

1. Marta plans to buy 2 baseballs and 1 baseball glove. Is $50 enough to bring to the store or does Marta need to bring

   $55? Explain your answer. _____

   _____

   _____

THE SPORTS COVE
Baseball...........6.50
Baseball
glove..............37.99
Baseball
cap................13.79

Read each statement. Decide whether you agree (A) or disagree (D). Place a checkmark in the appropriate column and then justify your reasoning.

Multiply and Divide Fractions			
Statement	A	D	Why?
An addition sentence for $3 \times \frac{4}{5}$ is $\frac{4}{5} + \frac{4}{5} + \frac{4}{5}$.			
A whole number can be written as a fraction with a denominator of 1.			
To multiply two fractions, multiply only the denominators.			
To multiply mixed numbers, multiply the whole numbers then multiply the fractions.			
Any two numbers with a sum of 1 are called reciprocals.			
To divide by a fraction, multiply by its reciprocal.			

When Will You Use This?

Here are a few examples of how fractions are used in the real world.

Activity 1 Describe a situation where you have had a large amount of something that you need to separate into smaller portions. What method did you use to separate it?

_____

Activity 2 Go online at **connectED.mcgraw-hill.com** to read the graphic novel *Buying Birdseed*. How many pounds of birdseed are in each original bag?

_____

Dwayne, Elisa & Luis in **Buying Birdseed**

So, how many small bags of birdseed can we make?

Try the Quick Check below.
Or, take the Online Readiness Quiz.

Quick Review   5.NF.1, 5.NF.2

Example 1

Estimate $4\frac{5}{6} + 1\frac{1}{8}$.

Think: $\frac{5}{6} > \frac{1}{2}$.
So, $4\frac{5}{6}$ rounds to 5.

$4\frac{5}{6} + 1\frac{1}{8}$

Think: $\frac{1}{8} < \frac{1}{2}$.
So, $1\frac{1}{8}$ rounds to 1.

$5 + 1 = 6$

So, $4\frac{5}{6} + 1\frac{1}{8}$ is about 6.

Example 2

Add $5\frac{9}{10} + 3\frac{3}{4}$. Write in simplest form.

$5\frac{9}{10} \rightarrow 5\frac{18}{20}$ Rename the fractions using the LCD, 20.

$+ 3\frac{3}{4} \rightarrow 3\frac{15}{20}$

$8\frac{33}{20}$ or $9\frac{13}{20}$

So, $5\frac{9}{10} + 3\frac{3}{4}$ is equal to $9\frac{13}{20}$.

Quick Check

Estimate with Fractions Estimate each sum or difference.

1. $6\frac{7}{8} + 5\frac{1}{4} \approx$ _____

2. $3\frac{1}{7} + 8\frac{1}{9} \approx$ _____

3. $12\frac{1}{5} - 5\frac{5}{6} \approx$ _____

Show your work.

Add and Subtract Fractions Add or subtract. Write in simplest form.

4. $\frac{9}{10} + \frac{3}{5} =$ _____

5. $7\frac{2}{3} - 3\frac{1}{7} =$ _____

6. $9\frac{7}{8} - 2\frac{5}{6} =$ _____

7. Suppose a plant grew $4\frac{1}{2}$ inches one week and $2\frac{3}{8}$ inches the next week.

   How many inches did it grow during both weeks?

   _____

Which problems did you answer correctly in the Quick Check?
Shade those exercise numbers below.

① ② ③ ④ ⑤ ⑥ ⑦

Estimate Products of Fractions

Nature A wildlife preserve has 16 tigers, about $\frac{1}{3}$ of which are male. The tigers are represented by the counters below.

Essential Question

WHAT does it mean to multiply and divide fractions?

Content Standards
Preparation for 6.NS.1

MP Mathematical Practices
1, 3, 4, 5

1. Can you separate the counters into three equal groups? Explain.

   _____

   _____

2. Fill in the multiples of 3 on the number line. Place a dot at 16.

3. What multiple of 3 is closest to 16? ☐

4. Arrows jump from 0 to 3 to 6. Continue the pattern. How many

   jumps are from 0 to your answer in Exercise 3? ☐ jumps

5. About how many tigers in the preserve are male? Explain.

   _____

   _____

Which MP Mathematical Practices did you use?
Shade the circle(s) that applies.

① Persevere with Problems   ⑤ Use Math Tools

② Reason Abstractly   ⑥ Attend to Precision

③ Construct an Argument   ⑦ Make Use of Structure

④ Model with Mathematics   ⑧ Use Repeated Reasoning

Estimate Using Compatible Numbers

You have already used compatible numbers to estimate quotients. You can also use compatible numbers to estimate products.

Examples

1. **Estimate $\frac{1}{4} \times 13$.**

Find a multiple of 4 close to 13. 12 and 4 are compatible numbers since $12 \div 4 = 3$.

> **Method 1** **Use a model.**
>
> Divide the bar representing 12 into 4 sections.
>
> Each section is $\frac{1}{4}$ of 12, or 3.
>
> Label each section 3.

> **Method 2** **Use compatible numbers.**
>
> $\frac{1}{4} \times 13 \approx \frac{1}{4} \times 12$
>
> ≈ 3
>
> So, $\frac{1}{4} \times 13$ *is about* 3.

Compatible Numbers

Use compatible numbers to estimate the product of a fraction and a whole number. Find compatible numbers using the denominator of the fraction and the whole number.

2. **Estimate $\frac{2}{5}$ of 11.**

Find a multiple of 5 close to 11. 10 and 5 are compatible numbers since $10 \div 5 = 2$.

$\frac{1}{5} \times 11 \approx \frac{1}{5} \times 10$

≈ 2

If $\frac{1}{5}$ of 10 is 2, then

$\frac{2}{5}$ of 10 is 2×2, or $\boxed{}$.

So, $\frac{2}{5} \times 11$ *is about* $\boxed{}$.

 Show your work.

Got it? **Do these problems to find out.**

a. $\frac{1}{5} \times 16$ b. $\frac{5}{6} \times 13$ c. $\frac{3}{4}$ of 23

a. _____

b. _____

c. _____

Estimate by Rounding to 0, ½, or 1

Words If the numerator of a fraction between 0 and 1 is almost as large as the denominator, round up. If the numerator is much smaller than the denominator, round down.

Examples $\frac{7}{8}$ rounds to 1; $\frac{1}{8}$ rounds to 0

Rounding fractions can help you find products of fraction factors.

Examples

Tutor

3. Estimate $\frac{1}{3} \times \frac{7}{9}$.

Dots are placed at $\frac{1}{3}$ and $\frac{7}{9}$.

Round to 0, ½, or 1. $\frac{1}{3}$ is about $\frac{1}{2}$ and $\frac{7}{9}$ is about 1.

$\frac{1}{3} \times \frac{7}{9} \longrightarrow \frac{1}{2} \times 1$

$\frac{1}{2} \times 1 = \frac{1}{2}$

So, $\frac{1}{3} \times \frac{7}{9}$ *is about* $\frac{1}{2}$.

Number line showing $\frac{1}{3}$ and $\frac{7}{9}$ with marks at 0, ½, 1.

4. Estimate $\frac{1}{9} \times \frac{4}{5}$.

Place dots on the number line at $\frac{1}{9}$ and $\frac{4}{5}$.

Round to 0, ½, or 1. $\frac{1}{9}$ is about 0 and $\frac{4}{5}$ is about 1.

$\frac{1}{9} \times \frac{4}{5} \longrightarrow 0 \times 1 = \boxed{}$

So, $\frac{1}{9} \times \frac{4}{5}$ *is about* $\boxed{}$.

Number line with marks at 0, ½, 1.

Show your work.

Got it? Do these problems to find out.

Estimate each product.

d. $\frac{5}{8} \times \frac{9}{10}$ **e.** $\frac{5}{6} \times \frac{9}{10}$ **f.** $\frac{5}{6}$ of $\frac{1}{9}$

d. _____

e. _____

f. _____

Example

5. Estimate the area of the flower bed.

Round each mixed number to the nearest whole number.

$$14\frac{7}{8} \times 6\frac{1}{8} \quad \rightarrow \quad 15 \times 6 = 90$$

So, the area *is about* 90 square feet.

$6\frac{1}{8}$ ft

$14\frac{7}{8}$ ft

Guided Practice

Estimate each product. Use a bar diagram if needed. (Examples 1–4)

1. $\frac{1}{8} \times 15 \approx$ _____

2. $\frac{2}{5}$ of $26 \approx$ _____

 Show your work.

3. $\frac{1}{5} \times \frac{8}{9} \approx$ _____

4. $6\frac{2}{3} \times 4\frac{1}{5} \approx$ _____

5. A border is made of $32\frac{2}{3}$ bricks that are $1\frac{1}{6}$ feet long. About how long is the border? (Example 5)

_____

_____

6. A kitchen measures $24\frac{1}{6}$ feet by $9\frac{2}{3}$ feet. Estimate the area of the kitchen. (Example 5)

_____

_____

Rate Yourself!

How confident are you about estimating the products of fractions? Color the square that applies.

☐ ☐ ☐ ☐ ☐

For more help, go online to access a Personal Tutor.

 Tutor

7. **Building on the Essential Question** Why is estimating products of fractions useful? _____

_____

_____

_____

Independent Practice

Go online for Step-by-Step Solutions

Estimate each product. Use a bar diagram if needed. (Examples 1–4)

1. $\frac{1}{4} \times 21 \approx$ _____

Show your work.

2. $\frac{5}{7}$ of $22 \approx$ _____

3. $\frac{5}{7} \times \frac{1}{9} \approx$ _____

4. $4\frac{1}{3} \times 2\frac{3}{4} \approx$ _____

5. Cyrus is inviting 11 friends over for pizza. He would like to have enough pizza so each friend can have $\frac{1}{4}$ of a pizza. About how many pizzas should he order? (Example 5)

_____

6. Hakeem's front porch measures $9\frac{3}{4}$ feet by 4 feet. Estimate the area of his front porch. (Example 5)

_____

7. **MP Use Math Tools** Refer to the graphic novel frame for Exercises a–b.

I hope 5 bags is enough.

a. If each bag holds $3\frac{3}{4}$ pounds, estimate how many pounds of birdseed Elisa, Luis, and Dwayne purchased. _____

_____

b. Suppose each bag costs $14.99. Estimate the total cost of 5 bags.

_____

Estimate the area of each rectangle.

8.

$5\frac{3}{4}$ ft

$8\frac{1}{8}$ ft

9

$3\frac{1}{4}$ in.

$9\frac{5}{8}$ in.

10. **STEM** Seattle, Washington, received rain on $\frac{7}{10}$ of the days in a recent month. If this pattern continues, about how many days would it *not* rain in 90 days? _____

_____

_____

🔥 H.O.T. Problems Higher Order Thinking

11. **MP** **Justify Conclusions** By what fraction would you multiply $8\frac{1}{2}$ so that the product is about 5? Explain your reasoning. _____

_____

_____

12. **MP** **Persevere with Problems** Determine which point on the number line could be the graph of the product of the numbers graphed at C and D.

0   M   N C R D   1

Explain your reasoning. _____

_____

_____

13. **MP** **Reason Inductively** Is the best estimate for the area of the rectangle at the right 2 square yards or 6 square yards? Explain your reasoning.

$2\frac{2}{3}$ yd

$1\frac{5}{6}$ yd

_____

_____

_____

Extra Practice

Estimate each product. Use a bar diagram if needed.

14. $\frac{1}{5} \times 26 \approx$ _5_

Homework Help ➤

$26 \approx 25; \frac{1}{5} \times \overset{5}{\cancel{25}} = \frac{5}{1}$ or 5

15. $\frac{1}{3}$ of $41 \approx$ _14_

$41 \approx 42; \frac{1}{3} \times \overset{14}{\cancel{42}} = \frac{14}{1}$ or 14

16. $\frac{1}{6}$ of $17 \approx$ _____

17. $\frac{2}{9}$ of $88 \approx$ _____

18. $\frac{2}{3} \times 10 \approx$ _____

19. $\frac{3}{8} \times 4 \approx$ _____

Estimate the area of the each rectangle.

20. _____

$3\frac{1}{5}$ ft

$5\frac{7}{9}$ ft

21. _____

$1\frac{1}{3}$ cm

$7\frac{7}{8}$ cm

22. Tara would like to finish $\frac{2}{5}$ of her book by next Friday. If the book has 203 pages, about how many pages does she need to read? _____

23. Javier is organizing his movie collection. He discovers that $\frac{5}{8}$ of his movies are action movies. If he has 46 movies, about how many are action movies? _____

24. **MP** **Justify Conclusions** Marco has a collection of 38 state quarters. If $\frac{3}{5}$ of his quarters are dated 2005, what is the approximate value of the quarters from 2005? Explain your answer to a classmate.

_____

25. The table shows the number of students in grades 6–8 who went to a local museum. Of these, between one half and three fourths packed their lunch. Which of the following ranges could represent the number of students who packed their lunch?

Students Visiting the Museum	
Grade	Number of Students
6	45
7	48
8	40

☐ less than 70   ☐ between 70 and 100   ☐ between 100 and 130   ☐ more than 130

26. Estimate the area of the rectangle at the right.

Area ≈ _____

$6\frac{1}{4}$ ft

$10\frac{5}{6}$ ft

Spiral Review

Round each fraction to 0, $\frac{1}{2}$, or 1. **4.NF.2**

27. $\frac{7}{12} \approx$ _____

28. $\frac{5}{6} \approx$ _____

29. $\frac{2}{11} \approx$ _____

30. $\frac{4}{9} \approx$ _____

31. Graph $\frac{3}{4}, \frac{1}{3}, \frac{1}{4}$, and $\frac{2}{3}$. **4.NF.2**

32. Graph $\frac{5}{10}, \frac{4}{5}, \frac{2}{5}$, and $\frac{7}{10}$. **4.NF.2**

33. Jasper wants to paint one wall of his room. If the wall is 12 feet wide and 10 feet tall, what is the area of the wall? **4.MD.3** _____

Multiply Fractions and Whole Numbers

Vocabulary Start-Up

A commuter train travels back and forth but does not change the distance traveled. In mathematics, operations that follow the **Commutative Property** can be performed in any order. For example, addition and multiplication are commutative.

Draw a line to "Commutative" if the examples can be done in either order. Draw a line to "Not Commutative" if the order changes the outcome.

Commutative	$12 \div 6; 6 \div 12$
	tying your left shoe; tying your right shoe
	$5 \times 7; 7 \times 5$
Not Commutative	play a soccer game; change into your team uniform
	$15 + 5; 5 + 15$

 Essential Question

WHAT does it mean to multiply and divide fractions?

 Vocabulary

Commutative Property

Content Standards
Preparation for 6.NS.1

MP **Mathematical Practices**
1, 3, 4, 7

Real-World Link

Some morning routines can be done in any order. Sometimes, the order matters. Describe a situation when the order you perform two actions is important.

_____
_____
_____

Which MP Mathematical Practices did you use?
Shade the circle(s) that applies.

① Persevere with Problems
② Reason Abstractly
③ Construct an Argument
④ Model with Mathematics

⑤ Use Math Tools
⑥ Attend to Precision
⑦ Make Use of Structure
⑧ Use Repeated Reasoning

Multiply a Whole Number by a Fraction

Words Write the whole number as a fraction. Multiply the numerators and multiply the denominators.

Example $5 \times \frac{3}{4} = \frac{5}{1} \times \frac{3}{4}$ Write 5 as $\frac{5}{1}$.

$= \frac{5 \times 3}{1 \times 4}$ Multiply.

$= \frac{15}{4}$ or $3\frac{3}{4}$ Simplify.

Example

1. Find $2 \times \frac{2}{5}$.

Method 1 Use an area model.

Shade $\frac{2}{5}$ of each of the first two columns.

A total of $\frac{4}{5}$ has been shaded.

Shade $\frac{4}{5}$ on the third column.

Method 2 Use an equation.

Estimate $2 \times \frac{1}{2} = 1$

$2 \times \frac{2}{5} = \blacksquare$

$2 \times \frac{2}{5} = \frac{2}{1} \times \frac{2}{5}$ Write 2 as $\frac{2}{1}$.

$= \frac{2 \times 2}{1 \times 5}$ Multiply.

$= \frac{4}{5}$ Simplify.

Using either method, $2 \times \frac{2}{5}$ is $\frac{4}{5}$.

Check for Reasonableness $\frac{4}{5} \approx 1$ ✔

Got it? Do these problems to find out.

a. $6 \times \frac{2}{3}$ **b.** $9 \times \frac{1}{3}$ **c.** $4 \times \frac{1}{8}$

a. _____

b. _____

c. _____

Multiply a Fraction by a Whole Number

When multiplying whole numbers and fractions, the order of the factors does not change the product. So, $4 \times \frac{3}{5} = \frac{3}{5} \times 4$. This is an example of the Commutative Property.

Examples

2. Find $\frac{3}{5} \times 4$.

Method 1 Use an area model.

Shade $\frac{3}{5}$ of each of the 4 columns.

A total of $\frac{12}{5}$ or $2\frac{2}{5}$ has been shaded.

Method 2 Use an equation. **Estimate** $\frac{1}{2} \times 4 = 2$

$\frac{3}{5} \times 4 = \blacksquare$

$\frac{3}{5} \times 4 = \frac{3}{5} \times \frac{4}{1}$ Write 4 as $\frac{4}{1}$.

$\quad = \frac{3 \times 4}{5 \times 1}$ Multiply.

$\quad = \frac{12}{5}$ Simplify. Compare to the estimate.

$\quad = 2\frac{2}{5}$

Using either method, $\frac{3}{5} \times 4$ is $2\frac{2}{5}$.

3. Find $\frac{1}{4} \times 5$.

Estimate $\frac{1}{4} \times 4 = 1$

$\frac{1}{4} \times 5 = \frac{1}{4} \times \frac{5}{1}$ Write 5 as $\frac{5}{1}$.

$\quad = \frac{1 \times 5}{4 \times 1}$ Multiply.

$\quad = \frac{5}{4}$ or $1\frac{1}{4}$ Simplify.

Check for Reasonableness $1\frac{1}{4} \approx 1$ ✔

Got it? Do these problems to find out.

d. $\frac{1}{2} \times 3$ **e.** $\frac{2}{5} \times 4$ **f.** $\frac{3}{4} \times 5$

Renaming

To rename an improper fraction as a mixed number, divide the numerator by the denominator. Write the remainder as a fraction with the divisor as the denominator.

Show your work.

d. _____

e. _____

f. _____

Example

4. A sloth spends $\frac{4}{5}$ of its life asleep. If a sloth lives to be 28 years old, how many years does it spend asleep?

Find $\frac{4}{5}$ of 28.

Estimate $\frac{4}{5}$ of 30 is 24.

$\frac{4}{5} \times 28 = \frac{4}{5} \times \frac{28}{1}$ Write 28 as $\frac{28}{1}$.

$= \frac{4 \times 28}{5 \times 1}$ Multiply.

$= \frac{112}{5}$ or $22\frac{2}{5}$ Simplify. Compare to the estimate.

A sloth spends $22\frac{2}{5}$ years of its life asleep.

Check ✓

Guided Practice

Multiply. Write in simplest form. (Examples 1–3)

1. $10 \times \frac{4}{5} =$ _____

Show your work.

2. $2 \times \frac{3}{4} =$ _____

3. $\frac{3}{8} \times 11 =$ _____

4. $\frac{3}{7} \times 9 =$ _____

5. A cat spends $\frac{2}{3}$ of its life asleep. If a cat lives to be 15 years old, how many years did it spend asleep? (Example 4) _____

6. **Building on the Essential Question** How is the process used to multiply a fraction and a whole number similar to the process used to multiply two whole numbers?

_____

_____

_____

_____

Rate Yourself!

Are you ready to move on? Shade the section that applies.

YES   ?   NO

For more help, go online to access a Personal Tutor.

Tutor

FOLDABLES Time to update your Foldable!

Independent Practice

Go online for Step-by-Step Solutions

Multiply. Write in simplest form. (Examples 1–3)

1. $20 \times \frac{3}{4} =$ _____

2. $14 \times \frac{2}{7} =$ _____

3. $10 \times \frac{1}{5} =$ _____

4. $\frac{3}{4} \times 6 =$ _____

5. $\frac{2}{5} \times 11 =$ _____

6. $\frac{1}{4} \times 6 =$ _____

7. **STEM** The male Cuban tree frog is about $\frac{2}{5}$ the size of the female Cuban tree frog. The average size of the female Cuban tree frog is shown at the right. What is the size of the male Cuban tree frog? (Example 4) _____

6 in.

8. The Mississippi River is the second longest river in the United States, second only to the Missouri River. The Mississippi River is about $\frac{23}{25}$ the length of the Missouri River. If the Missouri River is 2,540 miles long, how long is the Mississippi River? (Example 4)

_____

_____

9. One evening, $\frac{2}{3}$ of Mrs. Thorne's students watched a reality television show. Of Mrs. Lombardo's students, $\frac{4}{5}$ watched the same reality show. Which teacher had more students that watched the reality show? Explain.

_____

_____

Teacher	Number of Students
Mrs. Thorne	36
Mrs. Lombardo	30
Mr. Hollern	28

10. **MP** **Persevere with Problems** The table shows where sixth grade students at Sharonton Middle School attended fifth grade. There are 156 sixth grade students. How many more students attended Sharonton Elementary than Deacon Elementary?

School	Fraction of Students
Sharonton Elementary	$\frac{1}{2}$
Deacon Elementary	$\frac{1}{4}$
Banyon Elementary	$\frac{1}{6}$
New Students	$\frac{1}{12}$

11. **MP** **Persevere with Problems** Students at Marzo Middle School were recently surveyed. The results reported $\frac{1}{4}$ of sixth grade students, $\frac{3}{10}$ of seventh grade students, and $\frac{2}{7}$ of eighth grade students plan a career in STEM. In which grade do the most students plan to have careers in STEM?

Grade	Total Students
6	152
7	160
8	147

_____

12. **MP** **Persevere with Problems** Refer to the table in Exercise 11. Suppose in the sixth grade, $\frac{3}{4}$ of the students want a career in either teaching or STEM. How many sixth grade students did not choose a career in teaching or STEM?

_____

 ## H.O.T. Problems   Higher Order Thinking

13. **MP** **Identify Structure** Write a real-world problem involving the multiplication of a fraction and a whole number with a product that is between 8 and 10. Then solve the problem.

_____

_____

_____

14. **MP** **Find the Error** Noah is finding $\frac{3}{4}$ of 8. Find his mistake and correct it.

_____

_____

_____

$$\frac{3}{4} \times 8 = \frac{24}{32}$$

15. **MP** **Persevere with Problems** Use the digits 2, 3, and 5 to create a fraction and a whole number with a product greater than 2.

_____

_____

16. **MP** **Reason Inductively** Jenny made five loaves of banana bread that had $\frac{1}{4}$ cup of oil in each loaf. After she was done baking, she had $\frac{5}{8}$ cup of oil remaining. How much oil did she have before baking?

_____

Name _____ My Homework _____

Extra Practice

Multiply. Write in simplest form.

17. $12 \times \frac{1}{3} =$ _4_____

 Homework Help

$\frac{\cancel{12}^4}{1} \times \frac{1}{\cancel{3}_1} = \frac{4}{1}$ or 4

18. $18 \times \frac{1}{3} =$ 6 _____

$\frac{\cancel{18}^6}{1} \times \frac{1}{\cancel{3}_1} = \frac{6}{1}$ or 6

19. $8 \times \frac{1}{4} =$ _____

20. $\frac{1}{5} \times 7 =$ _____

21. $\frac{3}{7} \times 8 =$ _____

22. $\frac{5}{6} \times 15 =$ _____

23. For a singing contest in which 42,000 votes were cast, the winner received $\frac{3}{5}$ of the votes. How many votes did the winner *not* receive?

 Show your work.

_____

24. **STEM** In a recent year, the weather was partly cloudy $\frac{2}{5}$ of the days. Assuming there are 365 days in a year, how many days were partly cloudy?

_____

25. **MP Model with Mathematics** Write a real-world problem that involves multiplying a fraction and a whole number. Solve the problem and use estimation to check for reasonableness.

_____

_____

26. Leonard used $\frac{2}{7}$ of his paycheck to pay his cell phone bill and put $\frac{1}{3}$ of his paycheck in savings. Based on the amount of money left, which of the following are amounts he could spend on a new video game?

Pay to the order of	1438
Leonard White	$63.00

Leonard White	1438
Time	Net Pay
10 hours	$63.00

☐ $20   ☐ $24   ☐ $28   ☐ $34

27. David used the measuring cup shown to measure milk for a batch of muffins. He needs to make 12 batches of muffins for a fund raiser. He bought one gallon of milk. Does he have enough milk for all 12 batches of muffins? Explain your answer. *(Hint: one gallon = 16 cups)*

Spiral Review

Multiply. 5.NBT.5.

28. $22 \times 13 =$ _____

29. $18 \times 11 =$ _____

30. $17 \times 9 =$ _____

31. Hayley's guitar lesson lasts $\frac{3}{4}$ hour. How many minutes does Hayley spend at her guitar lesson? Use the clock to help you find your answer.

5.NF.6 _____

32. Miguel has one foot of string. He cuts the string into fourths. How many inches is each piece of string? 5.MD.1 _____

Multiply Fractions

  Real-World Link Watch ▶

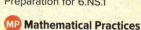 **Essential Question**

WHAT does it mean to multiply and divide fractions?

Content Standards
Preparation for 6.NS.1

MP Mathematical Practices
1, 3, 4, 5, 7

Reptiles A chameleon's body is about $\frac{1}{2}$ the length of its tongue. A certain chameleon has a tongue that is $\frac{2}{3}$ foot long.

Use an area model to show $\frac{1}{2}$ of $\frac{2}{3}$ or $\frac{1}{2} \times \frac{2}{3}$.

1. Divide the rectangle into 2 rows. Then divide it into 3 columns.

2. Shade a rectangle that is $\frac{1}{2}$ unit wide by $\frac{2}{3}$ unit long.

3. Refer to the model. The section that was shaded represents $\frac{1}{2} \times \frac{2}{3}$. What fraction represents $\frac{1}{2} \times \frac{2}{3}$? _____

4. What is the relationship between the numerators and denominators of the factors and the numerator and denominator of the product?

_____

_____

_____

 Which MP Mathematical Practices did you use?
Shade the circle(s) that applies.

① Persevere with Problems
② Reason Abstractly
③ Construct an Argument
④ Model with Mathematics

⑤ Use Math Tools
⑥ Attend to Precision
⑦ Make Use of Structure
⑧ Use Repeated Reasoning

Multiply Fractions

Work Zone

Words   Multiply the numerators and multiply the denominators.

Models

Numbers $\frac{2}{5} \times \frac{1}{2} = \frac{2 \times 1}{5 \times 2}$

Symbols $\frac{a}{b} \times \frac{c}{d} = \frac{a \times c}{b \times d}$, where b and d are not 0.

Example

Tutor

1. Find $\frac{1}{3} \times \frac{1}{4}$. Write in simplest form.

Method 1   Use a model.

Divide the rectangle into 4 rows. Then divide the rectangle into 3 columns.

Shade a section that is $\frac{1}{4}$ unit wide by $\frac{1}{3}$ unit long.

The section that is shaded represents $\frac{1}{4} \times \frac{1}{3}$, or $\frac{1}{12}$.

Method 2   Use an equation.

$\frac{1}{3} \times \frac{1}{4} = \blacksquare$

$\frac{1}{3} \times \frac{1}{4} = \frac{1 \times 1}{3 \times 4}$ Multiply the numerators.
Multiply the denominators.

$= \frac{1}{12}$ Simplify.

Show your work.

So, $\frac{1}{3} \times \frac{1}{4}$ is $\frac{1}{12}$.

Got it?   Do these problems to find out.

a. $\frac{1}{2} \times \frac{3}{5}$ b. $\frac{1}{3} \times \frac{3}{4}$ c. $\frac{2}{3} \times \frac{5}{6}$

a. _____

b. _____

c. _____

Simplify Before Multiplying

If the numerators and the denominators have a common factor you can simplify *before* you multiply. Remember that factors are two or more numbers that are multiplied together to form a product.

$$\frac{\overset{1}{\cancel{2}}}{3} \times \frac{5}{\cancel{6}_{\,3}} = \frac{1}{3} \times \frac{5}{3}$$

Think: $2 \div 2 = 1$
Think: $6 \div 2 = 3$

$$= \frac{5}{9}$$

Examples

> ### Simplifying
> When multiplying fractions, it is easier to find the answer if you simplify before multiplying.

2. Find $\frac{3}{4} \times \frac{5}{6}$.

Estimate $\frac{1}{2} \times 1 = \frac{1}{2}$

$$\frac{3}{4} \times \frac{5}{6} = \frac{\overset{1}{\cancel{3}} \times 5}{4 \times \cancel{6}_{\,2}}$$ Divide both the numerator and the denominator by 3.

$$= \frac{5}{8}$$ Simplify. Compare to the estimate.

Check for reasonableness $\frac{1}{2} \approx \frac{5}{8}$ ✔

3. Find $\frac{4}{9} \times 18$.

Estimate $\frac{1}{2} \times 18 = 9$

$$\frac{4}{9} \times 18 = \frac{4}{9} \times \frac{18}{1}$$ Write 18 as a fraction with a denominator of 1.

$$= \frac{4 \times \overset{2}{\cancel{18}}}{\underset{1}{\cancel{9}} \times 1}$$ Divide both the numerator and the denominator by 9.

$$= \frac{8}{1} \text{ or } 8$$ Simplify. Compare to the estimate.

Check for reasonableness $9 \approx 8$ ✔

> Show your work.

Got it? Do these problems to find out.

d. $\frac{3}{4} \times \frac{4}{9}$ **e.** $\frac{5}{6} \times \frac{9}{10}$ **f.** $\frac{3}{5} \times 10$

d. _____

e. _____

f. _____

4. Frank had $\frac{1}{2}$ of the lawn left to mow. On Saturday, he mowed $\frac{2}{3}$ of what was left. What fraction of the entire lawn did Frank mow on Saturday?

$\frac{1}{2} \times \frac{2}{3} = \frac{1 \times 2}{2 \times 3}$ Divide both the numerator and denominator by 2.

$= \frac{1}{3}$ Simplify.

So, Frank mowed $\frac{1}{3}$ of the lawn on Saturday.

Guided Practice

 Check ✓

Multiply. Write in simplest form. (Examples 1–3)

1. $\frac{1}{8} \times \frac{1}{2} =$ _____

2. $\frac{2}{3} \times \frac{4}{5} =$ _____

3. $\frac{4}{5} \times 10 =$ _____

Show your work.

4. $\frac{3}{4} \times 12 =$ _____

5. $\frac{3}{10} \times \frac{5}{6} =$ _____

6. $\frac{3}{5} \times \frac{5}{6} =$ _____

7. Rick has $\frac{1}{2}$ of a footlong sub left from yesterday. He ate $\frac{1}{3}$ of the leftover sandwich as a snack. What fraction of the entire sandwich did he eat as a snack? (Example 4)

_____

_____

8. **Building on the Essential Question** If two positive fractions are less than 1, why is their product also less than 1?

_____

_____

_____

Rate Yourself!

Are you ready to move on? Shade the section that applies.

I have a few questions.

I'm ready to move on.

I have a lot of questions.

For more help, go online to access a Personal Tutor. Tutor

FOLDABLES Time to update your Foldable!

Independent Practice

Go online for Step-by-Step Solutions

Multiply. Write in simplest form. (Examples 1–3)

1. $\frac{1}{3} \times \frac{2}{5} =$ _____

 Show your work.

2. $\frac{3}{4} \times \frac{5}{8} =$ _____

3 $\frac{2}{3} \times 4 =$ _____

4. $\frac{5}{6} \times 15 =$ _____

5. $\frac{2}{3} \times \frac{1}{4} =$ _____

6. $\frac{4}{9} \times \frac{3}{8} =$ _____

7 **Financial Literacy** Juanita spent $\frac{3}{4}$ of her allowance at the mall. Of the money spent at the mall, $\frac{1}{2}$ was spent on new earphones. What part of her allowance did Juanita spend on earphones? (Example 4)

_____

_____

8. A paint store has 35 gallons of paint in storage, $\frac{2}{5}$ of which are for outdoor use. The others are for indoor use. If each gallon costs $22, what is the total cost of the indoor paint in storage?

_____

_____

9 Homeroom 101 and Homeroom 102 share a hallway bulletin board. If Homeroom 101 uses $\frac{3}{5}$ of their half to display artwork, what fraction of the bulletin board is used to display Homeroom 101's artwork?

_____

_____

10. **MP** **Use Math Tools** Mr. Williams' physical education class lasts $\frac{7}{8}$ hour.
   a. How many minutes are spent warming up and cooling down?

   _____

   _____

   b. How many minutes are *not* spent on instruction? Explain.

   _____

   _____

Part of $\frac{7}{8}$-hour Class	
playing game	$\frac{1}{2}$
instruction	$\frac{1}{5}$
warm-up and cool-down	$\frac{3}{10}$

11. **MP** **Multiple Representations** Use the bar diagram.

a. **Words** Write a real-world problem represented by the

bar diagram. _____

_____

_____

b. **Models** Draw an area model to represent the situation.

c. **Words** Explain how you would solve your problem.

_____

_____

🔥 H.O.T. Problems Higher Order Thinking

12. **MP** **Reason Inductively** State whether each statement is *true* or *false*.
If the statement is *false*, provide a counterexample.

a. The product of two fractions that are each between 0 and 1 is also

between 0 and 1. _____

b. The product of a mixed number between 4 and 5 and a fraction

between 0 and 1 is always less than 4. _____

c. The product of two mixed numbers that are each between 4 and 5 is

between 16 and 25. _____

13. **MP** **Identify Structure** If the product of two positive fractions a and b

is $\frac{15}{56}$, find three pairs of possible values for a and b. _____

_____

14. **MP** **Persevere with Problems** Justify why $\frac{a}{b} \times \frac{b}{c} \times \frac{c}{d} \times \frac{d}{e}$ is equal to $\frac{a}{e}$

when b, c, d, and e are not zero. _____

_____

15. **MP** **Model with Mathematics** Write a word problem in which you multiply a
fraction by a number greater than 1. Estimate the product, then compare the

product to your estimate. _____

_____

_____

_____

_____

Extra Practice

Multiply. Write in simplest form.

16. $\frac{1}{8} \times \frac{3}{4} = \frac{3}{32}$ _____

$$\frac{1 \times 3}{8 \times 4} = \frac{3}{32}$$

Homework Help →

17. $\frac{2}{5} \times \frac{3}{7} = \frac{6}{35}$ _____

$$\frac{2 \times 3}{5 \times 7} = \frac{6}{35}$$

18. $\frac{3}{4} \times 2 =$ _____

19. $\frac{3}{8} \times 11 =$ _____

20. $\frac{3}{5} \times \frac{5}{7} =$ _____

21. $\frac{2}{5} \times \frac{5}{6} =$ _____

22. The bleachers at a football game are $\frac{7}{8}$ full, and $\frac{1}{2}$ of the fans in the bleachers are rooting for the home team. What fraction of the bleachers are filled with home-team fans? Justify your procedure.

_____

_____

23. The table shows the fraction of the votes that each candidate received. If 230 students voted, how many students voted for

each candidate? _____

_____

_____

Candidate	Fraction of Votes
Nyemi	$\frac{3}{5}$
Luke	$\frac{3}{10}$
Natalie	$\frac{1}{10}$

24. **MP** **Model with Mathematics** Alberto rode $\frac{5}{8}$ of the water rides at a water park. His sister, Reina, rode half of the rides that Alberto rode. What fraction of the water rides did Reina *not* ride? Support your answer with a model. _____

_____

25. **MP** **Justify Conclusions** Lee is making chocolate chip cookies and the recipe calls for $\frac{3}{4}$ cup of chocolate chips. If she wants to make $\frac{2}{3}$ of the recipe, what fraction of a cup of chocolate chips will she need? Explain.

_____

_____

26. The table shows the fraction of votes each candidate received for Student Council president. If 180 students voted, fill in the table to show the number of votes each student received.

Candidate	Fraction of Votes	Number of Votes
David	$\frac{1}{5}$	
Liam	$\frac{3}{10}$	
Marta	$\frac{1}{10}$	
Tiffany	$\frac{2}{5}$	

27. Four fifths of Terrence's text messages are to his friends. One half of those messages are to his friend Bianca. Terrence sends 120 text messages. How many text messages are to Bianca? ☐

Spiral Review

Multiply. 5.NBT.5

28. $12 \times 6 \times 9 =$ _____

29. $5 \times 22 \times 3 =$ _____

30. $15 \times 8 \times 11 =$ _____

31. Elise planted a row of flowers in an area with the dimensions shown at the right. What is the area of her flower garden? **5.NF.4b**

8 ft $\quad \frac{1}{2}$ ft

_____

_____

32. Without multiplying, determine whether the product of $5 \times \frac{4}{5}$ is located on the number line at point *A*, *B*, or *C*. Explain your reasoning.
5.NF.5b

_____

_____

_____

Multiply Mixed Numbers

 Watch ▶

Animals The eyeball of an Atlantic Giant Squid is about 12 times as large as the average human eyeball. The average human eyeball is $1\frac{1}{4}$ inches across. Use a bar diagram to compare the average size of a human eyeball to the average size of a Atlantic Giant Squid's eyeball.

------------- squid eyeball -----------------

$1\frac{1}{4}$

human
eyeball

 Essential Question

WHAT does it mean to multiply and divide fractions?

Content Standards
Preparation for 6.NS.1

MP **Mathematical Practices**
1, 3, 4, 5

1. Use the diagram above to compare the average size of the Atlantic Giant Squid's eyeball to the average size of the human eyeball. Use repeated addition.

2. Write a multiplication expression that shows the size of the Atlantic Squid's eyeball. _____

3. Write the multiplication expression from Exercise 2 using improper fractions. Multiply to find the size of the squid's eyeball.

_____

Which MP Mathematical Practices did you use?
Shade the circle(s) that applies.

① Persevere with Problems

② Reason Abstractly

③ Construct an Argument

④ Model with Mathematics

⑤ Use Math Tools

⑥ Attend to Precision

⑦ Make Use of Structure

⑧ Use Repeated Reasoning

Multiply a Fraction and a Mixed Number

To multiply a fraction and a mixed number, first write the mixed number as an improper fraction. Remember that when mixed numbers are written as improper fractions, the denominator does not change. Then multiply as with fractions.

$$2\frac{1}{2} \times \frac{1}{4} = \frac{5}{2} \times \frac{1}{4}$$

$$= \frac{5 \times 1}{2 \times 4}$$

$$= \frac{5}{8}$$

Examples

1. Find $\frac{1}{3} \times 1\frac{3}{4}$. **Write in simplest form.**

Estimate Use compatible numbers. $\frac{1}{2} \times 2 = 1$

$$\frac{1}{3} \times 1\frac{3}{4} = \frac{1}{3} \times \frac{7}{4} \qquad \text{Write } 1\frac{3}{4} \text{ as } \frac{7}{4}.$$

$$= \frac{1 \times 7}{3 \times 4} \qquad \text{Multiply.}$$

$$= \frac{7}{12} \qquad \text{Simplify. Compare to the estimate.}$$

2. Find $5\frac{1}{2} \times \frac{1}{3}$. **Write in simplest form.**

Estimate $\boxed{} \times \dfrac{\boxed{}}{\boxed{}} = \boxed{}$

$$5\frac{1}{2} \times \frac{1}{3} = \frac{\boxed{}}{\boxed{}} \times \frac{1}{3} \qquad \text{Write } 5\frac{1}{2} \text{ as an improper fraction.}$$

$$= \frac{\boxed{} \times}{\boxed{} \times} \qquad \text{Multiply.}$$

$$= \frac{\boxed{}}{\boxed{}} \text{ or } 1\frac{\boxed{}}{\boxed{}} \qquad \text{Simplify.}$$

Check for Reasonableness $1\frac{5}{6} \approx 2$ ✔

Show your work.

Got it? Do these problems to find out.

a. $\frac{2}{3} \times 2\frac{1}{2}$ 　　　　　**b.** $\frac{3}{8} \times 3\frac{1}{3}$

a. _____

b. _____

Multiply Mixed Numbers

To multiply two mixed numbers, write each mixed number as an improper fraction. Use the greatest common factor, or GCF, to simplify.

Examples

Tutor

3. Find $1\frac{7}{8} \times 3\frac{1}{3}$. Write in simplest form.

$$1\frac{7}{8} \times 3\frac{1}{3} = \frac{15}{8} \times \frac{10}{3}$$ Write $1\frac{7}{8}$ as $\frac{15}{8}$. Write $3\frac{1}{3}$ as $\frac{10}{3}$.

$$= \frac{\overset{5}{\cancel{15}}}{8} \times \frac{\overset{5}{\cancel{10}}}{\underset{1}{\cancel{3}}}$$ Divide 15 and 3 by their GCF, 3.
   Then divide 10 and 8 by their GCF, 2.

$$= \frac{25}{4}$$ Multiply the numerators and multiply the denominators.

$$= 6\frac{1}{4}$$ Simplify.

4. The Hoover Dam contains $4\frac{1}{2}$ million cubic yards of concrete. The Grand Coulee Dam, in Washington state, contains $2\frac{2}{3}$ times as much concrete. How much concrete does it contain?

Estimate $4 \times 3 = 12$

$$4\frac{1}{2} \times 2\frac{2}{3} = \frac{9}{2} \times \frac{8}{3}$$ Write the mixed numbers as improper fractions.

$$= \frac{\overset{3}{\cancel{9}}}{\underset{1}{\cancel{2}}} \times \frac{\overset{4}{\cancel{8}}}{\underset{1}{\cancel{3}}}$$ Divide 9 and 3 by their GCF, 3.
   Then divide 8 and 2 by their GCF, 2.

$$= \frac{3}{1} \times \frac{4}{1}$$ Multiply the numerators and multiply the denominators.

$$= \frac{12}{1} \text{ or } 12$$ Simplify.

There are 12 million cubic yards of concrete in the Grand Coulee Dam.

Check for Reasonableness $12 = 12$ ✔

> STOP **and Reflect**
>
> Is the product of two mixed numbers greater than or less than both the factors? Explain below.

Got it?   Do this problem to find out.

c. Mr. Wilkins is laying bricks to make a rectangular patio. The area he is covering with bricks is $15\frac{1}{2}$ feet by $9\frac{3}{4}$ feet. What is the area of the patio?

Show your work.

c. _____

Example

Tutor

5. Mr. Conrad's pecan pie recipe calls for $1\frac{3}{4}$ cups of pecans. He plans to make 8 pies for his family reunion. How many cups of pecans will Mr. Conrad need?

Estimate $2 \times 8 = 16$

$$1\frac{3}{4} \times 8 = \frac{7}{4} \times \frac{8}{1}$$ Write the mixed number as an improper fraction. Write the whole number as a fraction with a denominator of 1.

$$= \frac{7}{\overset{}{4}} \times \frac{\overset{2}{8}}{1}$$ Divide 8 and 4 by their GCF, 4.

$$= \frac{7}{1} \times \frac{2}{1}$$ Multiply the numerators and multiply the denominators.

$$= \frac{14}{1} \text{ or } 14$$ Simplify.

Check for Reasonableness $14 \approx 16$ ✓

Mr. Conrad will need 14 cups of pecans.

Guided Practice

Check ✓

Multiply. Write in simplest form. (Examples 1–3)

1. $\frac{1}{2} \times 2\frac{3}{8} =$ _____

 Show your work.

2. $1\frac{3}{4} \times 2\frac{4}{5} =$ _____

3. $1\frac{2}{3} \times 2\frac{4}{7} =$ _____

4. Melanie is training for a track meet. She ran $2\frac{1}{4}$ miles 5 times this week. How far did Melanie run this week?

(Examples 4 and 5) _____

5. **Building on the Essential Question** How do you multiply mixed numbers? _____

_____

_____

Rate Yourself!

☐ I understand how to multiply mixed numbers.

▶▶ Great! You're ready to move on!

☐ I still have some questions about multiplying mixed numbers.

⚅ No Problem! Go online to access a Personal Tutor.

Tutor

FOLDABLES Time to update your Foldable!

284 **Chapter 4** Multiply and Divide Fractions

Name _____   My Homework _____

Independent Practice

Go online for Step-by-Step Solutions

Multiply. Write in simplest form. (Examples 1–3)

1. $\frac{1}{2} \times 2\frac{1}{3} =$ _____

2. $1\frac{7}{8} \times \frac{4}{5} =$ _____

3. $\frac{7}{8} \times 3\frac{1}{4} =$ _____

4. $1\frac{2}{3} \times 1\frac{1}{4} =$ _____

5. $3\frac{3}{4} \times 2\frac{2}{5} =$ _____

6. $6\frac{2}{3} \times 3\frac{3}{10} =$ _____

7. A carp can travel at a speed of $3\frac{7}{10}$ miles per hour. At this rate, how far can a carp travel in $2\frac{1}{2}$ hours? (Example 4)

8. Juliette is making fruit salad. She purchased $9\frac{2}{3}$ ounces each of 6 different fruits. How many ounces of fruit did she purchase? (Example 5)

9. A waffle recipe calls for $2\frac{1}{4}$ cups of flour. If Chun wants to make $1\frac{1}{2}$ times the recipe, how much flour does he need? (Example 4)

10. **MP** **Model with Mathematics** Use the formula $d = rt$ to find the distance d a long-distance runner can run at a rate r of $9\frac{1}{2}$ miles per hour for time t of $1\frac{3}{4}$ hours.

11. **STEM** Earth is about $92\frac{9}{10}$ million miles from the Sun. Use the table shown.

   a. How far is Venus from the Sun? _____

   b. How far is Mars from the Sun? _____

   c. How far is Jupiter from the Sun? _____

   d. How far is Saturn from the Sun? _____

Planet	Approximate Number of Times as Far from the Sun as Earth
Venus	$\frac{3}{4}$
Mars	$1\frac{1}{2}$
Jupiter	$5\frac{1}{4}$
Saturn	$9\frac{1}{2}$

Multiply. Write in simplest form.

12. $\frac{3}{4} \times 2\frac{1}{2} \times \frac{4}{5} =$ _____

13. $\frac{1}{7} \times 5\frac{5}{6} \times 1\frac{1}{4} =$ _____

 H.O.T. Problems Higher Order Thinking

14. **MP** **Persevere with Problems** Analyze each product in the table.

First Factor		Second Factor		Product
$\frac{1}{2}$	×	$\frac{3}{4}$	=	$\frac{3}{8}$
1	×	$\frac{3}{4}$	=	$\frac{3}{4}$
$\frac{3}{2}$	×	$\frac{3}{4}$	=	$\frac{9}{8}$

a. Why is the first product less than $\frac{3}{4}$?

_____

_____

b. Why is the second product equal to $\frac{3}{4}$?

_____

_____

_____

c. Why is the third product greater than $\frac{3}{4}$?

_____

_____

_____

d. Based on your observations, make a conjecture about the product when one of the factors is a number less than 1.

_____

_____

15. **MP** **Use Math Tools** Without multiplying, determine whether the product of $2\frac{1}{2} \times \frac{2}{3}$ is located on the number line at point *A*, *B*, or *C*. Explain your reasoning. _____

_____

16. **MP** **Model with Mathematics** Give two examples of mixed numbers that when multiplied by $\frac{3}{4}$ give a product between $\frac{3}{4}$ and 1.

_____

Extra Practice

Multiply. Write in simplest form.

17. $\frac{3}{4} \times 2\frac{5}{6} = 2\frac{1}{8}$ _____

 Homework Help

$\frac{3}{4} \times 2\frac{5}{6} = \frac{3}{4} \times \frac{\overset{1}{17}}{\underset{2}{6}}$

$= \frac{1 \times 17}{4 \times 2}$

$= \frac{17}{8}$ or $2\frac{1}{8}$

18. $1\frac{4}{5} \times \frac{5}{6} = 1\frac{1}{2}$ _____

$1\frac{4}{5} \times \frac{5}{6} = \frac{\overset{3}{9}}{\underset{1}{5}} \times \frac{\overset{1}{5}}{\underset{2}{6}}$

$= \frac{3 \times 1}{1 \times 2}$

$= \frac{3}{2}$ or $1\frac{1}{2}$

19. $\frac{3}{10} \times 2\frac{5}{6} =$ _____

20. $3\frac{1}{5} \times 3\frac{1}{6} =$ _____

21. $4\frac{1}{2} \times 2\frac{5}{6} =$ _____

22. $3\frac{3}{5} \times 5\frac{5}{12} =$ _____

23. **MP** **Model with Mathematics** A reproduction of Claude Monet's *Water-Lilies* has dimensions $34\frac{1}{2}$ inches by $36\frac{1}{2}$ inches. Find the area of the painting.

_____

_____

24. A photograph is $5\frac{1}{3}$ inches wide. It is being enlarged to 3 times its original size. What is the width of the enlarged photograph?

_____

_____

25. Jalisa is making bracelets with leather bands. Each bracelet uses $7\frac{3}{4}$ inches of leather banding. She plans to make 4 bracelets. How many inches of leather banding will she need?

_____

_____

26. **MP** **Use Math Tools** Find examples of mixed numbers in a newspaper or magazine, on television, or on the Internet. Write a real-world problem in which you would multiply mixed numbers.

_____

_____

27. The picture below is enlarged to $4\frac{1}{2}$ times its original size. Fill in the boxes with the deminsions of the enlargement.

$3\frac{3}{4}$ in.

$6\frac{1}{4}$ in.

28. Ally is interested in the following classes. She wants to know the total time for each class before she decides to sign up. Complete the table to show the total time for each class.

Class	Days Offered	Time (h)	Duration	Total Time (h)
Cooking	M W F	$\frac{3}{4}$	3 weeks	
Dance	T R	$\frac{1}{2}$	10 weeks	
Pottery	M W	$2\frac{1}{4}$	8 weeks	
Sewing	T R F	1	6 weeks	
Wood Carving	W F	$1\frac{3}{4}$	5 weeks	

Ally decides on wood carving and wants to take another class. She plans to attend no more than 5 hours of class per week. Which additional class could she choose? Explain.

Spiral Review

Find each equivalent measurement. 5.MD.1

29. 1 foot = _____ inches

30. 1 gallon = _____ quarts

31. 1 yard = _____ feet

Convert Measurement Units

Real-World Link

 Watch ▶

Animals The table shows the approximate weights in tons of several large land animals. One ton is equivalent to 2,000 pounds. You can use a ratio table to convert each weight from tons to pounds.

Animal	Weight (T)
Grizzly bear	1
White rhinoceros	4
Hippopotamus	5
African elephant	8

Essential Question

WHAT does it mean to multiply and divide fractions?

Vocab **Vocabulary**

unit ratio
dimensional analysis

Content Standards
6.RP.3, 6.RP.3d

MP **Mathematical Practices**
1, 3, 4, 6

1. Complete the ratio table. The first two ratios are done for you. To produce equivalent ratios, multiply the quantities in each row by the same number.

Tons	1	4	5	8
Pounds	2,000	8,000		

2. Use the coordinate plane shown.

   a. Graph the ordered pairs (tons, pounds) from the table on the coordinate plane.

   b. Label the horizontal axis *Weight in Tons*.

   c. Label the vertical axis *Weight in Pounds*.

   d. Connect the points and describe the graph.

Which MP Mathematical Practices did you use?
Shade the circle(s) that applies.

① Persevere with Problems
② Reason Abstractly
③ Construct an Argument
④ Model with Mathematics
⑤ Use Math Tools
⑥ Attend to Precision
⑦ Make Use of Structure
⑧ Use Repeated Reasoning

Convert Larger Units to Smaller Units

Each relationship in the table can be written as a ratio. For example, you know that 1 yard = 3 feet. You can use the ratio $\frac{3 \text{ ft}}{1 \text{ yd}}$ to convert from yards to feet.

Customary Conversions			
Type of Measure	**Larger Unit**	⟶	**Smaller Unit**
Length	1 foot (ft)	=	12 inches (in.)
	1 yard (yd)	=	3 feet
	1 mile (mi)	=	5,280 feet
Weight	1 pound (lb)	=	16 ounces (oz)
	1 ton (T)	=	2,000 pounds
Capacity	1 cup (c)	=	8 fluid ounces (fl oz)
	1 pint (pt)	=	2 cups
	1 quart (qt)	=	2 pints
	1 gallon (gal)	=	4 quarts

Like a unit rate, a **unit ratio** is one in which the denominator is 1 unit. So, the ratio $\frac{3 \text{ ft}}{1 \text{ yd}}$ is a unit ratio.

Dimensional analysis is the process of including units of measurement as factors when you compute.

Example

Tutor

1. **Convert 20 feet to inches.**

Since 1 foot = 12 inches, the unit ratio is $\frac{12 \text{ in.}}{1 \text{ ft}}$.

$20 \text{ ft} = 20 \text{ ft} \times \frac{12 \text{ in.}}{1 \text{ ft}}$ Multiply by $\frac{12 \text{ in.}}{1 \text{ ft}}$.

$\quad\quad = 20 \text{ ft} \times \frac{12 \text{ in.}}{1 \text{ ft}}$ Divide out common units, leaving the desired unit, inches.

$\quad\quad = 20 \times 12 \text{ in.}$ Multiply.

$\quad\quad = 240 \text{ in.}$

So, 20 feet = 240 inches.

Got it? Do these problems to find out.

Complete.

a. 36 yd = ■ ft   **b.** $\frac{3}{4}$ T = ■ lb   **c.** $1\frac{1}{2}$ qt = ■ pt

---

Multiplying by 1

The ratio $\frac{3 \text{ ft}}{1 \text{ yd}}$ is equivalent to 1 because the numerator and the denominator represent the same amount.

Show your work.

a. _____

b. _____

c. _____

Example

Tutor

2. Marco mixes $\frac{1}{4}$ cup of fertilizer with soil before planting each bulb. How many fluid ounces of fertilizer does he use per bulb?

$\frac{1}{4}$ c = $\frac{1}{4}$ c × $\frac{8 \text{ fl oz}}{1 \text{ c}}$ Since 1 cup = 8 fluid ounces, multiply by $\frac{8 \text{ fl oz}}{1 \text{ c}}$. Then, divide out common units.

   = $\frac{1}{4}$ × 8 fl oz   Multiply.

   = 2 fl oz

So, 2 fluid ounces of fertilizer are used per bulb.

Got it? Do this problem to find out.

d. Jen runs $\frac{1}{8}$ of a mile before tennis practice. How many feet does she run before practice?

Show your work.

d. _____

Convert Smaller Units to Larger Units

Remember that the ratios $\frac{1 \text{ yd}}{3 \text{ ft}}$ and $\frac{3 \text{ ft}}{1 \text{ yd}}$ are equivalent. To convert from smaller units to larger units, choose the ratio that allows you to divide out the common units.

Example: 12 ft × $\frac{1 \text{ yd}}{3 \text{ ft}}$ ✓

Example

Tutor

3. Convert 15 quarts to gallons.

Since 1 gallon = 4 quarts, and quarts are smaller units than gallons, use the ratio $\frac{1 \text{ gal}}{4 \text{ qt}}$.

15 qt = 15 qt × $\frac{1 \text{ gal}}{4 \text{ qt}}$ Multiply by $\frac{1 \text{ gal}}{4 \text{ qt}}$.

   = 15 qt × $\frac{1 \text{ gal}}{4 \text{ qt}}$ Divide out common units, leaving the desired unit, gallons.

   = 15 × $\frac{1}{4}$ gal   Multiplying 15 by $\frac{1}{4}$ is the same as dividing 15 by 4.

   = 3.75 gal

e. _____

f. _____

Got it? Do these problems to find out.

e. 2,640 ft = ▪ mi   **f.** 100 oz = ▪ lb   **g.** 3 c = ▪ pt

g. _____

Example

 Tutor

4. Umeka needs $4\frac{1}{2}$ feet of fabric to make a costume for a play. How many yards of fabric does she need?

$$4\frac{1}{2} \text{ ft} = 4\frac{1}{2} \cancel{\text{ft}} \times \frac{1 \text{ yd}}{3 \cancel{\text{ft}}}$$ Since 1 yard = 3 feet, multiply by $\frac{1 \text{ yd}}{3 \text{ ft}}$. Then, divide out common units.

$$= \frac{\overset{3}{\cancel{9}}}{2} \times \frac{1}{\underset{1}{\cancel{3}}} \text{ yd}$$ Write $4\frac{1}{2}$ as an improper fraction. Then divide out common factors.

$$= \frac{3}{2} \text{ yd or } 1\frac{1}{2} \text{ yd}$$ Multiply.

So, Umeka needs $1\frac{1}{2}$ yards of fabric.

Guided Practice

Check ✓

Complete. (Examples 1 and 3)

1. $5\frac{1}{3}$ yd = _____ ft

2. $4\frac{1}{2}$ pt = _____ c

3. 12 qt = _____ gal

4. 28 in. = _____ ft

 Show your work.

5. A large grouper can weigh $\frac{1}{3}$ ton. How much does a large grouper weigh to the nearest pound? (Example 2) _____

6. The world's narrowest electric vehicle is about 35 inches wide. How wide is this vehicle to the nearest foot? (Example 4) _____

7. **Building on the Essential Question** How can you use ratios to convert units of measurement?

_____

_____

_____

Rate Yourself!

Are you ready to move on? Shade the section that applies.

YES   ?   NO

For more help, go online to access a Personal Tutor.

Tutor

Independent Practice

Go online for Step-by-Step Solutions eHelp

Complete. (Examples 1 and 3)

 Show your work.

1. 18 ft = _____ yd

2. 2 lb = _____ oz

3 6.5 c = _____ fl oz

4. 2 mi = _____ ft

5. 5,000 lb = _____ T

6. $2\frac{3}{4}$ qt = _____ pt

7 One of the largest pumpkins ever grown weighed about $\frac{3}{4}$ ton. How many pounds did the pumpkin weigh? (Example 2)

_____

8. A 40-foot power boat is for sale by owner. How long is the boat to the nearest yard? (Example 4)

_____

Show your work.

9. A 3-pound pork loin can be cut into 10 pork chops of equal weight. How many ounces is each pork chop? _____

_____

10. (MP) **Model with Mathematics** Will a 2-quart pitcher hold the entire recipe of citrus punch given at the right? Explain your reasoning.

_____

_____

_____

_____

Citrus Punch Drink

2 cups orange juice
2 cups grapefruit juice
$\frac{1}{4}$ cup apricot nectar
$\frac{1}{3}$ cup pineapple juice
4 cups ginger ale

11. **MP** **Multiple Representations** Use the graph at the right.

a. **Numbers** What does an ordered pair from this graph

represent? _____

_____

b. **Measurement** Use the graph to find the capacity in quarts of a 2.5-gallon container. Explain your reasoning.

_____

_____

c. **Unit Rate** What is the unit rate that converts gallons to quarts?

_____

d. **Expressions** Write an expression you could use to convert 2.5 gallons to quarts.

_____

🔥 H.O.T. Problems   Higher Order Thinking

12. **MP** **Model with Mathematics** Write a real-world problem in which you

would need to convert pints to cups. _____

_____

_____

MP **Persevere with Problems** Fill in each ◯ with <, >, or = to make a true sentence. Justify your answers.

13. 16 in. ◯ $1\frac{1}{2}$ ft

14. $8\frac{3}{4}$ gal ◯ 32 qt

_____

_____

15. **MP** **Persevere with Problems** Give two different measurements that are

equivalent to $2\frac{1}{2}$ quarts. _____

16. **MP** **Model with Mathematics** Write a real-world problem that can be

represented by the graph. _____

_____

_____

Extra Practice

Complete.

17. 72 oz = $4\frac{1}{2}$ lb

Homework Help →

$72 \text{ oz} = 72 \text{ oz} \times \frac{1 \text{ lb}}{16 \text{ oz}}$

$= \frac{\overset{9}{\cancel{72}}}{1} \times \frac{1}{\underset{2}{\cancel{16}}} \text{ lb}$

$= \frac{9}{2} \text{ or } 4\frac{1}{2}$

18. 4 gal = 16 qt

$4 \text{ gal} = 4 \text{ gal} \times \frac{4 \text{ qt}}{1 \text{ gal}}$

$= \frac{4}{1} \times \frac{4}{1} \text{ qt}$

$= \frac{16}{1} \text{ or } 16 \text{ qt}$

19. 3 c = _____ fl oz

20. $1\frac{1}{4}$ mi = _____ ft

21. 13 c = _____ pt

22. $3\frac{3}{8}$ T = _____ lb

23. Speed skiing takes place on a course that is $\frac{2}{3}$ mile long. How many feet long is the course?

Show your work. →

_____

24. A total of 35 pints of blood were collected at a local blood drive. How many quarts of blood were collected?

_____

25. **STEM** On Monday, it snowed a total of 15 inches. On Tuesday and Wednesday, it snowed an additional $4\frac{1}{2}$ inches and $6\frac{3}{4}$ inches, respectively. A weather forecaster says that over the last three days, it snowed over $2\frac{1}{2}$ feet. Is this a valid claim? Justify your answer.

_____

_____

MP **Be Precise** Complete the following statements.

26. If 16 c = 1 gal, then $1\frac{1}{4}$ gal = _____ c.

27. If 1,760 yd = 1 mi, then 880 yd = _____ mi.

28. If 36 in. = 1 yd, then $2\frac{1}{3}$ yd = _____ in.

29. The heights of 5 friends are shown in the table. Determine if each statement is true or false .

Friend	Height
Adrienne	65 in.
Eliott	$5\frac{2}{3}$ ft
Jaquie	$1\frac{2}{3}$ yd
Maria	$5\frac{1}{6}$ ft
Martin	$66\frac{1}{2}$ in.

    a. Maria is shorter than Jaquie.   ☐ True   ☐ False

    b. Martin is taller than Eliott.   ☐ True   ☐ False

    c. Jaquie is shorter than Eliott.   ☐ True   ☐ False

    d. Maria is taller than Adrienne.   ☐ True   ☐ False

30. Sort the measurements from least to greatest.

48 fl oz $10\frac{3}{4}$ c $\frac{1}{2}$ gal $2\frac{1}{4}$ qt $3\frac{1}{2}$ pt

	Measurement
Least	
Greatest	

Spiral Review

Divide. 5.NBT.6

31. 156 ÷ 4 = _____

32. 212 ÷ 8 = _____

33. 90 ÷ 12 = _____

34. David baked 78 cookies for a bake sale. He set aside 12 cookies to share with volunteers. The remaining cookies are bagged with 3 cookies in each bag. How many bags does David need? 4.OA.3 _____

35. Refer to the diagram of a living room. The doorway will not have a baseboard. How many feet of baseboard are needed to go around the room? 4.MD.3 _____

$12\frac{1}{4}$ ft $2\frac{5}{6}$ ft   10 ft

Draw a Diagram

Content Standards
6.NS.1

MP **Mathematical Practices**
1, 3, 4

Case #1 Traction Action

Manuel and his friends celebrated his birthday at the FunTimes game center. He spent $\frac{4}{7}$ of his money at the fun center on go-karts and now he has $15 left.

How much money did he spend on go-karts?

Understand What are the facts?

You know that Manuel spent $\frac{4}{7}$ of his money on go-karts. You need to determine how much money he spent on go-karts.

Plan What is your strategy to solve this problem?

He spent a fraction of his money. Draw a bar diagram.

Solve How can you apply the strategy?

Complete the bar diagram using information from the problem. Fill in the missing numbers to show the value of each section.

				5	5	5

|— — — Amount Spent on Go-Karts — — — —|— — — — $15 — — — — —|

$ [ ] ÷ [ ] = $ [ ]

So, each section represents $ [ ].

Manuel spent 4 × $ [ ] or $ [ ] on go-karts.

Check Does the answer make sense?

Four sevenths of $35 is 4 × [ ] or $ [ ]. ✔

Analyze the Strategy Tutor

MP **Justify Conclusions** Suppose Manuel had $9 left. How much money did he start with? Explain.

_____

_____

Case #2 Text Tally

Jeremy has $\frac{3}{5}$ as many saved text messages as Ria.

Jeremy has 24 saved text messages.

How many saved text messages do they have in all?

 Understand

Read the problem. What are you being asked to find?

I need to find _____ .

Underline key words and values in the problem. What information do you know?

Jeremy has $\dfrac{\boxed{}}{\boxed{}}$ as many saved texts as Ria.

Jeremy has $\boxed{}$ saved texts.

 Plan

Choose a problem-solving strategy.

I will use the _____ strategy.

 Solve

Solve the problem using your problem-solving strategy.

• Divide the tape diagram for Jeremy into 3 equal sections and the diagram for Ria into 5 equal sections.

• Jeremy has 24 messages. Fill in the boxes.

Jeremy $\boxed{}$

Ria $\boxed{}$

Ria has $\boxed{}$ × $\boxed{}$ = $\boxed{}$ saved texts.

So, Jeremy and Ria have $\boxed{}$ + $\boxed{}$, or $\boxed{}$ saved texts in all.

 Check

Use information from the problem to check your answer.

$\frac{3}{5}$ × $\boxed{}$ = _____

**Work with a small group to solve the following cases.
Show your work on a separate piece of paper.**

Collaborate

Case #3 Internet

Francesca spent 25 minutes on the Internet yesterday.

If this is $\frac{5}{6}$ of the time she spent on the computer, how long did she spend on the computer, but not on the Internet?

Case #4 Basketball

Mieko practiced shooting a basketball for $\frac{7}{10}$ of her total practice time. During the other time, she practiced dribbling.

If she practiced dribbling for 18 minutes, how many minutes did she practice shooting?

Case #5 Biking

Two bike riders, 175 miles apart, begin traveling toward each other at noon. One travels at 20 miles per hour, the other at 15 miles per hour. Also at noon, a fly begins flying between the riders, starting at the front of the slower bike. The fly travels at 20 miles per hour and can change direction without losing any time.

How far will the fly travel before the bicycles meet?

Use any strategy!

Case #6 Fruit

Use the table that shows the prices of different amounts of mixed fruit at the grocery store.

How much will 13 pounds of fruit cost?

Pounds	Cost ($)
2	4.50
4	9.00
6	13.50
8	18.00

Mid-Chapter Check

Vocabulary Check

1. **MP** **Be Precise** Define *Commutative Property*. Provide an example of an operation that is commutative. Provide an example of an operation which is not commutative. (Lesson 2)

_____

_____

_____

Skills Check and Problem Solving

Multiply. Write in simplest form. (Lessons 1–4)

2. $8 \times \frac{2}{5} =$ _____

3. $\frac{7}{8} \times \frac{2}{3} =$ _____

4. $4\frac{3}{4} \times 2\frac{1}{8} =$ _____

5. A new shirt costs $14.99. If the shirt is on sale for $\frac{1}{5}$ off its price, about how much would you save? (Lesson 1) _____

6. **MP** **Justify Conclusions** Corey needs 24 boards that are $47\frac{1}{2}$ inches long. (Lesson 5)

   a. How many feet of boards should he buy? Explain. _____

   _____

   b. If you can only buy 8-foot boards, how many should he buy? Explain.

   _____

   _____

   _____

7. **MP** **Persevere with Problems** What is the area of the picture and frame shown? Write your answer as a mixed number in simplest form.

   (Lesson 4) _____

$7\frac{1}{4}$ in.

$11\frac{2}{3}$ in.

Inquiry Lab

Divide Whole Numbers by Fractions

 Inquiry HOW can a bar diagram help you understand what it means to divide fractions?

Content Standards
6.NS.1

MP Mathematical Practices
1, 3, 4

Juan is building a set for the school musical. He has a 3-foot board that he needs to equally divide into $\frac{1}{2}$-foot pieces. How many pieces will he have after he cuts the board?

What do you know? _____

_____

_____

What do you need to find? _____

_____

Hands-On Activity 1

Step 1   Draw a model that represents the length of the board. Draw lines to separate the board into thirds. Each third represents one foot.

Step 2   Divide each foot into halves.

Step 3   Determine how many groups of $\frac{1}{2}$ are in 3. Circle the groups that are the size of the divisor $\frac{1}{2}$.

There are ☐ groups of $\dfrac{☐}{☐}$. So, $3 \div \frac{1}{2} = ☐$.

Check by multiplying: $☐ \times \dfrac{☐}{☐} = ☐$ ✔

Hands-On Activity 2

Find $4 \div \frac{2}{3}$.

Step 1   The model represents 4.

4

Step 2   Divide each whole into thirds.

Step 3   Circle groups of $\frac{2}{3}$ on the model.   *Think: How many groups of $\frac{2}{3}$ are in 4?*

There are ☐ groups of $\frac{2}{3}$. So, $4 \div \frac{2}{3} = $ ☐.

Check by multiplying: ☐ $\times \dfrac{\boxed{}}{\boxed{}} = $ ☐ ✔

Hands-On Activity 3

Find $2 \div \frac{3}{4}$.

Step 1   The model represents 2.

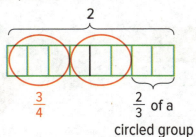

2

$\frac{3}{4}$ $\frac{2}{3}$ of a
circled group

Step 2   Divide each whole into _____.

Step 3   Determine how many groups of $\dfrac{\boxed{}}{\boxed{}}$ are in ☐.

Circle groups of $\dfrac{\boxed{}}{\boxed{}}$ on the model.

There are ☐ groups of $\frac{3}{4}$ and $\frac{2}{3}$ of a group left over. So, $2 \div \frac{3}{4} = \boxed{}\dfrac{\boxed{}}{\boxed{}}$.

Check by multiplying: $\boxed{}\dfrac{\boxed{}}{\boxed{}} \times \dfrac{\boxed{}}{\boxed{}} = $ ☐ ✔

Investigate

Collaborate

MP Model with Mathematics Work with a partner. Draw a diagram to find each quotient.

1. $3 \div \frac{1}{3} =$ _____

2. $2 \div \frac{1}{4} =$ _____

Show your work.

3. $6 \div \frac{2}{3} =$ _____

4. $4 \div \frac{1}{2} =$ _____

5. $3 \div \frac{3}{4} =$ _____

6. $4 \div \frac{3}{4} =$ _____

7. $5 \div \frac{2}{3} =$ _____

8. $2 \div \frac{4}{5} =$ _____

Analyze and Reflect

MP Use Math Tools Refer to the models you drew on the previous page. Work with a partner to complete the following table. The first one is done for you.

	Division Problem	Quotient	Is the Quotient Less Than or Greater Than the Dividend?
	$3 \div \frac{1}{3}$	9	greater
9.	$2 \div \frac{1}{4}$		
10.	$6 \div \frac{2}{3}$		
11.	$4 \div \frac{1}{2}$		
12.	$3 \div \frac{3}{4}$		
13.	$4 \div \frac{3}{4}$		

14. **MP Reason Inductively** When a whole number is divided by a fraction that is less than one, will the quotient be less than or greater than the whole number? Explain.

_____

_____

_____

Create

15. **MP Model with Mathematics** Write a story context that involves $4 \div \frac{4}{5}$. Solve the problem and multiply to check your answer.

_____

_____

_____

16. **Inquiry** HOW can a bar diagram help you understand what it means to divide

fractions? _____

_____

_____

Divide Whole Numbers by Fractions

Vocabulary Start-Up

 Vocab abc

Any two numbers with a product of 1 are called **reciprocals**.

Complete the table below by finding the reciprocal of $\frac{2}{3}$. Use the *guess, check, and revise* strategy. The first one is done for you.

Number	Product	Reciprocal
$\frac{1}{2}$	$\frac{1}{2} \times 2 = 1$	2
$\frac{2}{3}$	$\frac{2}{3} \times \dfrac{\square}{\square} = 1$	$\dfrac{\square}{\square}$

Describe the relationship between the numerator and the

denominator of a number and its reciprocal. _____

_____

 ## Real-World Link

Another name for reciprocal is *multiplicative inverse*. What are some words in everyday language that are similar to reciprocal or inverse?

_____

Pilots can fly in an *inverted* position, or upside down. How can you use the everyday meaning of *invert* to help you remember the mathematical meaning of multiplicative inverse, or reciprocal?

_____

_____

Essential Question

WHAT does it mean to multiply and divide fractions?

Vocab abc **Vocabulary**

reciprocals

Content Standards
6.NS.1

MP **Mathematical Practices**
1, 3, 4, 5

Which MP Mathematical Practices did you use?
Shade the circle(s) that applies.

① Persevere with Problems

② Reason Abstractly

③ Construct an Argument

④ Model with Mathematics

⑤ Use Math Tools

⑥ Attend to Precision

⑦ Make Use of Structure

⑧ Use Repeated Reasoning

Find Reciprocals

Dividing 3 by $\frac{1}{2}$ gives the same result as multiplying 3 by 2, which is the reciprocal of $\frac{1}{2}$. Any two numbers with a product of 1 are called reciprocals.

reciprocals

$$3 \div \frac{1}{2} = 6 \qquad 3 \times 2 = 6$$

same result

Examples

Tutor

1. **Find the reciprocal of $\frac{2}{3}$.**

   Since $\frac{2}{3} \times \frac{3}{2} = 1$, the reciprocal of $\frac{2}{3}$ is $\frac{3}{2}$.

2. **Find the reciprocal of $\frac{1}{8}$.**

   Since $\frac{1}{8} \times \frac{8}{1} = 1$, the reciprocal of $\frac{1}{8}$ is $\frac{8}{1}$ or 8.

3. **Find the reciprocal of 5.**

   Write the whole number as a fraction.

   $5 = \dfrac{\boxed{}}{1}$

   Find the missing factor.

   $\dfrac{\boxed{}}{\boxed{}} \times \dfrac{\boxed{}}{\boxed{}} = 1$

   The reciprocal of 5 is $\dfrac{\boxed{}}{\boxed{}}$.

Reciprocals

The examples suggest that you "invert" the fraction to find the reciprocal. That is, switch the numerator and denominator. You can use reciprocals to divide fractions.

Show your work.

a. _____

b. _____

c. _____

Got it? Do these problems to find out.

Find the reciprocal of each number.

a. $\frac{3}{5}$ 　　　　 b. $\frac{1}{3}$ 　　　　 c. 11

Divide by a Fraction

Words To divide a whole number by a fraction, multiply by its reciprocal.

Example $5 \div \frac{2}{3} = \frac{5}{1} \times \frac{3}{2}$

The division expression $5 \div \frac{2}{3}$ is read as *5 divided by two thirds.* You need to find how many two thirds are in 5.

Examples

Tutor

4. **Find $2 \div \frac{1}{3}$. Write in simplest form.**

2

Method 1 Use a model.

Model the dividend, 2.

Divide each whole into thirds.

Think: How many thirds are in 2?

$\frac{1}{3}$ $\frac{1}{3}$ $\frac{1}{3}$ $\frac{1}{3}$ $\frac{1}{3}$ $\frac{1}{3}$

Count the total number of sections.
There are 6 sections.

Method 2 Use an equation.

$2 \div \frac{1}{3} = \frac{2}{1} \times \frac{3}{1}$ Multiply by the reciprocal of $\frac{1}{3}$.

$= \frac{6}{1}$ or 6   Multiply the numerators.
                     Multiply the denominators.

5. **Find $7 \div \frac{2}{3}$. Write in simplest form.**

$7 \div \frac{2}{3} = \dfrac{\boxed{}}{\boxed{}} \times \dfrac{\boxed{}}{\boxed{}}$ Write the whole number as a fraction.
                                 Multiply by the reciprocal of $\frac{2}{3}$.

$\dfrac{\boxed{}}{\boxed{}} \times \dfrac{\boxed{}}{\boxed{}} = \dfrac{\boxed{}}{\boxed{}}$ Multiply the numerators.
                                 Multiply the denominators.

$\dfrac{\boxed{}}{\boxed{}} = \boxed{}\,\dfrac{\boxed{}}{\boxed{}}$ Simplify.

Show your work.

d. _____

e. _____

f. _____

Got it? **Do these problems to find out.**

d. $6 \div \frac{1}{3}$ **e.** $5 \div \frac{2}{3}$ **f.** $4 \div \frac{3}{4}$

Example

6. At summer camp, the duration of a field hockey game is $\frac{3}{4}$ hour. The camp counselors have set aside 6 hours for field hockey games. How many games can be played?

Divide 6 by three-fourths.

$6 \div \frac{3}{4} = \frac{6}{1} \times \frac{4}{3}$ Multiply by the reciprocal.

$= \frac{\overset{2}{\cancel{6}}}{1} \times \frac{4}{\underset{1}{\cancel{3}}}$ Divide 3 and 6 by the GCF, 3.

$= \frac{8}{1}$ or 8   Simplify.

So, 8 games can be played.

Guided Practice

Find the reciprocal of each number. (Examples 1–3)

1. $\frac{2}{3}$ _____

2. $\frac{1}{7}$ _____

3. 4 _____

Show your work.

Divide. Write in simplest form. (Examples 4 and 5)

4. $2 \div \frac{1}{3} =$ _____

5. $2 \div \frac{4}{5} =$ _____

6. $5 \div \frac{2}{7} =$ _____

7. A neighborhood development that is 4 acres is to be divided into $\frac{2}{3}$-acre lots. How many lots can be created?

(Example 6) _____

8. **Building on the Essential Question** Why does a whole number divided by a fraction less than one have a quotient greater than the whole number dividend?

_____

_____

_____

Rate Yourself!

How well do you understand dividing whole numbers by fractions? Circle the image that applies.

Clear   Somewhat Clear   Not So Clear

For more help, go online to access a Personal Tutor.

FOLDABLES Time to update your Foldable!

Independent Practice

Go online for Step-by-Step Solutions

Find the reciprocal of each number. (Examples 1–3)

1. $\frac{3}{5}$ _____

2. $\frac{1}{4}$ _____

3. 1 _____

Show your work.

Divide. Write in simplest form. (Examples 4 and 5)

4. $3 \div \frac{3}{4} =$ _____

5. $5 \div \frac{3}{4} =$ _____

6. $8 \div \frac{4}{7} =$ _____

7. $6 \div \frac{3}{5} =$ _____

8. $2 \div \frac{5}{8} =$ _____

9. $4 \div \frac{8}{9} =$ _____

10. Jamar has an 8-foot-long piece of wood that he wants to cut to build a step stool for his tree house. If each piece is going to be $\frac{5}{6}$ foot long, what is the greatest number of pieces he will be able to use? (Example 6)

_____

_____

11. The average adult horse needs $\frac{2}{5}$ bale of hay each day to meet dietary requirements. A horse farm has 44 bales of hay. How many horses can be fed in one day with 44 bales of hay? (Example 6)

_____

_____

12. **MP** **Justify Conclusions** Ethan ordered 4 sub sandwiches for a party. Each $\frac{1}{2}$ sandwich is one serving. Does he have enough to serve 7 friends? How much is leftover or how much more is needed? Explain. _____

_____

_____

_____

13. Chelsea has four hours of free time on Saturday. She would like to spend no more than $\frac{2}{3}$ of an hour on each activity. How many activities can she do during that time? Justify your procedure.

14. MP **Model with Mathematics** Find an example of dividing a whole number by a fraction in a newspaper or on the Internet. Write a real-world problem in which you would divide a whole number by a fraction.

🔥 H.O.T. Problems Higher Order Thinking

15. MP **Find the Error** Daniella is solving $\frac{8}{9} \div 4$. Find her mistake and correct it.

$$\frac{8}{9} \div 4 = \frac{8}{9} \times \frac{4}{1}$$
$$= \frac{32}{9} \text{ or } 3\frac{5}{9}$$

16. MP **Persevere with Problems** The Snack Shack is making a batch of trail mix. They use $9\frac{1}{3}$ pounds of granola, $9\frac{1}{3}$ pounds of mixed nuts, and $9\frac{1}{3}$ pounds of yogurt raisins to make the trail mix. They divide the mixture into 14 packages. How much is in each package? Explain.

17. MP **Reason Inductively** The bag of peanuts shown is on sale for $9.30. The Pep club wants to divide the larger bag into $\frac{3}{4}$-pound bags to sell at football games.

a. How much will it cost to fill one smaller bag?

b. How many pounds of peanuts will be remaining?

c. How many smaller bags can the club get from one 5-pound bag?

d. Suppose the club wants to sell 25 smaller bags. How many of the larger bags will they need to purchase? Explain.

Extra Practice

Find the reciprocal of each number.

18. $\frac{1}{10}$ 10

Homework
Help

$\frac{1}{10} \times \frac{10}{1} = 1$

The reciprocal is $\frac{10}{1}$ or 10.

19. $\frac{7}{9}$ _____

20. 8 _____

Divide. Write in simplest form.

21. $2 \div \frac{3}{5} = 3\frac{1}{3}$ _____

$2 \div \frac{3}{5} = \frac{2}{1} \times \frac{5}{3}$

$= \frac{10}{3}$ or $3\frac{1}{3}$

22. $5 \div \frac{5}{6} =$ _____

23. $3 \div \frac{5}{6} =$ _____

24. $10 \div \frac{5}{6} =$ _____

25. $4 \div \frac{5}{9} =$ _____

26. $6 \div \frac{2}{3} =$ _____

27. Turner has 6 pounds of pasta. Each time he makes dinner he uses $\frac{3}{4}$ pound of pasta. How many dinners can he make?

_____

28. **MP** **Use Math Tools** Rafael took 4 pumpkin pies to a family gathering. If he divides each pie into six equal-size slices, how many slices can

he serve? _____

29. Lenora is using the recipe shown to make salad dressing for a salad. She wants to make 20 batches of the recipe. How much vegetable oil and how much vinegar does she need?

Salad Dressing
$\frac{1}{3}$ cup vegetable oil
$\frac{1}{4}$ cup vinegar
salt and pepper
to taste

vegetable oil: [ ]

vinegar: [ ]

30. Derrick has $4 to play video games at the mall. Each game costs a quarter to play. Which is a correct method for determining the total number of games he can play? Select all that apply.

☐ Take the number of dollars he has and multiply it by 0.25.

☐ Take the number of dollars he has and divide it by 0.25.

☐ Take the number of dollars he has and multiply it by 4.

☐ Take the number of dollars he has and divide it by $\frac{1}{4}$.

Spiral Review

Find an equivalent fraction. **5.NF.5b**

31. $\frac{2}{3} = \frac{\boxed{}}{9}$

32. $\frac{3}{5} = \frac{\boxed{}}{20}$

33. $\frac{1}{4} = \frac{\boxed{}}{24}$

34. $\frac{5}{6} = \frac{\boxed{}}{18}$

35. $\frac{3}{4} = \frac{\boxed{}}{32}$

36. $\frac{4}{7} = \frac{\boxed{}}{28}$

37. The table shows how far four students walked in 5 minutes. How far did they walk together? **5.NF.1** _____

38. Sixty people can receive a piece of pizza if 5 pizzas are purchased. How many people can receive a slice of pizza if 7 pizzas are purchased?

4.OA.3 _____

	Distance (miles)
April	$\frac{3}{4}$
Ping	$\frac{1}{2}$
Hannah	$\frac{2}{3}$
Raj	$\frac{3}{4}$

Inquiry Lab
Divide Fractions

 HOW can using models help you divide one fraction by another fraction?

Content Standards
6.NS.1

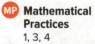 **Mathematical Practices**
1, 3, 4

Toby bought $\frac{8}{9}$ pound of trail mix from the grocery store. He wants to divide the trail mix into $\frac{2}{9}$-pound bags. How many bags can Toby make?

What do you know? _____

_____

What do you need to find? _____

_____

Hands-On Activity 1

To solve the problem use the division sentence $\frac{8}{9} \div \frac{2}{9}$. This shows how many groups of $\frac{\boxed{}}{\boxed{}}$ are in $\frac{8}{9}$.

Step 1 To make a bar diagram that represents the amount of Toby's trail mix, divide the bar into $\boxed{}$ sections.

Step 2 Shade $\boxed{}$ of the sections to represent $\frac{8}{9}$ pound.

Step 3 Circle each group of $\frac{2}{9}$ in the shaded section. Determine the number of equal groups of $\frac{2}{9}$.

There are _____ groups of $\frac{2}{9}$ in $\frac{8}{9}$. So, $\frac{8}{9} \div \frac{2}{9} = \boxed{}$.

Toby can make _____ bags of trail mix that each weigh $\frac{\boxed{}}{\boxed{}}$ pound.

Hands-On Activity 2

Draw a diagram to find $\frac{3}{4} \div \frac{3}{8}$.

Step 1   Rename so the fractions have common denominators. Since 8 is
a multiple of 4, rename the fraction $\frac{3}{4}$ as $\frac{\boxed{}}{\boxed{}}$.

Step 2   Draw a diagram with $\boxed{}$ sections and shade $\boxed{}$ of the sections
to represent $\frac{6}{8}$ pound.

Step 3   Circle each group of $\frac{3}{8}$ in the shaded section. Determine the
number of equal groups of $\frac{3}{8}$.

There are _____ groups of $\frac{3}{8}$ in $\frac{6}{8}$.

So, $\frac{3}{4} \div \frac{3}{8} = \boxed{}$.

Hands-On Activity 3

Draw a diagram to find $\frac{2}{3} \div 2$.

Step 1   Draw a diagram and shade the sections to represent $\frac{\boxed{}}{\boxed{}}$.

Step 2   Divide the shaded sections into _____ equal groups.

Step 3   Write the fraction that names each group. $\frac{\boxed{}}{\boxed{}}$

So, $\frac{2}{3} \div 2 = \frac{\boxed{}}{\boxed{}}$.

Investigate

Collaborate

MP **Model with Mathematics** Work with a partner. Draw a diagram to find each quotient.

1. $\frac{6}{7} \div \frac{2}{7} =$ _____

2. $\frac{4}{5} \div \frac{2}{5} =$ _____

 Show your work.

3. $\frac{6}{7} \div \frac{3}{7} =$ _____

4. $\frac{8}{10} \div \frac{2}{5} =$ _____

5. $\frac{3}{4} \div \frac{1}{2} =$ _____

6. $\frac{5}{6} \div \frac{2}{3} =$ _____

7. $\frac{4}{7} \div 2 =$ _____

8. $\frac{12}{13} \div 3 =$ _____

Analyze and Reflect

MP Use Math Tools Work with a partner to complete the table. The first one is done for you.

	Division Expression	Quotient	Multiplication Sentence
	$\frac{4}{5} \div \frac{1}{5}$	4	$\frac{1}{5} \times 4 = \frac{4}{5}$
9.	$\frac{8}{9} \div 8$		
10.			$\frac{3}{4} \times \frac{1}{2} = \frac{3}{8}$
11.	$\frac{6}{8} \div \frac{2}{8}$		
12.			$\frac{3}{7} \times 1 = \frac{3}{7}$
13.	$\frac{10}{11} \div 5$		
14.			$\frac{5}{9} \times 1 = \frac{5}{9}$

15. **MP Reason Inductively** Use the table to compare the divisor and dividend to the value of the quotient. When is the quotient greater than 1?

_____

_____

Create

16. **MP Make a Conjecture** Some quotients in the table are less than 1. Use the table to write a rule about when the quotient of two fractions will be less than 1.

_____

17. **MP Model with Mathematics** Write a story context that involves $\frac{6}{8} \div \frac{2}{8}$. Solve the problem and multiply to check your answer.

_____

_____

18. **Inquiry** HOW can using models help you divide one fraction by another fraction?

_____

_____

Divide Fractions

Real-World Link

Murals Three students are painting an art mural. The art mural is half painted.

1. Use the picture after Exercise 4. Divide the painted area into 3 equal parts.

2. Place an X over each part of the painted area. This represents the part each student has painted. Then divide the unpainted area into the same number of parts.

3. What fraction of the whole mural has each student painted? $\frac{\square}{\square}$

4. So, $\frac{1}{2} \div 3 = \frac{\square}{\square}$. It is also true that $\frac{1}{2} \times \frac{1}{3} = \frac{\square}{\square}$. Compare

and contrast the division problem and the multiplication problem.

_____

_____

_____

 Essential Question

WHAT does it mean to multiply and divide fractions?

Content Standards
6.NS.1

MP **Mathematical Practices**
1, 2, 3, 4, 5, 7, 8

Which **MP** **Mathematical Practices** did you use?
Shade the circle(s) that applies.

① Persevere with Problems

② Reason Abstractly

③ Construct an Argument

④ Model with Mathematics

⑤ Use Math Tools

⑥ Attend to Precision

⑦ Make Use of Structure

⑧ Use Repeated Reasoning

Divide by a Fraction

Words   To divide by a fraction, multiply by its reciprocal.

Example   **Numbers**   **Algebra**

$$\frac{5}{6} \div \frac{2}{3} = \frac{5}{6} \times \frac{3}{2}$$ $$\frac{a}{b} \div \frac{c}{d} = \frac{a}{b} \times \frac{d}{c}, \text{ where } b, c, \text{ and } d \neq 0$$

Work Zone

Example

Tutor

1. Find $\frac{1}{2} \div \frac{1}{3}$. **Write in simplest form.**

Method 1   **Use a model.**

Model the dividend, $\frac{1}{2}$.

Divide the whole into thirds.

$\frac{1}{2} \div \frac{1}{3}$ means how many thirds are in $\frac{1}{2}$.

Count the number of $\frac{1}{3}$ sections that are in $\frac{1}{2}$.

$\frac{1}{2}$ of another third

> **Greatest Common Factor (GCF)**
>
> The GCF of two or more numbers is the product of all common prime factors.
> $4 = 2 \times 2$
> $8 = 2 \times 2 \times 2$
> The GCF is 2×2 or 4.

Method 2   **Use an equation.**

$$\frac{1}{2} \div \frac{1}{3} = \blacksquare$$

$$\frac{1}{2} \div \frac{1}{3} = \frac{1}{2} \times \frac{3}{1}$$ Multiply by the reciprocal, $\frac{3}{1}$.

$$= \frac{3}{2} \text{ or } 1\frac{1}{2}$$ Multiply the numerators.
Multiply the denominators.

So, $\frac{1}{2} \div \frac{1}{3} = 1\frac{1}{2}$.

Check by multiplying: $\frac{3}{2} \times \frac{1}{3} = \frac{1}{2}$ ✔

Show your work.

Got it? Do these problems to find out.

a. $\frac{1}{4} \div \frac{3}{8}$ b. $\frac{2}{3} \div \frac{3}{8}$ c. $\frac{5}{6} \div \frac{1}{3}$

a. _____

b. _____

c. _____

Example

Tutor

2. **Write a story context for** $\frac{2}{3} \div \frac{1}{6}$. **Use a model to solve.**

Mariska has $\frac{2}{3}$-pound of sunflower seeds. Each day, she feeds the cardinals in her yard $\frac{1}{6}$ pound of seeds. For how many days will she be able to feed the cardinals?

Model $\frac{2}{3}$. The whole is divided into six sections. Count the number of $\frac{1}{6}$ sections.

So, Mariska can feed the cardinals for 4 days.

Got it? **Do this problem to find out.**

d. Write a story context for $\frac{3}{4} \div \frac{1}{8}$. Use a model to solve.

d. _____

Divide a Fraction by a Whole Number

When you divide a fraction by a whole number, rewrite the whole number as a fraction. Then divide as with fractions.

Example

Tutor

3. **Find** $\frac{5}{7} \div 10$. **Write in simplest form.**

$$\frac{5}{7} \div 10 = \frac{5}{7} \div \frac{10}{1}$$ Write the whole number as a fraction with a denominator of 1.

$$= \frac{5}{7} \times \frac{1}{10}$$ Multiply by the reciprocal.

$$= \frac{\overset{1}{5} \times 1}{7 \times \underset{2}{10}}$$ Divide 5 and 10 by their GCF, 5.

$$= \frac{1}{14}$$ Multiply the numerators.
Multiply the denominators.

Got it? **Do these problems to find out.**

e. $\frac{8}{9} \div 4$ **f.** $\frac{4}{5} \div 8$ **g.** $\frac{12}{13} \div 4$

e. _____

f. _____

g. _____

Tutor

Example

4. Ramón is making party favors. He is dividing $\frac{3}{4}$ pound of almonds into 12 packages. Write and solve an equation to find how many pounds of almonds are in each package.

To find the number of pounds in each package, solve the equation $\frac{3}{4} \div 12 = \blacksquare$.

$$\frac{3}{4} \div 12 = \frac{3}{4} \times \frac{1}{12}$$ Multiply by the reciprocal, $\frac{1}{12}$.

$$= \frac{\overset{1}{\cancel{3}} \times 1}{4 \times \underset{4}{\cancel{12}}}$$ Divide 3 and 12 by their GCF, 3.

$$= \frac{1}{16}$$ Multiply the numerators.
Multiply the denominators.

There will be $\frac{1}{16}$ pound of almonds in each package.

Guided Practice

 Check ✓

Divide. Write in simplest form. Check by multiplying. (Examples 1 and 3)

1. $\frac{1}{4} \div \frac{1}{2} =$ _____

 Show your work.

2. $\frac{5}{6} \div \frac{2}{3} =$ _____

3. $\frac{1}{8} \div 3 =$ _____

4. Write a story context for $\frac{2}{3} \div \frac{5}{6}$. Use a model to solve.
(Example 2)

_____

_____

_____

5. A neighborhood garden that is $\frac{2}{3}$ of an acre is to be

divided into 4 equal-size sections. Write and solve an
equation to find the size of each section. (Example 4)

_____

6. 📧 **Building on the Essential Question** How is the
process used to divide fractions similar to the process
used to multiply fractions?

_____

_____

_____

Rate Yourself!

How confident are you about
dividing fractions? Shade the
ring on the target.

**For more help, go online to
access a Personal Tutor.**

FOLDABLES Time to update your Foldable!

Independent Practice

Go online for Step-by-Step Solutions

Divide. Write in simplest form. Check by multiplying. (Examples 1 and 3)

1. $\frac{1}{8} \div \frac{1}{2} =$ _____

 Show your work.

2. $\frac{3}{4} \div \frac{2}{3} =$ _____

3. $\frac{3}{4} \div 9 =$ _____

4. $\frac{1}{6} \div \frac{4}{7} =$ _____

5. $\frac{1}{3} \div 8 =$ _____

6. $\frac{1}{3} \div \frac{5}{6} =$ _____

7. Write a story context for $\frac{5}{6} \div \frac{1}{12}$. Use a model to solve. (Example 2)

_____
_____
_____
_____

Write and solve an equation. (Example 4)

8. A piece of licorice is to be cut into 10 equal-size pieces. If the length of the piece of licorice is $\frac{2}{3}$ yard, how long will each piece of licorice be?

_____
_____

9. (MP) **Use Math Tools** To tie-dye one T-shirt, $\frac{3}{8}$ cup of dye is needed. The table shows the number of cups of each color of dye in Mr. Galvez's art class. How many T-shirts can be made using only orange dye?

_____
_____

Amount of Dye	
Color	Number of Cups
red	12
orange	$\frac{3}{4}$

10. (MP) **Reason Abstractly** Carlota has $\frac{3}{4}$ ton of mulch she is going to divide evenly among 5 flower beds. How much mulch will each flower bed contain?

_____

11. **MP** **Reason Abstractly** Complete the Venn diagram to compare and contrast the division and multiplication problems.

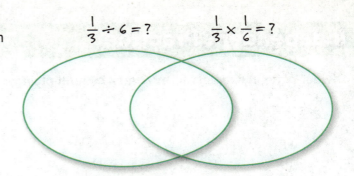

$$\frac{1}{3} \div 6 = ? \qquad \frac{1}{3} \times \frac{1}{6} = ?$$

🔥 H.O.T. Problems Higher Order Thinking

12. **MP** **Identify Structure** Find two positive fractions with a quotient of $\frac{5}{6}$. Give the equivalent multiplication sentence.

_____

13. **MP** **Identify Repeated Reasoning** Is the quotient $\frac{2}{3} \div \frac{1}{2}$ greater than or less than 1? Is the quotient of $\frac{1}{2} \div \frac{2}{3}$ greater than or less than 1? Explain your reasoning. _____

_____

14. **MP** **Persevere with Problems** Complete the steps to demonstrate why you multiply by the reciprocal when dividing fractions. Find $\frac{1}{4} \div \frac{3}{8}$.

Step 1 Rewrite it as $\dfrac{\frac{1}{4}}{\frac{3}{8}}$.

Step 2 Multiply the numerator and the denominator by the

reciprocal of $\dfrac{\square}{\square}$. $\dfrac{\frac{1}{4}}{\frac{3}{8}} = \dfrac{\frac{1}{4} \times \dfrac{\square}{\square}}{\frac{3}{8} \times \dfrac{\square}{\square}}$

Step 3 Simplify the denominator. $\dfrac{\frac{1}{4} \times \frac{8}{3}}{\square}$

Step 4 Simplify the fraction. $\dfrac{\square}{\square} \times \dfrac{\square}{\square}$

15. **MP** **Reason Inductively** In cooking, 1 drop is equal to $\frac{1}{6}$ of a dash. If a recipe calls for $\frac{2}{3}$ of a dash, write an expression that would give the number of drops that are needed.

_____

Extra Practice

Divide. Write in simplest form. Check by multiplying.

16. $\frac{1}{2} \div \frac{2}{3} = \underline{\frac{3}{4}}$

$$\frac{1}{2} \div \frac{2}{3} = \frac{1}{2} \times \frac{3}{2}$$
$$= \frac{3}{4}$$
$$\frac{3}{4} \times \frac{2}{3} = \frac{6}{12} \text{ or } \frac{1}{2} \checkmark$$

Homework Help →

17. $\frac{1}{5} \div 4 = \underline{\frac{1}{20}}$

$$\frac{1}{5} \div 4 = \frac{1}{5} \times \frac{1}{4}$$
$$= \frac{1}{20}$$
$$\frac{1}{20} \times \frac{4}{1} = \frac{4}{20} \text{ or } \frac{1}{5} \checkmark$$

18. $\frac{2}{5} \div \frac{3}{4} = \underline{\hspace{2cm}}$

19. $\frac{2}{7} \div 2 = \underline{\hspace{2cm}}$

20. $\frac{1}{5} \div \frac{5}{7} = \underline{\hspace{2cm}}$

21. $\frac{1}{4} \div \frac{3}{5} = \underline{\hspace{2cm}}$

22. Write a story context for $\frac{1}{4} \div \frac{1}{8}$. Use a model to solve.

_____

_____

Write and solve an equation.

23. A relay race is $\frac{1}{10}$ kilometer long. Four athletes will run an equal distance to complete the relay. How far does each athlete run?

_____

_____

24. Jalisa is using $\frac{5}{6}$ yard of ribbon to make bows for her party favors. Jalisa needs to make 6 bows. What is the length of the ribbon used for each bow?

_____

_____

25. Reaner Recycling shreds $\frac{7}{8}$ ton of aluminum each day. The machines can shred $\frac{1}{24}$ ton aluminum per cycle. How many cycles will be needed to shred the aluminum?

_____

_____

26. **MP** **Reason Abstractly** Reaner Recycling collected $\frac{7}{4}$ ton of aluminum last Saturday. If $\frac{7}{8}$ ton of aluminum can be shredded each day, how many days will it take to process what was collected on Saturday?

_____

27. You have 60 CD cases that you would like to store on the shelf shown. If each CD case is $\frac{3}{8}$ inch wide, is there enough room on the shelf for the CD cases? Explain your reasoning.

← 24 in. →

28. Which of the following numbers that when divided by $\frac{1}{2}$, give a result less than $\frac{1}{2}$? Select all that apply.

☐ $\frac{3}{8}$ ☐ $\frac{3}{16}$ ☐ $\frac{6}{25}$ ☐ $\frac{1}{3}$

Spiral Review

Find the greatest common factor of each pair of numbers. 4.OA.4

29. 4 and 8 _____

30. 6 and 3 _____

31. 12 and 8 _____

32. 6 and 8 _____

33. 12 and 16 _____

34. 9 and 15 _____

35. The set department has a 5 foot board. They cut 2 segments that are $1\frac{1}{2}$ foot each. How much of the board is left? 4.NF.3d

_____

36. The Sanchez family is building the dog pen shown. What is the area of the dog pen? 4.MD.3

8 ft

4 ft

_____

_____

Divide Mixed Numbers

Extreme Geography The deepest point in Earth's oceans is the Mariana Trench, which is located $6\frac{4}{5}$ miles beneath the ocean's surface. The average depth of Earth's oceans is $2\frac{1}{2}$ miles. By contrast, the highest elevation of Earth is Mt. Everest, which is about $5\frac{1}{2}$ miles high.

1. Write a division expression to find how many times as deep the Mariana Trench is than the average depth of the ocean.

Mariana Trench   Average Ocean Depth

2. Write a division expression to find how many times as tall Mt. Everest is than the average depth of the ocean.

Mount Everest   Average Ocean Depth

3. Rewrite the mixed number $6\frac{4}{5}$ as an improper fraction.

4. Rewrite the mixed number $5\frac{1}{2}$ as an improper fraction.

Which MP Mathematical Practices did you use?
Shade the circle(s) that applies.

① Persevere with Problems   ⑤ Use Math Tools

② Reason Abstractly   ⑥ Attend to Precision

③ Construct an Argument   ⑦ Make Use of Structure

④ Model with Mathematics   ⑧ Use Repeated Reasoning

Divide a Mixed Number by a Fraction

Dividing mixed numbers is similar to dividing fractions. To divide mixed numbers, write the mixed numbers as improper fractions and then divide as with fractions.

Examples

1. Find $1\frac{3}{4} \div \frac{2}{5}$.

Estimate $2 \div \frac{1}{2} = 4$

$$1\frac{3}{4} \div \frac{2}{5} = \frac{7}{4} \div \frac{2}{5}$$ Write the mixed number as an improper fraction.

$$= \frac{7}{4} \times \frac{5}{2}$$ Multiply by the reciprocal.

$$= \frac{35}{8} \text{ or } 4\frac{3}{8}$$ Simplify.

Check for Reasonableness $4\frac{3}{8} \approx 4$ ✔

. . . .

2. Find $3\frac{3}{4} \div \frac{4}{5}$.

Estimate $\boxed{} \div \boxed{} = \boxed{}$

$$3\frac{3}{4} \div \frac{4}{5} = \frac{\boxed{}}{\boxed{}} \div \frac{4}{5}$$ Write the mixed number as an improper fraction.

$$= \frac{\boxed{}}{\boxed{}} \times \frac{\boxed{}}{\boxed{}}$$ Multiply by the reciprocal.

$$= \frac{\boxed{}}{\boxed{}} \text{ or } \boxed{}\frac{\boxed{}}{\boxed{}}$$ Simplify.

Check for Reasonableness $\boxed{}\frac{\boxed{}}{\boxed{}} \approx \boxed{}$ ✔

Show your work.

a. _____

b. _____

c. _____

Got it? Do these problems to find out.

Divide. Write in simplest form. Check by multiplying.

a. $2\frac{3}{8} \div \frac{1}{4}$ b. $2\frac{1}{2} \div \frac{3}{7}$ c. $5\frac{5}{8} \div \frac{3}{4}$

Divide by a Mixed Number

To divide a mixed number by another mixed number, change both mixed numbers to improper fractions. Remember to simplify before you multiply.

Examples

3. Find $5\frac{1}{2} \div 2\frac{1}{2}$.

Estimate $6 \div 3 = 2$

$$5\frac{1}{2} \div 2\frac{1}{2} = \frac{11}{2} \div \frac{5}{2}$$ Write mixed numbers as improper fractions.

$$= \frac{11}{2} \times \frac{2}{5}$$ Multiply by the reciprocal.

$$= \frac{11}{\overset{}{2}} \times \frac{\overset{1}{2}}{5}$$ Divide 2 and 2 by the GCF, 2.

$$= \frac{11}{5} \text{ or } 2\frac{1}{5}$$ Simplify. Compare to the estimate.

STOP and Reflect

How is dividing two mixed numbers similar to dividing two fractions?

4. Find $4\frac{2}{3} \div 1\frac{3}{4}$.

Estimate $\boxed{} \div \boxed{} = \boxed{} \dfrac{\boxed{}}{\boxed{}}$

$$4\frac{2}{3} \div 1\frac{3}{4} = \frac{\boxed{}}{\boxed{}} \div \frac{\boxed{}}{\boxed{}}$$ Write the mixed numbers as an improper fractions.

$$= \frac{\boxed{}}{\boxed{}} \times \frac{\boxed{}}{\boxed{}}$$ Multiply by the reciprocal. Divide by the GCF.

$$= \frac{\boxed{}}{\boxed{}} \text{ or } \boxed{} \dfrac{\boxed{}}{\boxed{}}$$ Simplify.

Check for Reasonableness $\boxed{}\dfrac{\boxed{}}{\boxed{}} \approx \boxed{}\dfrac{\boxed{}}{\boxed{}}$ ✔

Got it? Do these problems to find out.

d. $4\frac{1}{5} \div 2\frac{1}{3}$ **e.** $8 \div 2\frac{1}{2}$ **f.** $1\frac{5}{9} \div 2\frac{1}{3}$

Show your work.

d. _____

e. _____

f. _____

Example

Watch ▶   Tutor 💬

5. The average adult male Giant Panda weighs about $1\frac{1}{5}$ times as much as the average adult female. If the average weight of a male Giant Panda is 330 pounds, how much does the average female Giant Panda weigh?

To find the average weight, solve the equation $330 \div 1\frac{1}{5} = \blacksquare$.

$330 \div 1\frac{1}{5} = \dfrac{330}{1} \div \dfrac{6}{5}$ Write the mixed number as an improper fraction.

$= \dfrac{330}{1} \times \dfrac{5}{6}$ Multiply by the reciprocal.

$= \dfrac{\overset{55}{\cancel{330}}}{1} \times \dfrac{5}{\underset{1}{\cancel{6}}}$ Divide 330 and 6 by their GCF, 6.

$= \dfrac{275}{1}$ or 275   Simplify.

So, the average female Giant Panda weighs about 275 pounds.

Guided Practice

Check ✓

Divide. Write in simplest form. Check by multiplying. (Examples 1–4)

1. $3\frac{1}{2} \div \frac{1}{2} =$ _____

2. $2\frac{2}{3} \div 1\frac{1}{6} =$ _____

3. $6\frac{2}{3} \div 2\frac{6}{7} =$ _____

Show your work.

4. A box of snack-size cracker packs weighs $28\frac{1}{2}$ ounces. Each snack pack weighs $4\frac{3}{4}$ ounces. How many snack packs are in the box? (Example 5)

_____

5. The soccer team has $16\frac{1}{2}$ boxes of wrapping paper left to sell. If each of the 12 players sells the same amount, how many boxes should each player sell? (Example 5)

_____

6. ⓠ **Building on the Essential Question** How do you divide mixed numbers? _____

_____

Rate Yourself!

☐   I understand how to divide mixed numbers.

▶▶ Great! You're ready to move on!

☐   I still have some questions about dividing mixed numbers.

⏸ No Problem! Go online to access a Personal Tutor.   Tutor 💬

FOLDABLES Time to update your Foldable!

Independent Practice

Go online for Step-by-Step Solutions

Divide. Write in simplest form. Check by multiplying. (Examples 1–4)

1. $4\frac{1}{6} \div 10 =$ _____

Show your work.

2. $6\frac{1}{2} \div \frac{3}{4} =$ _____

3 $3\frac{3}{4} \div 5\frac{5}{8} =$ _____

4. The length of a kitchen wall is $24\frac{2}{3}$ feet long. A border will be placed along the wall of the kitchen. If the border comes in strips that are each $1\frac{3}{4}$ feet long, how many strips of border are needed? (Example 5)

_____

_____

5 Jay is cutting a roll of biscuit dough into slices that are $\frac{3}{8}$ inch thick. If the roll is $10\frac{1}{2}$ inches long, how many slices can he cut? (Example 5)

_____

_____

6. **MP** **Be Precise** Refer to the graphic novel frame below for Exercises a–c.

We bought five $3\frac{3}{4}$ pound bags of bird seed.

Bird eed

a. What is the total weight of the birdseed they bought? _____

b. If each smaller bag contains $1\frac{1}{2}$ pounds, how many bags can they make?

_____

c. Will there be any birdseed left over? Explain. _____

_____

7. (MP) **Identify Structure** Complete the steps in dividing mixed numbers.

Divide Mixed Numbers
Is one of the numbers a whole number?

Yes

No

Write the whole number as a _____ .

Write the mixed number(s) as _____ fractions.

Write the _____ of the divisor.

Multiply as with _____ . Simplify.

🔥 H.O.T. Problems Higher Order Thinking

8. (MP) **Which One Doesn't Belong?** Select the expression that has a quotient greater than 1. Explain your reasoning.

$4\frac{2}{3} \div 5\frac{1}{4}$ $3\frac{1}{8} \div 2\frac{2}{5}$ $1\frac{6}{7} \div 2\frac{1}{3}$ $5\frac{3}{4} \div 7\frac{3}{8}$

_____

_____

9. (MP) **Persevere with Problems** Without dividing, explain whether $5\frac{1}{6} \div 3\frac{5}{8}$ is greater than or less than $5\frac{1}{6} \div 2\frac{2}{5}$. _____

_____

_____

_____

10. (MP) **Reason Inductively** Without doing the math, determine which is greater, $40 \cdot \frac{1}{4}$ or $40 \div \frac{1}{4}$. Explain your reasoning.

_____

_____

_____

Extra Practice

Divide. Write in simplest form. Check by multiplying.

11. $5\frac{1}{2} \div 2 = \underline{\quad 2\frac{3}{4} \quad}$

$$5\frac{1}{2} \div 2 = \frac{11}{2} \div \frac{2}{1}$$
$$= \frac{11}{2} \times \frac{1}{2}$$
$$= \frac{11}{4} \text{ or } 2\frac{3}{4}$$

 Homework Help

$$\frac{11}{\underset{2}{\cancel{4}}} \times \frac{\overset{1}{\cancel{2}}}{1} = \frac{11}{2} \text{ or } 5\frac{1}{2} \checkmark$$

12. $3 \div 4\frac{1}{2} = \underline{\qquad}$

13. $6 \div 2\frac{1}{4} = \underline{\qquad}$

14. $7\frac{4}{5} \div \frac{1}{5} = \underline{\qquad}$

15. $6\frac{1}{2} \div 3\frac{1}{4} = \underline{\qquad}$

16. $8\frac{3}{4} \div 2\frac{1}{6} = \underline{\qquad}$

17. $3\frac{3}{5} \div 1\frac{4}{5} = \underline{\qquad}$

18. $6\frac{3}{5} \div 2\frac{3}{4} = \underline{\qquad}$

19. $4\frac{2}{3} \div 2\frac{2}{9} = \underline{\qquad}$

20. **STEM** A human has 46 chromosomes. This is $5\frac{3}{4}$ times the number of chromosomes of a fruit fly. Write a division expression to find how many chromosomes a fruit fly has.

_____

_____

21. **MP** **Reason Abstractly** How many $\frac{3}{8}$-pound bags of trail mix can be made from $6\frac{3}{8}$ pounds of trail mix? Write a division expression.

_____

22. Natasha is setting tiles along the baseboard in her bathroom. One side of the bathroom is $18\frac{3}{4}$ feet. Each tile is $1\frac{1}{2}$ feet long. How many tiles does she need for this section? _____

Power Up! Test Practice

23. You have a shelf that holds $25\frac{1}{2}$ pounds. How many $1\frac{1}{4}$–pound books can the shelf hold? Explain your response.

24. Sort the expressions listed at the right into the appropriate bins.

Quotient Less Than 1	Quotient Equal to 1	Quotient Greater Than 1

$7\frac{2}{3} \div 5\frac{3}{8}$	$6\frac{2}{3} \div 6\frac{8}{12}$
$3\frac{2}{5} \div 6\frac{3}{4}$	$5\frac{1}{6} \div 4\frac{2}{5}$
$4\frac{1}{4} \div 4\frac{1}{4}$	$2\frac{1}{4} \div 2\frac{1}{2}$
$12\frac{4}{5} \div 8\frac{1}{4}$	$8\frac{3}{8} \div 8\frac{2}{9}$

Spiral Review

Multiply. Write in simplest form. 5.NF.4

25. $\frac{3}{4} \times 1 =$ _____

26. $\frac{3}{7} \times 2 =$ _____

27. $\frac{1}{2} \times \frac{1}{2} =$ _____

28. $\frac{1}{2} \times \frac{1}{4} =$ _____

29. $\frac{2}{5} \times \frac{1}{4} =$ _____

30. $\frac{3}{4} \times \frac{2}{3} =$ _____

31. Anna is planting corn on her farm. What is the area of cornfield? 5.NF.6

_____

$\frac{1}{2}$ mi

$\frac{5}{6}$ mi

21ST CENTURY CAREER
in Culinary Arts

Pastry Chef

Are you creative in the kitchen? Do you enjoy sharing your creations with others? If so, a career as a pastry chef might be perfect for you. Pastry chefs can have different responsibilities, such as creating desserts and dessert menus for restaurants; baking breads, pies, and cakes for neighborhood bakeries; or producing pastries and baked goods for grocery stores. Pastry chefs are not only artistic, but they are also precise and understand mathematics and the chemistry of the ingredients that they are using.

College & Career READINESS

Is This the Career for You?

Are you interested in a career as a pastry chef? Take some of the following courses in high school to get you started in the right direction.

◆ Algebra
◆ Chemistry
◆ Culinary Operations
◆ Food Science Technology

Find out how math relates to a career in Culinary Arts.

You be the Pastry Chef!

Use the information in the recipes below to solve each problem.

1. A chef is making only half of the frosting recipe. How much cream of tartar is needed? How much granulated sugar is needed? _____

2. The petits fours recipe is doubled. How much milk is needed for the cake? _____

3. If the frosting recipe is tripled, how much water is needed? Justify your procedure.

   _____

   _____

4. For a wedding, a pastry chef is increasing the cake recipe by $4\frac{1}{2}$ times. How much butter is needed? flour? _____

5. The recipe is increased to make $1\frac{3}{4}$ times the number of petits four than the original recipe. How much vanilla extract is needed for the cake and the frosting? Justify your procedure.

   _____

   _____

Petits Fours Frosting

3 cups granulated sugar

$\frac{1}{4}$ teaspoon cream of tartar

$1\frac{1}{2}$ cups water

1 cup powdered sugar, sifted

$\frac{1}{2}$ teaspoon vanilla extract

$\frac{1}{2}$ teaspoon almond extract

Petits Fours Cake

$\frac{1}{4}$ cup butter, melted

1 cup shortening

1 cup granulated sugar

1 teaspoon vanilla extract

$1\frac{1}{3}$ cups all-purpose flour

2 teaspoons baking powder

$\frac{1}{2}$ teaspoon salt

$\frac{2}{3}$ cup milk

3 egg whites

Career Project

It's time to update your career portfolio! Use the Internet or another source to research a career as a pastry chef. Write a paragraph that summarizes your findings.

_____

_____

_____

_____

_____

List the strengths you have that would help you succeed in this career.

- _____
- _____
- _____
- _____
- _____

Chapter Review

Vocabulary Check

Fill in the blank with the correct vocabulary term. Then circle the word that completes the sentence in the word search.

1. A number that has a whole number part and a fraction part is a _____.

2. The _____ is the greatest of the common factors of two or more numbers.

3. The product of a number and its _____ is one.

4. The number above the fraction bar is the _____.

5. The number below the fraction bar is the _____.

6. A _____ is a number that represents part of a whole or part of a set.

7. A _____ is a unit rate with a denominator of 1 unit.

8. A fraction with a numerator that is greater than or equal to the denominator is an _____.

9. _____ are numbers that are easy to divide mentally.

10. A fraction in which the GCF of the numerator and the denominator is 1 is written in _____.

```
N W V B C H H R O D M D C V P H H P R M R M
R O T C A F N O M M O C T S E T A E R G E I
K I I O S P Y K J M L B I Y Z S Q O M R B X
H D O T M J C G T P H J S C D B F E E H M E
D F E Z C R T Q P A I G Y Q Y T Q C P V U D
G E H C Y A B H S J G V G S S N I D E Q N N
X U N I P F R A C C O O N E V P B N C F E U
E M J O C M W F N S P V L D R X I O B O L M
H N G K M S N U R V T P T O P D A I B L B B
X O W T G I I O K E M N C V U L I T Y Z I E
B Z T F W M N X I I P A Y B W E F A G N T R
A X S Y K Y C A S T L O I W Z X M R K L A T
T E J K Q O L P T X C F R I S T K T L A P K
R O T A R E M U N O M A Q P G N C I Y X M T
Y H W D N X J S M X R X R U M O G N B R O I
M F F T B D Y N A P R G I F J I M U F I C A
```

Use Your FOLDABLES

Use your Foldable to help review the chapter.

Tape here

Tab 3	**Multiply and Divide Fractions**
Tab 2	
Tab 1	

Example

Example

fraction x mixed number

mixed number ÷ fraction

Got it?

The problems below may or may not contain an error. If the problem is correct, write a "✓" by the answer. If the problem is not correct, write an "X" over the answer and correct the problem.

1. $13 \times \frac{1}{3} = 4\frac{2}{3}$

 The first one is done for you. $13 \times \frac{1}{3} = \frac{13}{3}$ or $4\frac{1}{3}$

2. $16 \times \frac{5}{6} = 19\frac{1}{5}$

3. $35 \times \frac{3}{7} = 15$

4. $\frac{5}{8} \div \frac{3}{4} = \frac{15}{32}$

5. $3\frac{2}{3} \div \frac{5}{6} = 4\frac{2}{5}$

6. $2\frac{2}{3} \div 1\frac{1}{4} = 2\frac{2}{3}$

Power Up! Performance Task

Road Trip

The Johanssens are traveling across the country to visit relatives. Mr. Johanssen set the cruise control and drove at a constant speed for $622\frac{1}{2}$ miles. He took breaks, but remained at an average constant speed when driving. Some of the data related to the trip appear in the table below.

Time (h)	3	4	5	6	9	c
Distance (mi)	$186\frac{3}{4}$	a	$311\frac{1}{4}$	b	$560\frac{1}{4}$	$622\frac{1}{2}$

Write your answers on another piece of paper. Show all of your work to receive full credit.

Part A
What is Mr. Johanssen's rate of speed in miles per hour? Express your answer as a fraction or a mixed number.

Part B
Using the answer from Part A, find the missing values in the table. Express your answer as a fraction or a mixed number.

Part C
Graph the ordered pairs from the completed table on a coordinate plane. How far will the Johanssens have traveled after 15 hours?

Part D
The following year, Mr. Johanssen rents a vehicle for the same trip. The speedometer in this vehicle registers speeds in kilometers per hour instead of miles per hour. He wants to drive at a constant rate of 70 miles per hour. What is the maximum speed in kilometers per hour that he can drive to not exceed a rate of 70 miles per hour? Express your answer as a mixed number. Recall that 1 mile is approximately equal to $1\frac{3}{5}$ kilometers.

Reflect

 Answering the Essential Question

Use what you learned about multiplying and dividing fractions to complete the graphic organizer.

Essential Question

WHAT does it mean to multiply and divide fractions?

Operation	Dividend and Divisor	Is the answer less than or greater than the dividend? Provide an example.
multiply	whole number by whole number	
multiply	fraction by fraction	
divide	whole number by whole number	
divide	fraction by fraction	

 Answer the Essential Question. WHAT does it mean to multiply and divide fractions?

Chapter 5
Integers and the Coordinate Plane

Essential Question

HOW are integers and absolute value used in real-world situations?

Content Standards
6.NS.5, 6.NS.6, 6.NS.6a, 6.NS.6b, 6.NS.6c, 6.NS.7, 6.NS.7a, 6.NS.7b, 6.NS.7c, 6.NS.7d, 6.NS.8

MP Mathematical Practices
1, 2, 3, 4, 5, 7, 8

Math in the Real World

Rappelling Two friends rappel 350 feet down into a canyon. Their starting position is represented by 0 on the number line. Their ending position can be represented by −350.

Graph −350 on the number line below.

FOLDABLES® Study Organizer

1 Cut out the Foldable on page FL11 of this book.

2 Place your Foldable on page 418.

3 Use the Foldable throughout this chapter to help you learn about integers.

Vocabulary

absolute value	positive integer
bar notation	quadrants
integer	rational number
negative integer	repeating decimal
opposites	terminating decimal

Review Vocabulary

Using a graphic organizer can help you to remember important vocabulary terms. Fill in the graphic organizer below for the word *decimal*.

Decimal

Definition

Math Example

Real World Example

Place a checkmark below the face that expresses how much you know about each concept. Then scan the chapter to find a definition or example of it.

🙁 I have no clue.   😐 I've heard of it.   😀 I know it!

Integers and the Coordinate Plane

Concept	🙁	😐	😀	Definition or Example
absolute value				
comparing integers				
graphing integers on a number line				
graphing ordered pairs				
ordering rational numbers				
rational numbers				

When Will You Use This?

Here are a few examples of how integers are used in the real world.

Activity 1 Does anyone in your family own a digital camera? If so, how much data does the memory card hold? Use the Internet to determine how many pictures your memory card might hold.

_____

_____

_____

Activity 2 Go online at **connectED.mcgraw-hill.com** to read the graphic novel **Photo-Math**. According to Daniella, how many megabytes of data does the memory card hold? _____

Noah, Amanda, Luis, and Daniella in **Photo-Math**

How many pictures have we taken so far?

Well, you and I took 19 pictures during school.

Are You Ready?

Try the Quick Check below.
Or, take the Online Readiness Quiz.

Quick Review

4.NF.2, 5.NBT.3b

Example 1

Replace the ◯ with <, >, or = to make a true statement.

1.6 ◯ 1.3

1.3   1.6

1   1.2   1.4   1.6   1.8   2.0

Since 1.6 is to the right of 1.3, 1.6 > 1.3.

Example 2

Replace the ◯ with <, >, or = to make a true statement.

$\frac{2}{5}$ ◯ $\frac{7}{10}$

$\frac{2}{5}$ $\frac{7}{10}$

0         $\frac{1}{2}$ 1

Since $\frac{2}{5}$ is less than $\frac{1}{2}$ and $\frac{7}{10}$ is greater than $\frac{1}{2}$, $\frac{2}{5} < \frac{7}{10}$.

Quick Check

Compare Decimals Replace each ◯ with <, >, or = to make a true statement.

Show your work.

1. 4.8 ◯ 4.80

2. 7.7 ◯ 7.5

3. 1.2 ◯ 2.1

Compare Fractions Replace each ◯ with <, >, or = to make a true statement.

4. $\frac{2}{11}$ ◯ $\frac{9}{10}$ =

5. $\frac{3}{5}$ ◯ $\frac{1}{4}$ =

6. $\frac{2}{3}$ ◯ $\frac{4}{6}$ =

7. Jahan bought $\frac{2}{3}$ pound of peanuts and $\frac{1}{4}$ pound of walnuts. Did Jahan buy more peanuts or more walnuts?

How Did You Do?

Which problems did you answer correctly in the Quick Check?
Shade those exercise numbers below.

① ② ③ ④ ⑤ ⑥ ⑦

Inquiry Lab
Integers

Content Standards
6.NS.5, 6.NS.6, 6.NS.6c

MP **Mathematical Practices**
1, 3, 4

 Inquiry **HOW can positive and negative values be represented?**

In coastal regions, some animals live above sea level and other animals live in the ocean. A sea star can be found at an ocean depth of two feet. How can you represent an ocean depth of two feet?

What do you know? _____

_____

What do you need to find? _____

_____

Hands-On Activity

Sea level can be represented with the number 0.

To represent a location above sea level, use a positive number. A positive number can be written with or without a positive sign, such as 5 or +5.

To represent a location below sea level, use a negative number. A negative number is written with a negative sign, such as −5.

$$-5 \quad -4 \quad -3 \quad -2 \quad -1 \quad 0 \quad 1 \quad 2 \quad 3 \quad 4 \quad 5$$

Write a number to represent an ocean depth of two feet.

Step 1 Determine if a positive sign or a negative sign should be used.

Since the location is below, or less than, sea level, use a

_____ sign.

Step 2 Determine which number to use.

Use the number ☐ to represent two feet.

So, the number ☐ represents an ocean depth of two feet.

Investigate

Work with a partner. Write the correct number to represent each location in relationship to sea level. The first one is done for you. Then draw a number line to represent each number.

	Animal	Elevation (ft)	Above or Below Sea Level	Number
	Fiddler Crab	3	above sea level	+3
1.	Eagle's Nest	75	above sea level	
2.	Dolphin	10	below sea level	
3.	Spider Crab	375	below sea level	
4.	Blue Heron	4	above sea level	

Analyze and Reflect

5. **MP Reason Inductively** What negative number is the same distance from 0 as the number +4? Explain. Graph both numbers on the number line below.

_____

_____

Create

6. **MP Model with Mathematics** Write about a real-world situation that can be described using the number −6. Describe what the number 0 would represent in the situation. What would the number 6 represent?

_____

_____

_____

7. **Inquiry** HOW can positive and negative values be represented?

_____

_____

Integers and Graphing

 Real-World Link Watch ▶

Money The bar graph shows the amount of money remaining in the clothing budgets of four students at the end of one month. A value of —$2 means that someone overspent the budget and owes his or her parents 2 dollars.

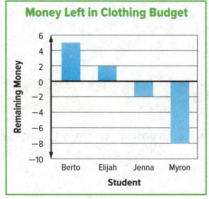

Money Left in Clothing Budget

Remaining Money

Berto   Elijah   Jenna   Myron

Student

 Essential Question

HOW are integers and absolute value used in real-world situations?

Vocab **Vocabulary**

integer
negative integer
positive integer

Content Standards
6.NS.5, 6.NS.6, 6.NS.6a, 6.NS.6c

MP **Mathematical Practices**
1, 3, 4, 5, 7

1. What number represents owing 8 dollars? ☐

2. What number represents having 5 dollars left? ☐

3. Who has the most money left? Who owes the most? Explain.

_____

_____

_____

Which MP Mathematical Practices did you use?
Shade the circle(s) that applies.

① Persevere with Problems   ⑤ Use Math Tools

② Reason Abstractly   ⑥ Attend to Precision

③ Construct an Argument   ⑦ Make Use of Structure

④ Model with Mathematics   ⑧ Use Repeated Reasoning

$15.00

Use Integers to Represent Data

Positive whole numbers, their opposites, and zero are called **integers**. To represent data that are less than a 0, you can use **negative integers**. A negative integer is written with a — sign. Data that are greater than zero are represented by **positive integers**.

| negative integers | Zero is neither negative nor positive. | positive integers |

$$-6 \quad -5 \quad -4 \quad -3 \quad -2 \quad -1 \quad 0 \quad +1 \quad +2 \quad +3 \quad +4 \quad +5 \quad +6$$

Opposites are numbers that are same distance from zero in opposite directions.

Examples

Write an integer for each situation. Explain the meaning of zero in each situation.

1. **a 10-yard loss**

Because it represents a loss, the integer is −10. In football, the integer 0 represents no yards lost or no yards gained.

2. **4 inches of rain above normal**

Because it represents above, the integer is 4. In this situation, the integer 0 represents the normal amount of rain.

3. **a $48 deposit into a savings account**

Because it represents an increase, the integer is [ ].

In this situation, the integer 0 represents _____

_____.

> **Zero**
>
> The number zero can have different meanings based on real-world context. Sometimes zero represents an amount that does not change. Zero can also be used to represent real-world ideas, such as sea level.

Show your work.

Got it? Do these problems to find out.

Write an integer for each situation. Explain the meaning of zero in each situation.

a. a gain of $2 a share   **b.** 10 degrees below zero

a. _____

b. _____

Graph Integers

Integers and sets of integers can be graphed on a horizontal or vertical number line. To graph a point on the number line, draw a point on the number line at its location. A set of integers is written using braces, such as {2, −9, 0}.

Examples

4. **Graph −7 on a number line.**

Draw a number line. Then draw a dot at the location that represents −7.

$$\overset{\bullet}{\underset{-10\ -9\ -8\ -7\ -6\ -5\ -4\ -3\ -2\ -1\ \ 0\ \ 1\ \ 2\ \ 3\ \ 4\ \ 5\ \ 6\ \ 7\ \ 8\ \ 9\ 10}{\longleftrightarrow}}$$

5. **Graph the set of integers {−4, 2, −1} on a number line.**

Draw a number line. Then draw a dot at the location of each integer.

6. **Graph the set of integers {0, 2, −3} on a number line.**

Draw a number line. Then draw a dot at the location of each integer.

Got it? Do these problems to find out.

Graph each set of integers on a number line.

c. {−3, 0, −2, 4}

d. {8, −6, −9, 5}

Show your work.

d. _____

Example

7. Alaina and her dad played golf on four different days. The data set {−1, +1, −3, +2} shows Alaina's scores in relation to par. Graph the scores. Explain the meaning of zero in this situation.

Draw a number line. Then draw a dot at the location of each golf score.

$$\overleftrightarrow{\;\;-5\;\;-4\;\;-3\;\;-2\;\;-1\;\;\;0\;\;\;1\;\;\;2\;\;\;3\;\;\;4\;\;\;5\;\;}$$

The integer 0 represents par.

Guided Practice

Check ✓

Write an integer for each situation. Explain the meaning of zero in each situation. (Examples 1–3)

1. 15-yard gain _____

2. loss of 2 hours _____

Graph each integer or set of integers on a number line. (Examples 4–6)

3. −2

$$\overleftrightarrow{\;\;\;|\;\;\;\;|\;\;\;\;|\;\;\;\;|\;\;\;\;|\;\;\;\;|\;\;\;}$$

4. {−1, 1, 0}

$$\overleftrightarrow{\;\;\;|\;\;\;\;|\;\;\;\;|\;\;\;\;|\;\;\;\;|\;\;\;\;|\;\;\;}$$

5. The data set {+5, 0, −15, +20} shows the number of points Delaney scored on each hand of a card game. Graph the scores. Explain the meaning of zero in this situation. (Example 7)

_____

6. 🄔 **Building on the Essential Question** How can you use integers to represent data?

_____

_____

_____

_____

Rate Yourself!

How confident are you about integers and graphing? Check the box that applies.

🙁   😐   🙂

☐ ☐ ☐ ☐ ☐

For more help, go online to access a Personal Tutor.

Tutor

Independent Practice

Go online for Step-by-Step Solutions

Write an integer for each situation. Explain the meaning of zero in each situation. (Examples 1–3)

1. 3 miles below sea level _____

2. earning $45 _____

3. moving back 5 spaces on a game board _____

_____

Graph each integer or set of integers on a number line. (Examples 4–6)

4. −5

5. {2, −3, 0, 1}

6. The data set {+4, −1, −2, 0} shows a change in number of state representatives for four states after the last census. Graph the change in number of representatives. Explain the meaning of zero in this situation. (Example 7)

_____

7. **MP Use Math Tools** The table shows the record low temperatures for several states. Graph the temperatures on a number line.

Record Low Temperature by State (°F)				
AL	**AK**	**CT**	**NJ**	**VA**
−27	−29	−32	−34	−30

8. **MP Use Math Tools** The table shows the number of points earned for each action in a video game. While playing the video game, Kevin fell in water, jumped over a rock, touched a cactus and climbed a mountain. Graph the number of points he earned for each action on the number line.

Action	Points
fall in water	−10
walk over a bridge	+5
climb mountain	+10
jump over rock	+5
walk through quicksand	−15
touch cactus	−15

9. **MP** **Model with Mathematics** Complete the graphic organizer by writing words or symbols used to represent positive and negative integers.

Positive Integer	Negative Integer
•	•
•	•
•	•
•	•

🔥 H.O.T. Problems Higher Order Thinking

10. **MP** **Persevere with Problems** A football team receives the ball on their own 10 yard line.

   a. They make a gain of 15 yards in the first play. What yard line is the ball on?

   _____

   b. What represents zero in this situation? Explain.

   _____

11. **MP** **Justify Conclusions** The temperature outside is 15°F. If the temperature drops 20°, will the outside temperature be represented by a positive or negative integer? Explain your reasoning.

   _____
   _____
   _____

12. **MP** **Identify Structure** Describe the characteristics of each set of numbers that make up the set of integers.

   _____
   _____
   _____

13. **MP** **Reason Inductively** Explain how to find the distance between −2 and 3 on a number line.

   _____
   _____
   _____
   _____

Extra Practice

Write an integer for each situation. Explain the meaning of zero in each situation.

14. 13° below zero <u>−13; The integer 0 represents zero degrees.</u>

15. spending $25 _____

16. 13-yard gain _____

Graph each integer or set of integers on a number line.

17. −8

18. {0, −3, 1, −1}

19. {−1, 1, −2}

20. {3, −5, 4, −1}

21. {4, −2, 2}

22. The data set {+3, −1, −2, +1} shows the moves a player made forward or backward in a board game. Graph the moves the player made. Explain the meaning of zero in this situation.

_____

23. **MP** **Model with Mathematics** The table shows the overnight low temperatures for 5 days in Minneapolis. Graph the temperatures on a number line.

Overnight Low Temperatures (°F)				
1	−1	3	−6	0

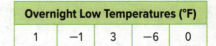

24. Jackson owes his sister Monica $15. Explain the meaning of zero in this situation.

25. On Monday Kennedy spent $2 on lunch. On Tuesday she spent $1 on a snack. On Wednesday, her sister gave her $3. Graph the integers on the number line.

Spiral Review

Fill in each ◯ with < or > to make the inequality true. **4.NBT.2**

26. 26 ◯ 22

27. 11 ◯ 13

28. 2.5 ◯ 3

29. 44 ◯ 4.4

30. 15 ◯ 6.8

31. 1.8 ◯ 1.9

32. Ally bought $\frac{1}{12}$ pound of cashews and $\frac{5}{6}$ pound of granola. Plot the fractions on the number line. Which quantity is greater? Explain. **4.NF.3d**

_____

33. The number of raffle tickets the student council sold over three days is shown in the table. How many total raffle tickets did they sell? **4.NBT.4**

_____

Day	Tickets Sold
Wednesday	35
Thursday	23
Friday	46

Inquiry Lab
Absolute Value

 inquiry **HOW can a number line help you find two integers that are the same distance from zero?**

Content Standards
6.NS.5, 6.NS.7, 6.NS.7c, 6.NS.7d
MP **Mathematical Practices**
1, 2, 3, 5

Several hot air balloons were flying at the same height. The dashed line below represents their starting point. Which two balloons moved the same distance but in opposite directions?

Hands-On Activity

In the diagram below, +8 means Balloon A climbed 8 feet and −10 means Balloon B moved down 10 feet.

Use the diagram to compare the distance each balloon moved.

Step 1 Complete the chart to compare the distance each balloon moved from the dashed line.

Balloon	Integer	Direction	Distance Moved (ft)
C	0	none	0
D	+10		
E	−15		

Step 2 Determine which two balloons moved the same distance away from the dashed line.

So, Balloon ☐ and Balloon ☐ moved ☐ feet from the dashed line.

Investigate

MP Use Math Tools Use the number line to determine the distance between each integer and zero.

Show your work.

$$-5 \quad -4 \quad -3 \quad -2 \quad -1 \quad 0 \quad 1 \quad 2 \quad 3 \quad 4 \quad 5$$

1. -2 _____

2. $+3$ _____

Work with a partner to complete the table. The first one is done for you.

Integer	Distance Between Integer and Zero	Opposite Integer	Distance Between Opposite Integer and Zero
3	3	–3	3
3. 7			
4. –11			
5. –13			
6. –21			

Analyze and Reflect

7. MP Reason Inductively What can you conclude about the distance from zero for both an integer and its opposite?

_____

Create

On Your Own

8. MP Model with Mathematics The movement of Balloon B in the Activity was represented by the number −10. Write a number to represent the starting point of the balloons. How is this number shown on the diagram?

_____

9. Inquiry HOW can a number line help you find two integers that are the same distance from zero?

_____

_____

Absolute Value

Vocabulary Start-Up

The distance between a number and 0 on the number line is called its **absolute value**.

1. Each mark on the number line indicates one yard. Draw a tree three yards west of the house. Draw a mailbox three yards east of the house.

West   East

2. The distance between the house and the tree is _____ the distance between the house and the mailbox.

3. The tree and the mailbox are in _____ directions from the house.

4. How does the number line above help you to understand absolute value? _____

_____

_____

Essential Question

HOW are integers and absolute value used in real-world situations?

 Vocabulary

absolute value
opposites

Content Standards
6.NS.6, 6.NS.6a, 6.NS.7, 6.NS.7c, 6.NS.7d

MP **Mathematical Practices**
1, 2, 3, 4

Real-World Link

5. **Errands** Jesse leaves home and walks 4 blocks west to the grocery store to buy milk then returns home. He then walks another 4 blocks east to the Post Office. Compare the distance and the direction of Jesse's house and the Post Office from the grocery store.

whew...

Which **MP** **Mathematical Practices** did you use?
Shade the circle(s) that applies.

① Persevere with Problems

⑤ Use Math Tools

② Reason Abstractly

⑥ Attend to Precision

③ Construct an Argument

⑦ Make Use of Structure

④ Model with Mathematics

⑧ Use Repeated Reasoning

Find Opposites

Positive numbers, such as 2, are graphed to the right (or above) zero on a number line. Negative numbers, such as −2, are graphed to the left (or below) zero on a number line.

Opposites are numbers that are the same distance from zero in opposite directions. Since 0 is not negative nor positive, 0 is its own opposite. The opposite of the opposite of a number, is the number itself. For example, the opposite of the opposite of 3, −(−3), is 3.

−2 is 2 units to the left of zero.

2 is 2 units to the right of zero.

Examples

Tutor

1. **Find the opposite of −5.**

Method 1 **Use a number line.**

Draw a number line and graph −5.

−5 is 5 units to the left of 0. The integer 5 is 5 units to the right of 0.

So, 5 is the opposite of −5.

Method 2 **Use symbols.**

The integer −5 uses the negative symbol.

The opposite of a negative symbol is a positive symbol.

So, the opposite of −5 is +5, or 5.

· ·

2. **Find the opposite of the opposite of 4.**

The opposite of 4 is −4. The opposite of −4 is 4.

So, 4 is the opposite of the opposite of 4.

Show your work.

Got it? **Do these problems to find out.**

a. _____

b. _____

a. What is the opposite of 3?

b. What is the opposite of the opposite of −2?

Absolute Value

Words The absolute value of a number is the distance between the number and zero on a number line.

Model

4 units   4 units

−6 −5 −4 −3 −2 −1 0 1 2 3 4 5 6

Symbols $|4| = 4$ The absolute value of 4 is 4.

$|-4| = 4$ The absolute value of −4 is 4.

The integers −4 and 4 are each 4 units from 0, even though they are on opposite sides of 0. |−4| is read *absolute value of negative four*.

Absolute Value

Since distance cannot be negative, the absolute value of a number is always positive or zero.

Examples Tutor

3. Evaluate $|-7|$.

7 units

−8 −7 −6 −5 −4 −3 −2 −1 0 1 2 3

The graph of −7 is 7 units from 0 on the number line.

So, $|-7| = 7$.

4. Evaluate $|5| + |-6|$.

$|5| + |-6| = 5 + |-6|$ The absolute value of 5 is 5.

$= 5 + 6$ The absolute value of −6 is 6.

$= 11$ Simplify.

5. Evaluate $|-7| - |3|$.

$|-7| - |3| = \boxed{} - \boxed{}$ Find the absolute value of −7 and 3.

$= \boxed{}$ Simplify.

 Show your work.

Got it? Do these problems to find out.

c. $|14|$ **d.** $|-9| + |3|$ **e.** $|-8| - |-2|$

c. _____

d. _____

e. _____

Example

6. A seagull is flying 25 feet above sea level. Nevaeh is diving 15 feet below sea level. What is the distance between Nevaeh and the seagull?

The expression |25| describes the seagull's distance above sea level. The expression |−15| describes Nevaeh's distance below sea level.

To find the distance, add the absolute values.

$|25| + |-15| = 25 + |-15|$ The absolute value of 25 is 25.

$\qquad\qquad = 25 + 15$ The absolute value of −15 is 15.

$\qquad\qquad = 40$ Add.

So, the total distance is 40 feet.

Guided Practice

Check ✓

1. What is the opposite of 0? (Example 1)

Show your work.

_____

2. What is the opposite of the opposite of 6? (Example 2)

_____

Evaluate each expression. (Examples 3–5)

3. $|-5| =$ _____

4. $|20| - |-3| =$ _____

5. $|-16| + |-12| =$ _____

6. A game show contestant lost 15 points. He answered another question incorrectly and lost another 15 points. How many total points has he lost? (Example 6)

_____

7. **Building on the Essential Question** How can absolute value help you to understand the size of a quantity? Give an example. _____

_____

_____

Independent Practice

Go online for Step-by-Step Solutions

Find the opposite of each integer. (Example 1)

1. 6 _____

2. −3 _____

3. 0 _____

Find the opposite of the opposite of each integer. (Example 2)

4. 12 _____

5. −9 _____

6. −17 _____

Evaluate each expression. (Examples 3–5)

7. $|-14| =$ _____

8. $|31| - |-1| =$ _____

9. $|-15| + |-6| =$ _____

10. Jayson spent $18 on a shirt. Then he spent $24 on a pair of pants. What is the total amount he spent? (Example 6) _____

11. Lilly saw a jelly fish at 6 feet below sea level. She saw a bright blue fish at 10 feet below sea level. What is the distance between the blue fish and the jelly fish? (Example 6) _____

12. **STEM** The table shows the melting points of various elements. Is the absolute value of the melting point of neon greater than or less than the absolute value of the melting point of hydrogen? _____

Element	Melting Point (°C)
Hydrogen	−259
Neon	−248
Oxygen	−218

13 **STEM** The surface of Jupiter is made of colorful clouds created by various chemicals in the atmosphere. The temperature at the top of the clouds is −230°F. The temperature below the clouds is 70°F. Which temperature has the lower absolute value? _____

MP **Reason Abstractly** Evaluate each expression.

14. $-|3| =$ _____

15. $|5 + 9| =$ _____

16. $|17 - 8| =$ _____

17. **MP** **Find the Error** Mei is evaluating an expression using absolute value. Find her mistake and correct it.

$$|-|4|| = -14$$

18. **MP** **Which One Doesn't Belong?** Identify the phrase that *cannot* be described by the same absolute value as the other three. Explain your reasoning.

a loss of 8 pounds	8 miles above sea level	giving away $8	18° below normal

MP **Persevere with Problems** Determine whether each statement is *always, sometimes,* or *never* true. Explain.

19. The absolute value of a positive integer is a negative integer.

20. If *a* and *b* are integers and *a* > *b*, then |*a*| > |*b*|.

21. **MP** **Reason Abstractly** Explain why the absolute value of a number is never negative.

22. **MP** **Reason Abstractly** Explain why an account balance less than −40 dollars represents a debt greater than 40 dollars.

23. **MP** **Reason Inductively** Is −*n* *always, sometimes,* or *never* a positive number? Explain your reasoning.

Extra Practice

Find the opposite of each integer.

24. −2 2 _____

Homework Help ➡

25. 15 _____

26. 42 _____

Find the opposite of the opposite of each integer.

27. 9 _____

28. 0 _____

29. −8 _____

Evaluate each expression.

30. $|18| =$ _____

31. $|0| =$ _____

32. $|25| =$ _____

33. $|2| + |{-13}| =$ _____

34. $|{-20}| - |17| =$ _____

35. $|{-16}| - |5| =$ _____

36. The balance of Bryce's account is $16. Jada's account is $5 overdrawn. What is the difference between their account balances?

_____

37. A football team lost 3 yards on their first play and 6 yards on their second play. How many total yards did they lose?

_____

38. The table shows the lowest elevations for several states. Is the absolute value of the lowest elevation of California greater than or less than the absolute value of the lowest elevation of Illinois?

_____

State	Lowest Elevation (ft)
Oklahoma	289
Illinois	279
Kentucky	257
California	−282

MP **Reason Abstractly** Evaluate each expression.

39. $-|{-10}| =$ _____

40. $|13 - 6| =$ _____

41. The table shows the freezing point of different liquids. What liquid's freezing point has the greatest absolute value?

◻ water

◻ acetic acid

◻ linseed oil

◻ acetone

Liquid	Freezing Point (°F)
Water	32
Acetic Acid	62
Linseed Oil	−4
Acetone	−94

42. Let $x = -1$, $y = -2$, and $z = 1$. Use the variables x, y, and z to complete the following number sentences.

a. $|\boxed{}| < |\boxed{}|$

b. $|\boxed{}| > |\boxed{}|$

c. $|\boxed{}| = |\boxed{}|$

Spiral Review

Fill in each ◯ with >, <, or = to make a true statement. **4.NBT.2**

43. 69.23 ◯ 69.25

44. 171.10 ◯ 171.09

45. 47.74 ◯ 47.740

46. Part of a sauce recipe is shown. If all the ingredients are mixed together, how much sauce will be made? **5.NF.1**

_____

6 teaspoons Worcestershire Sauce
$1\frac{1}{2}$ teaspoons seasoned salt
$\frac{1}{2}$ teaspoon onion powder
$\frac{3}{4}$ teaspoon garlic powder

Show your work.

47. Caroline's soccer practice starts at a quarter after 4 P.M. and ends at 5 P.M. How many minutes does her soccer practice last? **4.MD.2**

_____

Compare and Order Integers

 Real-World Link

Winter Fairbanks is located in interior Alaska. The average temperature for several months is shown on the thermometer.

1. The average temperature for December is −6.5°F and the average temperature for March is 11°F. Label December and March on the thermometer.

2. Which months have a greater average temperature than February? _____

3. Which months have a lower average temperature than November? _____

4. Complete the inequality to compare the temperatures of November and February.

   3 > ⬚

Essential Question

HOW are integers and absolute value used in real-world situations?

Content Standards
6.NS.7, 6.NS.7a, 6.NS.7b, 6.NS.7d

MP **Mathematical Practices**
1, 2, 3, 4, 5

°Fahrenheit

November: 3°
February: −4°
January: −10°

Which MP Mathematical Practices did you use?
Shade the circle(s) that applies.

① Persevere with Problems
② Reason Abstractly
③ Construct an Argument
④ Model with Mathematics

⑤ Use Math Tools
⑥ Attend to Precision
⑦ Make Use of Structure
⑧ Use Repeated Reasoning

Compare Integers

To compare integers, you can compare the signs as well as the magnitude, or size, of the numbers. Greater numbers are graphed farther to the right.

Compare the signs.

Positive numbers are greater than negative numbers. So, $2 > -3$.

Compare the position on the number line.

Since -2 is farther to the right, $-2 > -3$.

Example

Fill in the ◯ with <, >, or = to make a true sentence.

1. $12 \bigcirc -4$

Graph 12 and -4 on a number line. Then compare.

Since 12 is to the right of -4, $12 > -4$.

Got it? Do these problems to find out.

a. $-3 \bigcirc -5$ **b.** $-5 \bigcirc 0$ **c.** $6 \bigcirc -1$

Example

2. Justin has a score of -4 on the Trueville Trivia game. Desiree's score is -5. Write an inequality to compare the scores. Explain the meaning of the inequality.

$-4 > -5$ -4 is farther to the right on a number line than -5.

Since $-4 > -5$, Justin has a higher score than Desiree.

Got it? Do this problem to find out.

d. The temperature on Tuesday was 2°F. The temperature on Wednesday was -2°F. Write an inequality to compare the temperatures. Explain the meaning of the inequality.

Absolute Value

Although -5 is the lesser value in Example 2, it represents the greater point deficit.

$|-5| > |-4|$

Show your work.

d. _____

Order Integers

You can use a number line to order a set of integers. Integers can be ordered from least to greatest or from greatest to least.

Example

Tutor

3. Order the set {−9, 6, −3, 0} from least to greatest.

| Method 1 | Use a number line.

Graph the numbers on a number line.

The order from left to right is −9, −3, 0, and 6.

| Method 2 | Compare signs and values.

Compare negative numbers. Then compare positive numbers.

The negative integers are −9 and −3. −9 < −3

The integer 0 is neither positive nor negative.

The positive integer is 6.

So, the order from least to greatest is −9, −3, 0, and 6.

> **Absolute Value**
> Since absolute value is always positive, it is not used to compare and order integers.

Got it? Do these problems to find out.

Show your work.

e. Order the set {−4, 3, 11, −25} from greatest to least.

f. Order the set {−18, 30, 12, −6, 3} from least to greatest.

e. _____

f. _____

Example

4. **STEM** The table shows the lowest elevation for several continents. Order the elevations from least to greatest.

First, graph each integer. Then, write the integers as they appear on the number line from left to right.

Continent	Lowest Elevation (m)
Africa	−156
Asia	−418
Australia	−12
Europe	−28
North America	−86
South America	−105

A number line labeled from −500 to 0 in increments of 50.

The elevations from least to greatest are −418, −156, −105, −86, −28, and −12.

Guided Practice

Check ✓

Fill in each ◯ with <, >, or = to make a true statement. *(Example 1)*

1. 17 ◯ 31

2. −6 ◯ −10

3. −83 ◯ −38

4. Andrew and his father are scuba diving at −38 feet and Tackle Box Canyon has an elevation of −83 feet. Write an inequality to compare the depths. Explain the meaning of the inequality. *(Example 2)*

5. **STEM** The daily low temperatures in Kate's hometown last week were 2°C, −9°C, −18°C, −6°C, 3°C, 0°C, and −7°C. Order the temperatures from greatest to least. *(Examples 3 and 4)*

6. **Q** **Building on the Essential Question** How can symbols and absolute value help you to order sets of integers?

Rate Yourself!

How confident are you about comparing and ordering integers? Shade the ring on the target.

I'm on target.

I need help.

For more help, go online to access a Personal Tutor.

FOLDABLES Time to update your Foldable!

Independent Practice

Go online for Step-by-Step Solutions

Fill in each ⬭ **with <, >, or = to make a true statement.** (Example 1)

1. −2 ⬭ −4

2. 1 ⬭ −3

3. 5 ⬭ 0

4. Amy is building a house. The basement floor is at −15 feet. The roof of the house is above the ground 25 feet. Write an inequality to compare the heights. Explain the meaning of the inequality. (Example 2)

_____

_____

5 The low temperature in Anchorage, Alaska, one day was −9°F. On the same day, the low temperature in Flagstaff, Arizona, was 26°F. Write an inequality to compare the temperatures. Explain the meaning of the inequality. (Example 2)

_____

_____

Order each set of integers from least to greatest. (Example 3)

6. {15, 17, 21, 6, 3}

_____

7 {−55, 143, 18, −79, 44, 101}

_____

8. The table indicates Xavier's cell phone use over the last four months. Positive values indicate the number of minutes he went over his allotted time, and negative values indicate the number of minutes he was under. Arrange the months from least to most minutes used. (Example 4)

_____

Month	Time (min)
February	−156
March	12
April	0
May	−45

9. **MP** **Use Math Tools** Refer to the table and the following information. The apparent magnitude of an object measures how bright the object appears to the human eye. A negative magnitude identifies a brighter object than a positive magnitude.

a. Which object appears the brightest to the human eye?

_____

b. Order the objects from the brightest to the faintest.

_____

_____

c. Find the least apparent magnitude of this data set.

_____

Object	Approximate Apparent Magnitude
100-Watt Bulb	−19
Alpha Centauri	4
Andromeda Galaxy	0
Full Moon	−13
Sun	−27
Venus	−5

10. **MP** **Justify Conclusions** Refer to the graphic novel frame below for exercises a–c.

a. The memory card holds 65,536 kilobytes. If each picture is about 760 kilobytes, about how many more pictures can they take? _____

b. Write an inequality to compare the number of pictures taken during school to the number of pictures taken after school. _____

c. Explain the meaning of the inequality. _____

 ## H.O.T. Problems Higher Order Thinking

11. **MP** **Model with Mathematics** Write a real-world situation to explain the inequality −$15 < $7. _____

_____

12. **MP** **Reason Abstractly** Explain why −11 is less than −7, but |−11| is greater than |−7|. _____

_____

_____

13. **MP** **Persevere with Problems** Order the fractions $-\frac{1}{2}, \frac{5}{2}, -\frac{12}{4}, \frac{1}{6}$, and $\frac{7}{8}$ from least to greatest. _____

14. **MP** **Persevere with Problems** Find all integers that make $|n| < 3$ a true statement. Then graph the integers on the number line. _____

Extra Practice

Fill in each ◯ **with <, >, or = to make a true statement.**

15. −2 ⟨<⟩ 4

 Homework Help → *Since −2 is to the left of 4, −2 < 4.*

```
←——+——+——+——+——+——+——→
  −6  −4  −2   0   2   4   6
```

16. −6 ◯ 3

17. −3 ◯ 2

18. The elevation of Driskill Mountain, Louisiana, is 163 meters above sea level. Death Valley has an elevation of −86 meters. Write an inequality to compare the elevations. Explain the meaning of the inequality.

_____

19. Yvonne owes her sister $25. Michael's checking account balance is −$20. Write an inequality to compare the amounts. Explain the meaning of the inequality.

_____

_____

Order each set of integers from least to greatest.

20. {14, 1, 6, 23, 7, 5}

_____

21. {−221, 63, 54, −89, −71, −10}

_____

22. Gary, Sindhu, and Beth are all waiting for their trains to arrive. Gary's train leaves at 5 minutes before noon, Sindhu's leaves at 25 minutes after noon, and Beth's leaves 5 minutes before Sindhu's train. Order the three by who will leave first.

_____

23. **MP** **Use Math Tools** Use the bar graph and the information below. The bar graph gives the scores of four golfers (A, B, C, and D). The numbers indicate scores above and below par.

a. Order the scores on a number line.

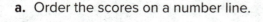
```
←——+——+——+——+——+——+——+——+——+——+——+——+——→
  −5  −4  −3  −2  −1   0   1   2   3   4   5   6   7
```

Golf Scores

Score Above or Below Par / Player

b. Which player had the worst score? Explain your answer.

_____

24. Verónica (V) was 12 minutes early to class, Deshawn (D) was right on time, and Kendis (K) was 3 minutes late. Which time line represents the students' arrival to class?

25. The table shows the scores at the end of a game of miniature golf. The integer 0 represents par. Sort the players from fewest shots taken to most shots taken.

Player	Score
Cristian	−6
Bailey	2
Liam	−3
Marisol	5

	Player
Fewest Shots	
Most Shots	

Spiral Review

Write each fraction as a decimal. **5.NF.5b**

26. $\frac{3}{4} =$ _____

27. $\frac{1}{5} =$ _____

28. $\frac{3}{20} =$ _____

29. The table shows the heights of Sonya's siblings. Who is taller, Frieda or Julio? Compare their heights using the symbol >. **4.NBT.2**

Member	Height (ft)
Frieda	$5\frac{1}{4}$
Julio	$5\frac{5}{6}$

_____

30. Kristen and Mitchell were given the same math assignment. Kristen completes 0.8 of her work in class. Mitchell completes 0.75 of his work during class. Who has more homework remaining after class? **5.NBT.3**

_____

Work Backward

Content Standards
6.NS.5

MP **Mathematical Practices**
1, 3

Case #1 Hit the Slopes!

Marissa and her family are on a ski trip at Mount Washington in New Hampshire. They returned from the slopes at 6 P.M. By 9 P.M., the temperature had fallen 18° to the day's low temperature of −8°F.

What was the temperature at 6 P.M.?

Understand *What are the facts?*

- By 9 P.M., the temperature had fallen 18°.
- The day's low temperature was −8°F.

Plan *What is your strategy to solve this problem?*

Work backward from the low temperature at 9 P.M. Use a thermometer diagram to find the temperature at 6 P.M.

Solve *How can you apply the strategy?*

Start at ☐ °F. Shade the thermometer ☐ degrees

to find the temperature at 6 P.M.

So, the temperature at 6 P.M. was _____.

Check *Does the answer make sense?*

−8°F is 8 degrees away from 0°F. 18 − 8 = 10

So, add to ☐ to 0°F. Since 0 + ☐ = ☐, the answer is reasonable.

Analyze the Strategy Tutor

MP **Justify Conclusions** The high temperature was 36°F. How far away from

−8°F is 36°F? Explain. _____

_____

Case #2 Get Ready, Get Set, Go

The table shows the amount of time it takes Henry to do different activities before going to soccer practice.

If he needs to be at practice at 8:15 A.M., what time should he wake up in the morning to get to the soccer field?

Activity	Time (minutes)
Travel to field from home	15
Eating breakfast	35
Changing into uniform	10
Checking E-mail messages	20

Understand

Read the problem. What are you being asked to find?

I need to find _____.

Underline key words and values. What information do you need to know?

The table shows the time it takes Henry to do each activity. It takes him ▢ minutes to get to the field, ▢ minutes to eat, ▢ minutes to change for practice, and ▢ minutes to check his email.

Plan

Choose a problem-solving strategy.

I will use the _____ strategy.

Solve

Use your problem-solving strategy to solve the problem.

8:15 A.M. − 20 min = _____ A.M.   7:45 A.M. − 35 min = _____ A.M.

7:55 A.M. − 10 min = _____ A.M.   7:10 A.M. − 15 min = _____ A.M.

So, Henry should wake up at _____.

Check

Use information from the problem to check your answer.

Begin at _____ A.M. and add the minutes from the table.

▢ + ▢ + ▢ + ▢ = ▢ minutes

_____ A.M. plus ▢ minutes is 8:15 A.M.

**Work with a small group to solve the following cases.
Show your work on a separate piece of paper.**

Case #3 Sea Level

Mr. Ignacio went diving along the coral reef in Oahu.
He descended 12 meters below sea level. The
difference between this point on the coral reef and the
highest point on the island, Mount Ka'ala, is 1,232 meters.
Two-fifths of the way up the mountain is a ranger station.

How far above sea level is the ranger station?

_____

Case #4 Cameras

Adamo saved 13 pictures on his digital camera for a total of
12,021.1 KB. He deleted 32 pictures for a total of 29,590.4 KB.

If there are now 108 pictures, how many
kilobytes of storage did he use at the
beginning?

_____

Case #5 Mystery Number

A number is multiplied by 4, and then 6 is added to the product. The
result is 18.

What is the number?

_____

Case #6 Ladders

You are standing on the middle rung of a ladder.

If you first climb up 3 rungs, then down 5 rungs, and then up
10 rungs to get onto the top rung, how many rungs are on the
ladder?

_____

Use any
strategy!

Mid-Chapter Check

Vocabulary Check

1. **MP** **Be Precise** Define *negative integer*. Give an example of a negative integer and then give its opposite. (Lesson 1)

   _____

   _____

2. Fill in the blank in the sentence below with the correct term. (Lesson 2)

   The _____ of the numbers −4 and 4 is 4.

Skills Check and Problem Solving

MP **Model with Mathematics** Graph each set of integers on a number line. (Lesson 1)

3. {−4, −6, 0, 3}

4. {2, −3, 1, −1}

Evaluate each expression. (Lesson 2)

5. $|{-12}| =$ _____

6. $|{-4}| + |{-10}| =$ _____

7. $|9| + |{-2}| =$ _____

8. $|13| - |{-5}| =$ _____

9. $|{-16}| - |{-2}| =$ _____

10. $|{-15}| + |{-7}| =$ _____

11. Hailey, Priya, and Shetal are auditioning for the same role. Hailey auditions at 10 minutes before four, Priya auditions 30 minutes before Hailey, and Shetal auditions at 5 minutes before four. Order the three by who will audition first. (Lesson 3) _____

12. **MP** **Persevere with Problems** The tables shows the overnight low temperatures for a four-day period. The low temperature on Wednesday was between the second and third highest temperatures of the ones shown, when placed in order. Give a possible temperature for Wednesday. (Lesson 3)

   _____

Temperature (°F)	
Thursday	−8
Friday	7
Saturday	18
Sunday	−11

Inquiry Lab

Number Lines

Content Standards
6.NS.6, 6.NS.6c, 6.NS.7

MP Mathematical Practices
1, 3, 4

 Inquiry **HOW can you use a number line to model and compare positive and negative rational numbers?**

Marcus and Silvio are at the beach. Marcus builds a sandcastle 0.6 meter high. Silvio digs a hole in the sand 0.8 meter deep.

Hands-On Activity 1

Just as you can graph integers on a number line, you can graph positive and negative fractions and decimals. Recall that positive numbers are to the right of zero on the number line and negative numbers are to the left of zero.

Step 1 Complete the number line from −1 to 1, with increments of 0.2.

Step 2 The sandcastle is above sea level. Its height is *greater than zero* on the number line, so draw a dot at ☐ to represent the sandcastle.

Step 3 The hole is below sea level. Its depth is *less than zero* on the number line. So draw a dot at ☐ to represent the hole.

MP **Model with Mathematics** Work with a partner. Graph each number on a number line.

1. −2.4

show your work.

2. 0.1

3. −4.5

4. −6.8

Hands-On Activity 2

Tools

Graph $-\frac{3}{4}$ on a number line.

Step 1   Model $-\frac{3}{4}$ using fraction tiles. Draw a number line from −1 to 0.

Since the denominator of the fraction is ☐ , divide your number

line into ☐ equal parts.

| $\frac{1}{4}$ | $\frac{1}{4}$ | $\frac{1}{4}$ | $\frac{1}{4}$ |

−1   0

Step 2   Each mark on the number line represents $\dfrac{☐}{☐}$. Label the number

line with $-\frac{3}{4}$, $-\frac{2}{4}$, and $-\frac{1}{4}$.

−1   0

Step 3   Draw a dot to graph $-\frac{3}{4}$ on the number line above.

Investigate

MP **Model with Mathematics** Work with a partner. Graph each number on a number line.

5. $-\dfrac{4}{5}$

show your work.

6. -5.75

7. $\dfrac{7}{10}$

8. $-\dfrac{3}{8}$

9. 8.75

10. $-\dfrac{3}{10}$

11. $-\dfrac{5}{12}$

Work with a partner to complete the table. The first one is done for you.

Number	Positive or Negative	Greater Than or Less Than Zero	Left or Right of 0 on the Number Line
−3.5	negative	<	left
12. $+\frac{4}{5}$			
13. $-\frac{1}{3}$			
14. $+0.3$			

15. **MP Reason Inductively** Which number is greater, 0.3 or −0.7? Explain.

_____

_____

16. **MP Reason Inductively** Jacyln thinks that $-\frac{1}{2}$ is greater than $\frac{1}{4}$ because it is farther from zero on the number line. Is her thinking correct? Explain.

_____

_____

_____

Create
On Your Own

17. **MP Model with Mathematics** Write a real-world problem that involves a positive and a negative value. Then graph the values used in the problem on a number line. Compare the values.

_____

_____

_____

18. **Inquiry** HOW can you use a number line to model and compare positive and negative rational numbers?

_____

_____

Terminating and Repeating Decimals

Vocabulary Start-Up

Any number that can be written as a fraction is called a **rational number**. Every rational number can be written as either a **terminating decimal** or a **repeating decimal**.

Draw lines from each word to its matching statement.

| terminating decimal | | the decimal form of a rational number; 0.33333... |

| repeating decimal | | the decimal form of a rational number which has a repeating digit of zero; 0.625 |

 Essential Question

HOW are integers and absolute value used in real-world situations?

 Vocabulary

rational number
terminating decimal
repeating decimal
bar notation

Content Standards
Preparation for 6.NS.6c and 6.NS.7a

MP **Mathematical Practices**
1, 2, 3, 4, 7, 8

 ## Real-World Link

Party Favors Jude is buying fruit snacks for party favors. He asks the cashier for a half pound of fruit snacks.

1. Express one half as a fraction.

_____

2. Write the decimal that represents half a pound.

_____

3. Suppose Jude wanted to buy one third of a pound. What decimal would the scale show?

_____

Which **MP** Mathematical Practices did you use?
Shade the circle(s) that applies.

① Persevere with Problems

② Reason Abstractly

③ Construct an Argument

④ Model with Mathematics

⑤ Use Math Tools

⑥ Attend to Precision

⑦ Make Use of Structure

⑧ Use Repeated Reasoning

Key Concept ▶ Rational Numbers

Watch

Words Rational numbers can be written as fractions.

Algebra $\frac{a}{b}$, where a and b are integers and $b \neq 0$.

Model

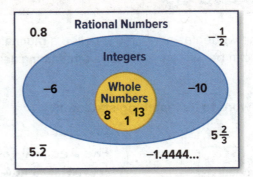

Work Zone

Fractions, terminating and repeating decimals, percents, and integers are all rational numbers. Every rational number can be expressed as a decimal by dividing the numerator by the denominator.

Rational Number	Repeating Decimal	Terminating Decimal
$\frac{3}{10}$	0.300...	0.3
$\frac{4}{5}$	0.800...	0.8
$\frac{5}{6}$	0.833...	does not terminate

To indicate the number pattern that repeats indefinitely, use bar notation. **Bar notation** is a bar placed over the digits that repeat.

$0.545454... = 0.\overline{54}$ $\qquad\qquad$ $0.583333... = 0.58\overline{3}$

Example

Tutor

1. Write $\frac{5}{12}$ as a decimal.

$$
\begin{array}{r}
0.4166 \\
12\overline{)5.000} \\
\end{array}
$$

Divide 5 by 12.

$\quad -\,4\,8$

$\quad\quad 20$

$\quad -\,12$

$\quad\quad\, 80$

$\quad\; -\,72$

$\quad\quad\quad 80$

$\quad\quad -\,72$

$\quad\quad\quad\quad 8$ $\quad$ The remainder will never be zero.

So, $\frac{5}{12} = 0.4166...$ or $0.41\overline{6}$.

Got it? Do these problems to find out.

Write each fraction as a decimal. Use bar notation if necessary.

a. $\dfrac{1}{6}$　　　　b. $\dfrac{8}{9}$　　　　c. $\dfrac{2}{11}$

Write a Negative Fraction as a Decimal

When writing negative fractions as decimals, the process is the same. Divide as with positive fractions. Write the negative sign in front of the decimal.

Examples

 Tutor

2. Write $-\dfrac{2}{9}$ as a decimal.

```
   0.222
9)2.000      Divide 2 by 9.
  −18
   20
  −18
   20
  −18
    2        The remainder will never be zero.
```

So, $-\dfrac{2}{9} = -0.222\ldots$ or $-0.\overline{2}$.

Repeating Decimals

When dividing, it is sometimes helpful to divide until the repeated pattern is shown at least three times.

3. Write $-2\dfrac{2}{3}$ as a decimal.

$-2\dfrac{2}{3}$ can be rewritten as $-\dfrac{8}{3}$.

The mixed number $-2\dfrac{2}{3}$ can be written as $-2.\overline{6}$.

```
  2.6...
3)8.0
 −6
  20
 −18
   2
```

Got it? Do these problems to find out.

Write each fraction as a decimal. Use bar notation if necessary.

d. $-\dfrac{1}{4}$　　　　e. $-\dfrac{5}{6}$　　　　f. $-2\dfrac{1}{6}$

d. _____

e. _____

f. _____

Example

4. Frankie made 34 out of 44 free throws this season. To the nearest thousandth, what is his free-throw average?

Using a calculator, divide 34 by 44.

34 ÷ 44 ENTER 0.77272727

To the nearest thousandth, his free-throw average is 0.773.

Got it? Do this problem to find out.

g. Of nine students surveyed, four said they prefer exercising in the morning rather than in the evening. Express this fraction as a decimal. Use bar notation if necessary.

Show your work.

g. _____

Guided Practice

Write each fraction as a decimal. Use bar notation if necessary. *(Examples 1–3)*

1. $\dfrac{7}{9}$ = _____

2. $-\dfrac{1}{33}$ = _____

3. $-2\dfrac{5}{6}$ = _____

4. $\dfrac{10}{15}$ = _____

5. $-\dfrac{4}{5}$ = _____

6. $1\dfrac{5}{9}$ = _____

7. Dana bought $\dfrac{2}{3}$ yard of fabric to make a new purse. Write the amount of fabric she used as a decimal. *(Example 4)*

_____

8. **Building on the Essential Question** How are repeating decimals used in real-world situations?

_____

_____

_____

_____

Independent Practice

Go online for Step-by-Step Solutions

Write each fraction as a decimal. Use bar notation if necessary.
(Examples 1–3)

1. $\dfrac{7}{15} =$ _____

2. $\dfrac{8}{18} =$ _____

3. $-\dfrac{8}{12} =$ _____

Show your work.

4. $-\dfrac{6}{7} =$ _____

5. $3\dfrac{15}{44} =$ _____

6. $-2\dfrac{5}{22} =$ _____

7. Sarafina had 34 out of 99 hits when she was at bat during the softball season. What was her batting average? (Example 4)

_____

8. Shiv and his friends ate $3\dfrac{1}{6}$ pizzas. Write this amount as a decimal. (Example 4)

_____

Write each decimal as a fraction or mixed number in simplest form.

9. $-0.9 =$ _____

10. $-0.85 =$ _____

11. $-3.8 =$ _____

Evaluate each expression.

12. $\left| -2.3 \right| =$ _____

13. $\left| \dfrac{4}{13} \right| =$ _____

14. $\left| -8\dfrac{7}{11} \right| =$ _____

15. **STEM** There are over 2,700 species of snakes in the world. Over 600 species are venomous. Write the fraction of species that are *not* venomous as a decimal. _____

16. **MP Justify Conclusions** The ratio of the distance around a circle to the distance across a circle through its center is represented by the number π. The number π is a decimal that does not repeat. The fraction $\dfrac{22}{7}$ is sometimes used as an estimate for π. Is $\dfrac{22}{7}$ a repeating decimal? Explain.

_____

17. (MP) **Reason Abstractly** Refer to the graphic novel frame below for Exercises a–b.

a. How many total photos were taken? _____

b. What fraction of the photos were taken after school? Write this fraction as a decimal. Round to the nearest thousandth. _____

🔥 H.O.T. Problems Higher Order Thinking

18. (MP) **Justify Conclusions** Name a number that is a rational number, but not an integer. Justify your response. _____

_____

19. (MP) **Persevere with Problems** Predict whether or not the decimal equivalent to $\frac{17}{36}$ is terminating. Explain your reasoning. Check your prediction with a calculator. _____

_____

20. (MP) **Which One Doesn't Belong?** Identify the decimal equivalent that does *not* have the same characteristic as the other three. Explain.

| $\frac{1}{12}$ | $\frac{2}{12}$ | $\frac{3}{12}$ | $\frac{4}{12}$ |

_____

21. (MP) **Identify Repeated Reasoning** Find the decimal equivalents for $\frac{1}{11}$, $\frac{2}{11}$, and $\frac{3}{11}$. Use the pattern to mentally find the decimal equivalents for $\frac{7}{11}$ and $\frac{8}{11}$.

_____

Extra Practice

Write each fraction as a decimal. Use bar notation if necessary.

22. $\frac{32}{75} =$ _0.42$\overline{6}$_

Homework Help →

$$
\begin{array}{r}
0.42\overline{6} \\
75\overline{)32.000} \\
-300 \\
\hline
200 \\
-150 \\
\hline
500 \\
-450 \\
\hline
50
\end{array}
$$

23. $\frac{3}{11} =$ _____

24. $-\frac{5}{8} =$ _____

25. $-\frac{7}{10} =$ _____

26. $2\frac{5}{7} =$ _____

27. $-1\frac{80}{99} =$ _____

28. Cris answered 61 out of 66 questions correctly on a test. What is his test average to the nearest thousandth? _____

Write each decimal as a fraction or mixed number in simplest form.

29. $-0.15 =$ _____

30. $-7.75 =$ _____

31. $-12.54 =$ _____

32. **MP Identify Repeated Reasoning** The table shows the decimal equivalent to fractions with a denominator of 7.

a. What do you notice about the pattern of the six repeated numbers?

_____

_____

Fraction	Decimal	Fraction	Decimal
$\frac{1}{7}$	$0.\overline{142857}$	$\frac{4}{7}$	$0.\overline{571428}$
$\frac{2}{7}$	$0.\overline{285714}$	$\frac{5}{7}$	$0.\overline{714285}$
$\frac{3}{7}$	$0.\overline{428571}$	$\frac{6}{7}$	$0.\overline{857142}$

b. Using the decimals, add the first half of each pattern to the numbers in the last half. For example, $\frac{1}{7} = 0.\overline{142857}$, so add $142 + 857$.

What pattern do you notice? _____

c. Using a calculator, try the same experiment with $\frac{5}{13}$. Is the result the same?

Justify your reasoning. _____

33. Write a decimal to represent the shaded portion of the figure.

34. Which of the following fractions are equivalent to $0.\overline{3}$?
Select all that apply.

☐ $\frac{1}{3}$　　　　☐ $\frac{3}{9}$　　　　☐ $\frac{3}{10}$　　　　☐ $\frac{11}{33}$

Spiral Review

Fill in each ◯ with < or > to make a true statement. 4.NBT.2

35. 4,556 ◯ 4,565

36. 8,698 ◯ 8,689

37. 47,872 ◯ 47,871

38. 26,525 ◯ 26,522

39. 1,123,004 ◯ 1,123,040

40. 5,776,050 ◯ 5,775,005

41. The table shows the number of miles Katie walked for two weeks.
Compare the distances using the < symbol. 5.NBT.3b

Week	Number of Miles
1	5.78
2	5.691

42. The table shows the amount of different colored paints in a bin in art
class. Compare the amount of blue and orange paint using the > symbol.

5.NBT.3

Color	Number of Ounces
Blue	47.362
Green	47.637
Orange	47.394
Yellow	47.583

Compare and Order Rational Numbers

Real-World Link

Insects The lengths of several common types of insects are shown in the table.

1. Which of the insects is the longest?

_____

2. Shade each fraction strip to represent the lengths of a fire ant and a housefly. Which is longer, the fire ant or housefly?

_____

Insect	Length (in.)
Green June Beetle	$\frac{3}{4}$
Cricket	$\frac{1}{1}$
Fire ant	$\frac{1}{3}$
Firefly	$\frac{3}{4}$
Housefly	$\frac{1}{4}$
Japanese beetle	$\frac{1}{2}$
Mosquito	$\frac{5}{8}$

Essential Question

HOW are integers and absolute value used in real-world situations?

Content Standards
6.NS.6, 6.NS.6c, 6.NS.7, 6.NS.7a, 6.NS.7b

MP Mathematical Practices
1, 2, 3, 4, 5, 7

$\frac{1}{4}$	$\frac{1}{4}$	$\frac{1}{4}$	$\frac{1}{4}$
$\frac{1}{3}$	$\frac{1}{3}$	$\frac{1}{3}$	

3. How many of the insects are longer than 0.5 inch?

_____

4. Order the lengths of a housefly, a Green June beetle, and a fire ant from the shortest to longest.

☐ ☐ ☐
☐ ☐ ☐ _____

Which MP Mathematical Practices did you use?
Shade the circle(s) that applies.

① Persevere with Problems

② Reason Abstractly

③ Construct an Argument

④ Model with Mathematics

⑤ Use Math Tools

⑥ Attend to Precision

⑦ Make Use of Structure

⑧ Use Repeated Reasoning

Compare Decimals and Fractions

Positive and negative rational numbers can be represented on a number line. You can use a number line to help you compare and order rational numbers.

Examples

Fill in each ◯ with **<, >, or = to make a true statement.**

1. −1.2 ◯ 0.8

Graph the decimals on a number line.

+—+—+—+—+—+—+—+—+—+—+—+—+—+—+—+—+—+
−1.4 −1.2 −1 −0.8−0.6−0.4−0.2 0 0.2 0.4 0.6 0.8 1 1.2 1.4

Since −1.2 is to the left of 0.8, −1.2 < 0.8.

- -

2. −1.40 ◯ −1.25

Graph the decimals on a number line.

Since −1.40 is below −1.25, −1.40 < −1.25.

- -

Least Common Multiple (LCM)

Find the LCM of 8 and 16.

8: 8 16 24

16: 16 32, 48

The LCM is 16.

3. $-\dfrac{3}{8}$ ◯ $-\dfrac{5}{16}$

Rename the fractions using the least common denominator.

$$-\frac{3}{8} = -\frac{3 \times 2}{8 \times 2} = -\frac{6}{16} \qquad\qquad -\frac{5}{16} = -\frac{5 \times 1}{16 \times 1} = -\frac{5}{16}$$

Since $-6 < -5$, $-\dfrac{6}{16} < -\dfrac{5}{16}$ and $-\dfrac{3}{8} < -\dfrac{5}{16}$.

Check ✔

Got it? Do these problems to find out.

a. 3.1 ◯ −3.7

b. −4.5 ◯ −4.49

c. $\dfrac{9}{16}$ ◯ $\dfrac{12}{16}$

d. $-\dfrac{7}{10}$ ◯ $-\dfrac{4}{5}$

Compare and Order Rational Numbers

To compare and order rational numbers, first write them in the same form.

Examples

Fill in each ◯ with <, >, or = to make a true statement.

4. -0.51 ◯ $-\dfrac{8}{15}$

Rename $-\dfrac{8}{15}$ as a decimal. Then graph both decimals on a number line.

$-0.5\overline{3}$ -0.51

$\begin{array}{cccccc} & & & & & \\ \hline -0.6 & -0.55 & -5 & -0.5 & -0.45 & -0.4 \end{array}$

$-\dfrac{8}{15} = -0.5\overline{3}$

Since -0.51 is to the right of $-0.5\overline{3}$ on the number line, $-0.51 > -\dfrac{8}{15}$.

5. Order the set $\left\{-2.46,\ -2\dfrac{22}{25},\ -2\dfrac{1}{10}\right\}$ from least to greatest.

Write $-2\dfrac{22}{25}$ and $-2\dfrac{1}{10}$ as decimals to the hundredths place.

$-2\dfrac{22}{25} = -2.88 \qquad -2\dfrac{1}{10} = -2.1$

-2.88 -2.46 -2.1

$\begin{array}{ccccc} & & & & \\ \hline -3.00 & -2.75 & -2.50 & -2.25 & -2.00 \end{array}$

Graph the decimals on the number line.

From least to greatest, the order is $-2\dfrac{22}{25},\ -2.46,$ and $-2\dfrac{1}{10}.$

Got it? Do these problems to find out.

Fill in each ◯ with <, >, or = to make a true statement.

e. $-3\dfrac{5}{8}$ ◯ -3.625

f. $\dfrac{3}{7}$ ◯ 0.413

g. Order the set $\left\{-7\dfrac{13}{20},\ -7.78,\ -7\dfrac{17}{100}\right\}$ from greatest to least.

STOP and **Reflect**

How could you represent that −8.3 feet is deeper than −5.7 feet? Explain.

Show your work.

g. _____

Example

6. Mr. Plum's science class is growing plants under different conditions. The table shows the difference from the average for some students' plants. Order the differences from least to greatest.

Student	Difference (in.)
Ricky	$3\frac{1}{4}$
Debbie	-2.2
Suni	1.7
Leonora	$-1\frac{7}{10}$

Express each number as a decimal.

Ricky's plant: $3\frac{1}{4} = 3.25$ Debbie's plant: -2.2

Suni's plant: 1.7 Leonora's plant: $-1\frac{7}{10} = -1.7$

From least to greatest, the differences are -2.2, $-1\frac{7}{10}$, 1.7, and $3\frac{1}{4}$.

Guided Practice

Fill in each ◯ with <, >, or = to make a true statement. (Examples 1–4)

1. $9.7 \bigcirc -10.3$

2. $\frac{5}{8} \bigcirc -\frac{3}{8}$

3. $-6.7 \bigcirc -6\frac{7}{10}$

4. $-\frac{5}{6} \bigcirc -0.94$

Show your work.

Order the following sets of numbers from least to greatest. (Example 5)

5. $\left\{-3\frac{1}{3}, 3.3, -3\frac{3}{4}, 3.5\right\}$ _____

6. $\left\{2.\overline{1}, -2.1, 2\frac{1}{11}, -2\right\}$ _____

7. Financial Literacy Steve recorded these amounts in his checkbook: $-\$6.50$, $\$7.00$, $-\$6.75$, and $\$7.25$. Order these amounts from least to greatest. (Example 6)

_____

8. **Building on the Essential Question** How can a number line help in ordering rational numbers?

_____

Rate Yourself!

Are you ready to move on? Shade the section that applies.

YES   ?   NO

For more help, go online to access a Personal Tutor.

FOLDABLES Time to update your Foldable!

Independent Practice

Go online for Step-by-Step Solutions

Fill in each ◯ with <, >, or = to make a true statement. (Examples 1–4)

1. $\frac{5}{4}$ ◯ $-\frac{1}{4}$

Show your work.

2. $-6\frac{1}{3}$ ◯ -6.375

3 $-\frac{3}{5}$ ◯ -0.6

4. $-9\frac{2}{7}$ ◯ -9.3

Order each set of numbers from least to greatest. (Example 5)

5 $\left\{2.8, -2\frac{3}{4}, 3\frac{1}{8}, -2.\overline{2}\right\}$ _____

6. $\left\{\frac{2}{3}, -0.6, 0.65, \frac{4}{5}\right\}$ _____

7. Financial Literacy The change in four stocks during a day are: $-4\frac{1}{2}$, 5.6, $-2\frac{3}{8}$, and 1.35.

Order the changes from least to greatest. (Example 6)

_____

8. **MP** **Multiple Representations** Consider the inequality $-3.5 < -1.5$.
   a. Words Write a real-world problem that could be represented by the inequality.

   _____

   b. Number Line Graph -3.5 and -1.5 on the number line.

   c. Symbols Use the symbol > to compare -3.5 and -1.5.

   _____

9 For a STEM competition, Julienne constructed a model rocket. The rocket can reach an average height of 545 feet. Find the differences between the average height and the actual heights reached. Then write them as positive and negative rational numbers. Order the differences from least to greatest.

Trials	Actual Height (ft)
1	534.2
2	556.4
3	554.0
4	535.3

_____

10. **MP** **Identify Structure** Fill in the diagram with appropriate numbers.

Terminating
Decimals

Negative
Numbers

H.O.T. Problems Higher Order Thinking

11. **MP** **Reason Inductively** Determine whether the following statement is *always*, *sometimes*, or *never* true. Give examples to justify your answer.
    If x and y are both greater than zero and x > y, then −x < −y.

12. **MP** **Justify Conclusions** Determine whether the fractions $-\frac{4}{5}$, $-\frac{4}{6}$, $-\frac{4}{7}$, and $-\frac{4}{8}$ are arranged in order from least to greatest. Explain.

13. **MP** **Reason Abstractly** Explain why −0.33 is greater than $-0.\overline{33}$.

14. **MP** **Persevere with Problems** Compare the set $\left\{-0.\overline{7}, -0.\overline{6}, -\frac{7}{9}, -\frac{2}{3}\right\}$. Explain your answer.

15. **MP** **Model with Mathematics** Write a real-world problem in which you would order a set of four rational numbers.

Extra Practice

Fill in each ◯ with <, >, or = to make a true statement.

16. -18.6 ⦗<⦘ -18.06

```
←—•——+——+——+——+——•——+——→
  -18.6   -18.4   -18.2   -18
```

17. -4.08 ◯ -4.7

18. $-\dfrac{3}{7}$ ◯ $-\dfrac{2}{5}$

19. -3.375 ◯ $-3\dfrac{4}{10}$

20. $-5\dfrac{1}{5}$ ◯ -5.2

21. $-8\dfrac{2}{5}$ ◯ -8.3

Order the following sets of numbers from least to greatest.

22. $\left\{ \dfrac{1}{8},\ -0.02\overline{5},\ 0.2,\ -\dfrac{1}{7} \right\}$ _____

23. $\left\{ 1.25,\ 1\dfrac{3}{4},\ 1.2\overline{5},\ 1\dfrac{1}{5} \right\}$ _____

24. **MP Reason Inductively** The average amount of time Brent spent in-line skating for one week was 34 minutes. During the next week, the difference between the average time and actual time spent skating was 4.2 minutes, $-5\dfrac{1}{3}$ minutes, $-2\dfrac{1}{2}$ minutes, and 3.75 minutes.

Order these differences from least to greatest. _____

Fill in each ◯ with <, >, or = to make a true statement.

25. $-4\dfrac{4}{5}$ ◯ $-4.\overline{7}$

26. $-3.2\overline{5}$ ◯ $-3.\overline{2}$

27. $-5.\overline{31}$ ◯ $-5.\overline{313}$

28. The table shows the profit or loss of the after-school snack stand.

   a. Write each profit as a positive number and each loss as a negative

      number. _____

   b. Order the numbers from least to greatest.

   _____

Day	Profit or Loss	($)
1	Profit	7.50
2	Loss	3.50
3	Loss	6.00
4	Profit	4.50

29. Refer to the number line. Determine if each inequality about the coordinates of the points is true or false.

a. $B > M$ ☐ True ☐ False

b. $F > B$ ☐ True ☐ False

c. $M < F$ ☐ True ☐ False

d. $F < M$ ☐ True ☐ False

30. Student Council's goal was to raise $50 each week for 4 weeks to have enough money for the school dance. The table shows the difference between the goal and the actual amount raised. Sort the amounts from least to greatest.

Week	1	2	3	4
Difference ($)	5.50	−6.25	7.80	−2.45

	Amount
Least	
Greatest	

Spiral Review

Graph the points on the coordinate plane. 5.G.1

31. $H(1, 6)$ **32.** $M(7, 0)$

33. $I(5, 8)$ **34.** $N(4, 9)$

35. $J(6, 3)$ **36.** $O(7, 5)$

37. $L(3, 1)$ **38.** $P(2, 2)$

39. Graph the point on the number line that represents $\frac{3}{10}$ and label it A. 4.NF.6

The Coordinate Plane

 Real-World Link Watch ▶

Maps The map shows the layout of a small town. The locations of buildings are described in respect to the town hall. Each unit on the grid represents one block.

1. Describe the location of the barber shop in relation to the town hall.

_____

_____

2. What building is located 7 blocks east and 5 blocks north of the

town hall? _____

_____

3. Violeta is at the library. Describe how many blocks and in what direction she should travel to get to the

supermarket. _____

_____

4. Town Hall and the bank are both located on the same vertical number line. The number 0 represents the location of Town Hall on the number line. What number represents the location of the bank? ☐

 Essential Question

HOW are integers and absolute value used in real-world situations?

Vocab **Vocabulary**

quadrants

Content Standards
6.NS.6, 6.NS.6b, 6.NS.6c

MP **Mathematical Practices**
1, 3, 4, 5, 7

Which MP Mathematical Practices did you use?
Shade the circle(s) that applies.

① Persevere with Problems
② Reason Abstractly
③ Construct an Argument
④ Model with Mathematics

⑤ Use Math Tools
⑥ Attend to Precision
⑦ Make Use of Structure
⑧ Use Repeated Reasoning

Identify Points and Ordered Pairs

A coordinate plane is formed when the *x*-axis and *y*-axis intersect at their zero points. The axes separate the coordinate plane into four regions called **quadrants**.

You can use the location on the plane or use the *x*-coordinates and *y*-coordinates to identify the quadrant in which a point is located.

Quadrant	*x*-coordinate	*y*-coordinate	Example
I	positive	positive	(2, 5)
II	negative	positive	(−2, 5)
III	negative	negative	(−2, −5)
IV	positive	negative	(2, −5)

Examples

1. Identify the ordered pair that names point *C*. Then identify the quadrant in which it is located.

Step 1 Start at the origin. Move right on the *x*-axis. The *x*-coordinate of point *C* is $1\frac{1}{2}$.

Step 2 Move up the *y*-axis. The *y*-coordinate is 1.

Point *C* is located at $\left(1\frac{1}{2}, 1\right)$. Both coordinates are positive.

So, point *C* is in Quadrant I.

2. Identify the point located at $\left(-1\frac{1}{2}, -1\right)$. Then identify the quadrant in which it is located.

Step 1 Start at the origin. Move left on the *x*-axis. The *x*-coordinate is $-1\frac{1}{2}$.

Step 2 Move down the *y*-axis. The *y*-coordinate is −1.

Point *B* is located at $\left(-1\frac{1}{2}, -1\right)$. Both coordinates are negative.

So, point *B* is in Quadrant III.

Ordered Pairs

A point located on the *x*-axis will have a *y*-coordinate of 0. A point located on the *y*-axis will have an *x*-coordinate of 0. Points located on an axis are not in any quadrant.

Got it? Do these problems to find out.

a. Refer to Example 1. Identify the ordered pair that names point *A*. Then identify the quadrant in which it is located.

b. Refer to Example 1. Identify the point located at (1, −2). Then identify the quadrant in which it is located.

a. _____

b. _____

Reflections on the Coordinate Plane

You can use what you know about number lines and opposites to compare locations on the coordinate plane. Consider the number line and coordinate plane below. The number line shows that −4 and 4 are opposites.

The coordinate plane shows that the points (−4, 0) and (4, 0) are the same distance from the *y*-axis in opposite directions. So, they are *reflected* across the *y*-axis. Notice that the *y*-coordinates did not change and that the *x*-coordinates are opposites.

Example

Tutor

3. **Name the ordered pair that is a reflection of (−3, 2) across the *x*-axis.**

To reflect across the *x*-axis, keep the same *x*-coordinate, −3, and take the opposite of the *y*-coordinate. The opposite of +2 is −2.

So, (−3, 2) reflected across the *x*-axis is located at (−3, −2).

Got it? Do these problems to find out.

Name the ordered pair that is a reflection of each point across the *x*-axis.

c. (1, −4) d. (−2, 5) e. (−3, −1)

c. _____

d. _____

e. _____

Show your work.

Example

4. Kendall is building a square fence. She places fence posts at the locations indicated on the grid. What is the location of the post that reflects (−4, 4) across the *y*-axis?

To reflect across the *y*-axis, keep the same *y*-coordinate, 4.

The opposite of the *x*-coordinate, −4, is 4.

So, (−4, 4) reflected across the *y*-axis is (4, 4).

Got it? Do this problem to find out.

f. Kendall also placed a fence post at (−4, −4). What is the location of the post that reflects (−4, −4) across the *y*-axis?

Show your work.

f. _____

Guided Practice

Identify the ordered pair that names each point or the name of each point. Then identify the quadrant in which it is located. (Examples 1 and 2)

1. *T*

_____

2. $\left(-1\frac{1}{2}, 0\right)$

_____

3. $\left(-2, 2\frac{1}{2}\right)$

_____

4. Refer to the diagram of a school. (Examples 3 and 4)

a. What is located at the reflection of (−3, −4) across the *y*-axis. What are the coordinates of this location?

_____

b. What is located at the reflection of the science labs across the *x*-axis? What are the coordinates of this

location? _____

5. **Building on the Essential Question** How are number lines and the coordinate plane related?

_____

_____

_____

_____

Rate Yourself!

Are you ready to move on? Shade the section that applies.

YES   ?   NO

For more help, go online to access a Personal Tutor.

Tutor

Independent Practice

Go online for Step-by-Step Solutions

Identify the ordered pair that names each point. Then identify the quadrant in which it is located. (Example 1)

1. R

2. G

3. B

_____

_____

_____

4. T

5 C

6. A

_____

_____

_____

show your work.

Identify the name of each point. Then identify the quadrant in which it is located. (Example 2)

7 (−2.5, 1.5)

8. (1, 1.5)

9. (0.5, −2.5)

_____

_____

_____

10. (2, −0.5)

11. (−0.5, 0)

12. (−1, −1.5)

_____

_____

_____

13. **MP** **Use Math Tools** Refer to the map of Wonderland Park. (Examples 3 and 4)

a. What is located closest to the origin?

_____

b. Liza is standing at (2, 4). What is located at the reflection of (2, 4) across the x-axis? What are the coordinates of this location?

_____

c. What is located at the reflection of (3, 1) across the y-axis? What are the coordinates of this

location? _____

d. The Pipeline Plunge is reflected across the x-axis. What are the coordinates of its new

location? _____

14. **MP Identify Structure** Fill in the graphic organizer below. Consider the point (−3, 2).

Action	Result
opposite of −3	
point (−3, 2) reflected across the y-axis	
point (−3, 2) reflected across the x-axis	

 H.O.T. Problems Higher Order Thinking

MP Persevere with Problems Without graphing, identify the quadrant(s) for which each of the following statements is true for any point (*x*, *y*). Justify your response.

15. The *x*- and *y*-coordinates have the same sign.

_____

_____

16. The *x*- and *y*-coordinates have opposite signs. _____

_____

17. **MP Reason Inductively** Does the order of the numbers in an ordered pair matter when naming a point? Can that point be represented by more than

one ordered pair? _____

_____

_____

18. **MP Model with Mathematics** A parallelogram is graphed on a coordinate plane so that two points are in the first quadrant and two points are in the third quadrant. What are possible coordinates of the vertices of the parallelogram?

_____

Extra Practice

Identify the ordered pair that names each point. Then identify the quadrant in which it is located.

19. *U*

(1, 3); I

Both numbers are positive so it is in the first quadrant.

20. *D*

_____

21. *S*

_____

22. *P*

_____

23. *J*

_____

24. *M*

_____

Identify the name of each point. Then identify the quadrant in which it is located.

25. $\left(-1\frac{1}{2}, \frac{1}{2}\right)$

_____

26. $\left(1, 1\frac{1}{2}\right)$

_____

27. $\left(\frac{1}{2}, -1\right)$

_____

28. $\left(1\frac{1}{2}, 0\right)$

_____

29. $\left(-1\frac{1}{2}, -1\frac{1}{2}\right)$

_____

30. $\left(-1, 1\frac{1}{2}\right)$

_____

31. **MP** **Model with Mathematics** Luke is making a model of a park. He has the basketball court drawn on his model.

a. The swing set is located at the reflection of point *B* across the *x*-axis. What ordered pair describes the location of the swing set?

_____

b. The slide is located at the reflection of point *C* across the *x*-axis. What ordered pair describes the location of the slide?

_____

c. A water fountain is located at the reflection of point *D* across the *y*-axis. What ordered pair describes the location of the water fountain?

_____

32. Identify the ordered pair that names each point.
Then identify the quadrant in which it is located.

a. *A* [ ]

b. *B* [ ]

c. *C* [ ]

d. *D* [ ]

e. *E* [ ]

33. Write the ordered pair that repesents the reflection of
point *J* across the *y*-axis.

[ ]

Spiral Review

Represent the set of numbers as decimals on the number line. 4.NF.6

34. $\left\{5\frac{3}{10}, 5\frac{1}{10}, 5\right\}$

35. $\left\{3\frac{1}{10}, 2\frac{7}{10}, 2\frac{9}{10}\right\}$

36. Draw a line of symmetry on the figure shown. 4.G.3

37. The table shows how many magazines three co-workers sold in one month.
How many magazines did they sell in total? 4.NBT.4

_____

Name	Number of Magazines
Julie	12
Dion	0
Calvin	7

Graph on the Coordinate Plane

Real-World Link

Watch ▶ | Tools

Scavenger Hunt Maria hid the clues to a scavenger hunt for her hiking club. Use the map to show where she hid the clues. Identify the location of each clue.

Essential Question

HOW are integers and absolute value used in real-world situations?

Content Standards
6.NS.6, 6.NS.6b, 6.NS.6c, 6.NS.8

MP Mathematical Practices
1, 2, 3, 4, 7

1. The first clue is hidden near a tree. What ordered pair describes its location?

_____

2. Maria hid the next clue at a location that is reflected across the *y*-axis from the tree. Where is it hidden?

_____

3. She then walks 3 blocks east and 2 blocks north to place the next clue. Where is it hidden?

_____

4. The next clue is at a location that is reflected across the *x*-axis from Clue 3. Where is it hidden?

_____

5. Maria then hid the next clue under a rock by the lake. How many blocks east did she walk to the lake?

_____

6. The final clue tells the hikers to walk 5 blocks north and three blocks east to find the prize. What ordered pair describes the location of the prize? _____

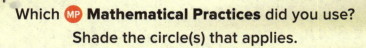

Which **MP** **Mathematical Practices did you use?**
Shade the circle(s) that applies.

① Persevere with Problems   ⑤ Use Math Tools

② Reason Abstractly   ⑥ Attend to Precision

③ Construct an Argument   ⑦ Make Use of Structure

④ Model with Mathematics   ⑧ Use Repeated Reasoning

Graph Ordered Pairs

To graph an ordered pair, draw a dot at the point that corresponds to the coordinates.

Examples

Tutor

1. Graph point *M* at (−3, 5).

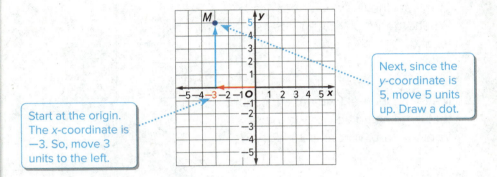

Next, since the *y*-coordinate is 5, move 5 units up. Draw a dot.

Start at the origin. The *x*-coordinate is −3. So, move 3 units to the left.

2. Graph point *N* at $\left(-2\frac{1}{2}, -3\frac{1}{2}\right)$.

The *x*-coordinate $-2\frac{1}{2}$ is between −2 and −3.

Start at the origin and move $2\frac{1}{2}$ units left.

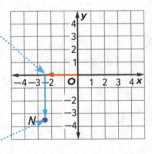

The *y*-coordinate $-3\frac{1}{2}$ is between −3 and −4.

Next, move $3\frac{1}{2}$ units down. Draw a dot.

Got it? Do these problems to find out.

Graph and label each point on the coordinate plane below.

a. *P*(−2, 4)

b. *Q*(0, −4)

c. $R\left(-\frac{1}{2}, -2\frac{1}{2}\right)$

d. *S*(4.5, 1)

Graph Reflections on the Coordinate Plane

You can graph points that are reflected across the *x*- and *y*-axes. Remember that points reflected across the *x*-axis will have the same *x*-coordinates and their *y*-coordinates will be opposites. Points reflected across the *y*-axis will have the same *y*-coordinates and their *x*-coordinates will be opposites.

Examples

3. **Graph A(2, −4). Then graph its reflection across the x-axis.**

Graph point *A*.

To reflect across the *x*-axis, keep the same *x*-coordinate, 2, and take the opposite of the *y*-coordinate.

The opposite of −4 is 4.

So, point *A* reflected across the *x*-axis is located at point *A′*(2, 4). Graph point *A′*.

4. **Graph B(−1.5, 3). Then graph its reflection across the y-axis.**

Graph point *B*.

To reflect across the *y*-axis, keep the same *y*-coordinate and take the opposite of the *x*-coordinate.

The opposite of −1.5 is 1.5.

So, point *B* reflected across the *y*-axis is point *B′*(1.5, 3).

Got it? **Do these problems to find out.**

e. Graph C(−1, 5). Then graph its reflection across the *x*-axis.

f. Graph $D\left(2, 3\frac{1}{2}\right)$. Then graph its reflection across the *y*-axis.

Example

5. Mr. Martin is using a coordinate plane to design a logo. He graphs points at (2, 4) and (2, −2). He reflects (2, −2) across the *y*-axis. Then he reflects the new point across the *x*-axis. What figure is Mr. Martin using for his logo?

Graph (2, 4) and (2, −2). (2, −2) reflected across the *y*-axis is (−2, −2).
Graph (−2, −2). (−2, −2) reflected across the *x*-axis is (−2, 2).
Graph (−2, 2).

So, the figure is a trapezoid.

Got it? Do this problem to find out.

g. _____

g. Ms. Shaull is drawing a map of the school. Her room is at (−3, 4) and the gym is at (3, 4). The library is a reflection of (3, 4) across the *x*-axis. This point is reflected across the *y*-axis to graph the office. What figure is graphed on the map?

Check ✓

Guided Practice

1. Use a coordinate plane to represent Jasmine's stone garden. Graph points $E(-1, -4)$ and $F\left(-3\frac{1}{2}, 4\right)$. Then reflect point *E* across the *y*-axis and point *F* across the *x*-axis. What is the shape of her stone garden? (Examples 1–5)

Show your work.

_____

2. **Building on the Essential Question** How can the coordinate plane be used to represent geometric figures?

_____

_____

Rate Yourself!

How confident are you about graphing on the coordinate plane? Check the box that applies.

□ □ □ □ □

For more help, go online to access a Personal Tutor.

Independent Practice

Go online for Step-by-Step Solutions

Graph and label each point on the coordinate plane to the right.
(Examples 1 and 2)

1. $T(0, 0)$

2. $D(2, 1)$

3. $K(-3.25, 3)$

4. $N\left(0, -1\frac{1}{2}\right)$

 $F(-4.5, 0)$

6. $A\left(-3\frac{1}{2}, -3\right)$

7. $L(2.5, -3.5)$

8. $S\left(4, 2\frac{1}{2}\right)$

9. Graph $U(3.5, -3)$ on the coordinate plane to the right.
Then graph its reflection across the x-axis. (Example 3)

10. Graph $B(-7, 6)$ on the coordinate plane on the right.
Then graph its reflection across the x-axis. (Example 3)

11. Graph $R(-2, 5)$ on the coordinate plane to the right.
Then graph its reflection across the y-axis. (Example 4)

12. Amelia is drawing a map of the park. She graphs the entrance at
$(2, -3)$. She reflects $(2, -3)$ across the y-axis. Then Amelia reflects
the new point across the x-axis. What figure is graphed on the map?
(Example 5)

_____

13. A point is reflected across the y-axis. The new point is located
at $(-4.25, -1.75)$. Write the ordered pair that represents the

original point. _____

14. **MP Model with Mathematics** A point is reflected across the x-axis.
The new point is $(-7.5, 6)$. What is the distance between the two points?

_____

15 On a coordinate plane, draw triangle *ABC* with vertices *A*(−1, −1), *B*(3, −1), and *C*(−1, 2). Find the area of the triangle in square units.

_____

16. The points (4, 3) and (−4, 0) are graphed on a coordinate plane. The point (4, 3) is reflected across the *x*- and *y*-axes. If all four points are connected, what figure is graphed?

_____

H.O.T. Problems Higher Order Thinking

17. **MP** **Identify Structure** Three vertices of a quadrilateral are (−1 −1), (1, 2), and (5, −1). What are the coordinates of two vertices that will form two different parallelograms? _____

MP **Persevere with Problems** Determine whether each statement is *sometimes*, *always*, or *never* true. Give an example or a counterexample.

18. When a point is reflected across the *y*-axis, the new point has a negative *x*-coordinate. _____

_____

_____

19. The point (*x*, *y*) is reflected across the *x*-axis. Then the new point is reflected across the *y*-axis. The location of the point after both reflections is (−*x*, −*y*). _____

_____

_____

20. The *x*-coordinate of a point that lies on the *x*-axis is negative.

_____

_____

_____

21. The *x*-coordinate of a point that lies on the *y*-axis is positive.

_____

_____

_____

Extra Practice

Graph and label each point on the coordinate plane to the right.

22. $B(-3, 4)$ *The x-coordinate is -3. The y-coordinate is 4.*

23. $D(-1.5, 2.5)$

24. $A\left(4\frac{3}{4}, -1\frac{1}{4}\right)$

25. $J\left(2\frac{1}{2}, -2\frac{1}{2}\right)$

26. $C(1, 4.5)$

27. $F(-4, -3.5)$

28. $G\left(3\frac{1}{2}, 3\right)$

29. $H\left(-3, -1\frac{1}{2}\right)$

30. Graph $N(1, -3)$ on the coordinate plane to the right. Then graph its reflection across the y-axis.

31. Graph $H(7, 8)$ on the coordinate plane on the right. Then graph its reflection across the x-axis.

32. Graph $F(-6, 5.5)$ on the coordinate plane to the right. Then graph its reflection across the x-axis.

33. Marcus is drawing a plan for his vegetable garden. He graphs one corner at $(-7.5, 2)$ and one corner at $(7.5, 2)$. He reflects $(-7.5, 2)$ across the x-axis. Then Marcus reflects the new point across the y-axis. What shape is the vegetable garden?

_____

34. A point is reflected across the x-axis. The new point is located at $(4.75, -2.25)$. Write the ordered pair that represents the original point.

_____

35. **MP** **Model with Mathematics** A point is reflected across the x-axis. The new point is $(5, -3.5)$. What is the distance between the two points?

_____

36. Draw triangle *XYZ* with vertices *X*(−3, 2), *Y*(4, −2), and *Z*(−3, −2) on the coordinate plane. Then find the area of the triangle in square units.

37. What are the coordinates of point *H* after it is reflected across the *x*-axis, and then reflected across the *y*-axis?

Spiral Review

Multiply. 4.NBT.5

38. $1 \times 1 \times 1 =$ _____

39. $3 \times 3 \times 3 =$ _____

40. $6 \times 6 \times 6 =$ _____

41. Use the geometric pattern below to find the number of unit squares in the next figure. 4.OA.5

42. Alexa saved a total of $210. Each week she saved the same amount of money. She has been saving for 7 weeks. How much money did Alexa save each week? 5.NBT.6

Inquiry Lab
Find Distance on the Coordinate Plane

 Inquiry **WHAT is the relationship between coordinates and distance?**

Content Standards
6.NS.8

MP **Mathematical Practices**
1, 3, 4

Taylor's house and school are each shown on the map. What is the distance between the two points?

What do you know? _____

_____

_____

What do you need to find? _____

_____

_____

Hands-On Activity 1

Find the distance between Taylor's house and the school.

Step 1   Find the coordinates of Taylor's house and the school.

Taylor's House: _____   School: _____

Step 2   Write the *x*-coordinates for each location in the table.

Location	x-coordinate
house	
school	

Step 3   Find the difference between the *x*-coordinates.

_____

Step 4   Draw a line connecting the points in the coordinate plane above. How many units are between the points? _____

How does this compare to your answer from Step 3?

_____

So, there are ☐ units between Taylor's house and the school.

Hands-On Activity 2

Find the distance between point *A* and point *B* on the coordinate plane.

Step 1 Find the coordinates of each point.

Point *A*: _____   Point *B*: _____

Step 2 Count the number of units between each *y*-coordinate and the *x*-axis.

Point	*y*-coordinate	Distance from *x*-axis
A		
B		

Step 3 To find the distance between the two points, add the distance from the *x*-axis to each point.

☐ + ☐ = ☐

Step 4 Draw a line connecting the points. How many

units are between the points? _____

How does this compare to your answer from Step 3?

_____

So, the distance between point *A* and point *B* is ☐ units.

Investigate

Collaborate

MP **Model with Mathematics** Work with a partner. Draw a line between each pair of points. Then find the distance between the points.

1. _____

2. _____

Investigate

Collaborate

MP **Model with Mathematics** Work with a partner. Plot each pair of points on the coordinate plane. Then find the distance between the points.

3. $C(-2, -4)$, $D(-2, -1)$ _____

 Show your work.

4. $E(-3, -1)$, $F(1, -1)$ _____

5. $G(1, -2)$, $H(3, -2)$ _____

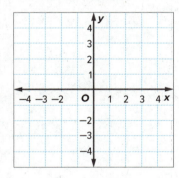

6. $K(3, -4)$, $L(3, 2)$ _____

7. $M(4, 1)$, $N(-1, 1)$ _____

8. $P(4, 4)$, $Q(4, 1)$ _____

Analyze and Reflect

Collaborate

With a partner to complete the table below. Use your answers from Exercises 3–6. The first one is done for you.

Exercise	Coordinates Used	Horizontal or Vertical Line?	Same or Different Quadrant?	Line Length
2	2 and −2	vertical	different	4 units
9.	3	and			
10.	4	and			
11.	5	and			
12.	6	and			

13. Compare your answers from Exercises 10 and 11. What is the relationship between the coordinates used and the length of each line?

_____

_____

_____

14. Name the coordinates of two points that have the same *x*-coordinates and are 8 units apart. _____

15. **MP** **Reason Inductively** Use absolute value to write a rule for determining the distance between two points on a coordinate plane that have the same *x*-coordinate. _____

_____

_____

Create

On Your Own

16. **MP** **Model with Mathematics** Write and solve a real-world problem that involves determining distance on a coordinate plane. _____

_____

_____

17. **Inquiry** WHAT is the relationship between coordinates and distance?

_____

_____

21ST CENTURY CAREER
in Art

Scientific Illustrator

If you are artistic and have a strong interest in science, you should think about a career as a scientific illustrator. Scientific illustrators combine their artistic abilities with their scientific backgrounds to draw scientifically accurate images. Karen Carr, a wildlife and natural history artist, has artwork in scientific publications, museums, and zoos. To draw animals that are extinct, she examines fossils, talks to scientists, and uses measurements and proportions from scientific literature.

College & Career READINESS

Is This the Career for You?

Are you interested in a career as a scientific illustrator? Take some of the following courses in high school.

- ◆ Algebra
- ◆ Biology
- ◆ Geometry
- ◆ Life/Figure Drawing
- ◆ Physics

Find out how math relates to a career in Art.

You be the Scientific Illustrator!

Use the information in the table to solve each problem. Write in simplest form.

1. Write the length and height of an Argentinosaurus as decimals. Use bar notation if necessary.

   _____

2. How much taller was a Velociraptor than a Microraptor? Write your answer as a decimal. _____

3. Which is greater, the height of the Argentinosaurus or the length of the Camptosaurus? _____

4. How much longer was a Camptosaurus than a Velociraptor? Plot your answer on the number line.

   0 1 2 3 4 5 6 7 8 9 10 11 12

5. Compare the heights of all four dinosaurs. Order them from least to greatest.

   _____

6. An artist is creating a mural in which a Microraptor is $1\frac{1}{2}$ times the actual size. What is the length of the dinosaur in the mural? _____

Dinosaur Measurements		
Dinosaur	**Length (ft)**	**Height (ft)**
Argentinosaurus	$114\frac{5}{6}$	$24\frac{1}{10}$
Camptosaurus	$16\frac{2}{5}$	$11\frac{4}{5}$
Microraptor	$2\frac{5}{8}$	$\frac{24}{25}$
Velociraptor	$5\frac{9}{10}$	$3\frac{7}{25}$

Career Project

It's time to update your career portfolio! Investigate the education and training requirements for a career as a scientific illustrator.

_____

_____

_____

_____

_____

What are some short-term goals you need to achieve to become a scientific illustrator?

- _____
- _____
- _____
- _____

Chapter Review

Vocabulary Check

Complete the puzzle by unscrambling the letters below to reveal words from the vocabulary list at the beginning of the chapter.

Complete each sentence using the vocabulary list at the beginning of the chapter.

1. A _____ is a number that can be written as a fraction.

2. A number that is less than zero is a _____.

3. A number that is greater than zero is a _____.

4. The _____ of a number is the distance between the number and zero on a number line.

5. The division of a _____ ends.

6. A decimal whose digits repeat in groups of one or more is

   a _____.

Use Your FOLDABLES

Use your Foldable to help review the chapter.

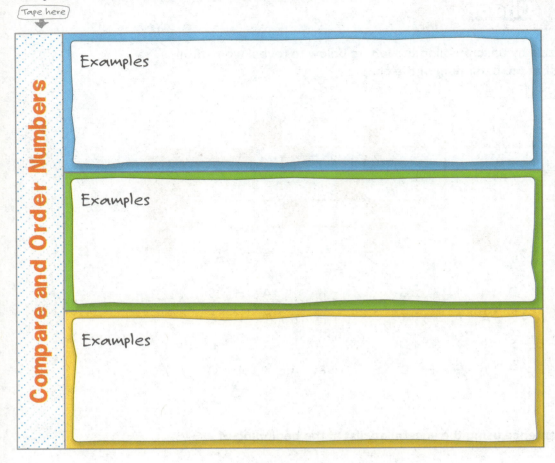

Tape here

Compare and Order Numbers

Examples

Examples

Examples

Got it?

Circle the correct term or number to complete each sentence.

1. The opposite of −4 is (−4, 4).

2. The distance of a number from 0 is its (opposite, absolute value).

3. The value listed first in an ordered pair is the (*x*-coordinate, *y*-coordinate).

4. The absolute value of 17 is (−17, 17).

5. (1.$\overline{25}$, 6.543) is a terminating decimal.

Power Up! Performance Task

City Grid

Maps are often represented on coordinate planes like the one shown below. Each unit on the grid represents one mile. Example: A building located at (0, 2) is 2 miles away from a building located at (0, 4).

Write your answers on another piece of paper. Show all of your work to receive full credit.

Part A

There are six buildings that need to be plotted on a city grid: City Hall (0, 1), Bobcat Stadium (3, 2), McKinley High School (−4, 1), Westfield Mall (4, −3), Roberts Middle School (−4, 3), and Grover's Mill (−3, −4). Place a point on the graph to represent each building and label it with the correct building name.

Part B

Which building is farther to the west: Roberts Middle School or the Grover's Mill? Explain your answer in words and by writing an inequality using the proper coordinates and either the greater than (<) or less than (>) sign.

Part C

All roads in the city travel either north and south or east and west. How many miles must the city's residents travel to go from the Bobcat Stadium to Grover's Mill? Explain. Your explanation should include absolute value.

Reflect

 Answering the Essential Question

Use what you learned about integers and the coordinate plane to complete the graphic organizer.

Essential Question

HOW are integers and absolute value used in real-world situations?

Vocabulary	Definition
integer	
absolute value	

Describe a real-world situation that can be represented by the absolute value of 27.

Describe a real-world situation that can be represented by the absolute value of −16?

Answer the Essential Question. HOW are integers and absolute value used in real-world situations?

_____

_____

UNIT PROJECT

Watch ▶

Get Out the Map! If you could go anywhere in the world, where would you go? This is your chance to explore someplace new. In this project you will:

- **Collaborate** with your classmates as you investigate a new place you would like to travel.
- **Share** the results of your research in a creative way.
- ℯ **Reflect** on how mathematical ideas can be represented.

By the end of this project, you will have a better understanding of how to use a map to plan the perfect adventure!

Collaborate

⏻ **Go Online** Work with your group to research and complete each activity. You will use your results in the Share section on the following page.

1. Think of someplace new that you would like to travel. Investigate the population of the area you choose as well as any interesting geographical features. Make a list of area attractions that you'd like to visit, such as museums, amusement parks, historical sites, national parks, and so on.

2. Research information about the climate for the area of your location in the time of year you are planning to travel. Important information could include the monthly rainfall, the average daily high and low temperatures, humidity levels, and wind speeds. Create a visual display to share your results.

3. Find lodging in the location you chose. Then, using maps and descriptions available on the Internet, plot three or four area attractions on a coordinate plane. Label each point.

4. Create a budget for each day of the trip showing the cost of the travel, lodging, daily activities, and food. Calculate the cost for your entire family, not just yourself.

5. Use an online map or GPS device to determine the actual distances you will need to travel to get from where you are staying to any attraction you plan to visit. Then find the total distance you will travel during the entire trip.

With your group, decide on a way to present what you have learned from each of the activities about the planning the perfect adventure. Some suggestions are listed below, but you can also think of other creative ways to present your information. Remember to show how you used math to complete each of the activities in this project!

- Create a travel brochure for your current location. Your objective is to increase the tourists for your town. The brochure should include each of the following: a detailed map, recommended restaurants, area attractions, and fun facts.
- Write a journal entry from the perspective of an early explorer who, 1,000 years ago, traveled to the location you chose. Then describe how technology makes planning, budgeting, and navigating at the same location much easier today.

Check out the note on the right to connect this project with other subjects.

 connect **with Science**

Use the Internet to research what technology was used by early explorers to navigate through unknown territories. Some questions to consider are:

- What tools were used to help explorers travel in the right direction?

- What constellations were used by the explorers and how were they used?

6. **ⓔ** **Answer the Essential Question** How can mathematical ideas be represented?

a. How did you use what you learned about computing with multi-digit numbers, and multiplying and dividing fractions to represent mathematical ideas in this project?

_____

_____

_____

b. How did you use what you learned about integers and the coordinate plane to represent mathematical ideas in this project?

_____

_____

_____

Go online for the eGlossary.

The eGlossary contains words and definitions in the following 13 languages:

Arabic	Cantonese	Hmong	Spanish	Urdu
Bengali	English	Korean	Tagalog	Vietnamese
Brazilian Portuguese	Haitian Creole	Russian		

English

Español

absolute value   The distance between a number and zero on a number line.

valor absoluto   Distancia entre un número y cero en la recta numérica.

acute angle   An angle with a measure greater than 0° and less than 90°.

ángulo agudo   Ángulo que mide más de 0° y menos de 90°.

acute triangle   A triangle having three acute angles.

triángulo acutángulo   Triángulo con tres ángulos agudos.

Addition Property of Equality   If you add the same number to each side of an equation, the two sides remain equal.

propiedad de adición de la igualdad   Si sumas el mismo número a ambos lados de una ecuación, los dos lados permanecen iguales.

algebra   A mathematical language of symbols, including variables.

álgebra   Lenguaje matemático que usa símbolos, incluyendo variables.

algebraic expression   A combination of variables, numbers, and at least one operation.

expresión algebraica   Combinación de variables, números y, por lo menos, una operación.

analyze   To use observations to describe and compare data.

analizar   Usar observaciones para describir y comparar datos.

angle   Two rays with a common endpoint form an angle. The rays and vertex are used to name the angle.

ángulo   Dos rayos con un extremo común forman un ángulo. Los rayos y el vértice se usan para nombrar el ángulo.

$\angle ABC$, $\angle CBA$, or $\angle B$

$\angle ABC$, $\angle CBA$ o $\angle B$

arithmetic sequence   A sequence in which the difference between any two consecutive terms is the same.

sucesión aritmética   Sucesión en la cual la diferencia entre dos términos consecutivos es constante.

Associative Property   The way in which numbers are grouped does not change the sum or product.

propiedad asociativa   La forma en que se agrupan tres números al sumarlos o multiplicarlos no altera su suma o producto.

average   The sum of two or more quantities divided by the number of quantities; the mean.

promedio   La suma de dos o más cantidades dividida entre el número de cantidades; la media.

Bb

bar notation   A bar placed over digits that repeat to indicate a number pattern that repeats indefinitely.

notación de barra   Barra que se coloca sobre los dígitos que se repiten para indicar el número de patrones que se repiten indefinidamente.

base   Any side of a parallelogram.

base   Cualquier lado de un paralelogramo.

base

base

base   One of the two parallel congruent faces of a prism.

base   Una de las dos caras paralelas congruentes de un prisma.

base
base

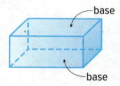

base
base

base   In a power, the number used as a factor. In 10^3, the base is 10. That is, $10^3 = 10 \times 10 \times 10$.

base   En una potencia, el número usado como factor. En 10^3, la base es 10. Es decir, $10^3 = 10 \times 10 \times 10$.

box plot   A diagram that is constructed using five values.

diagrama de caja   Diagrama que se construye usando cinco valores.

0 1 2 3 4 5 6 7 8 9 10

0 1 2 3 4 5 6 7 8 9 10

Cc

center   The given point from which all points on a circle are the same distance.

centro   Un punto dado del cual equidistan todos los puntos de un círculo o de una esfera.

circle   The set of all points in a plane that are the same distance from a given point called the center.

círculo   Conjunto de todos los puntos en un plano que equidistan de un punto dado llamado centro.

circle graph   A graph that shows data as parts of a whole. In a circle graph, the percents add up to 100.

Area of Oceans

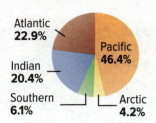

gráfica circular   Gráfica que muestra los datos como partes de un todo. En una gráfica circular los porcentajes suman 100.

Área de Superficie de los Océanos

circumference   The distance around a circle.

circumference

circunferencia   La distancia alrededor de un círculo.

circunferencia

cluster   Data that are grouped closely together.

agrupamiento   Conjunto de datos que se agrupan.

coefficient   The numerical factor of a term that contains a variable.

coeficiente   El factor numérico de un término que contiene una variable.

Commutative Property   The order in which numbers are added or multiplied does not change the sum or product.

propiedad commutativa   La forma en que se suman o multiplican dos números no altera su suma o producto.

compatible numbers   Numbers that are easy to use to perform computations mentally.

números compatibles   Números que son fáciles de usar para realizar computations mentales.

complementary angles   Two angles are complementary if the sum of their measures is 90°.

∠1 and ∠2 are complementary angles.

ángulos complementarios   Dos ángulos son complementarios si la suma de sus medidas es 90°.

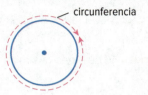

∠1 y ∠2 son complementarios.

composite figure   A figure made of triangles, quadrilaterals, semicircles, and other two-dimensional figures.

figura compuesta   Figura formada por triángulos, cuadriláteros, semicírculos y otras figuras bidimensionales.

congruent   Having the same measure.

congruente   Ques tienen la misma medida.

congruent figures Figures that have the same size and same shape; corresponding sides and angles have equal measures.

constant A term without a variable.

coordinate plane A plane in which a horizontal number line and a vertical number line intersect at their zero points.

corresponding sides The sides of similar figures that "match."

cubic units Used to measure volume. Tells the number of cubes of a given size it will take to fill a three-dimensional figure.

3 cubic units

figuras congruentes Figuras que tienen el mismo tamaño y la misma forma; los lados y los ángulos correspondientes con igual medida.

constante Un término sin una variable.

plano de coordenadas Plano en que una recta numérica horizontal y una recta numérica vertical se intersecan en sus puntos cero.

lados correspondientes Lados de figuras semejantes que coinciden.

unidades cúbicas Se usan para medir el volumen. Indican el número de cubos de cierto tamaño que se necesitan para llenar una figura tridimensional.

3 unidades cúbicas

Dd

data Information, often numerical, which is gathered for statistical purposes.

decagon A polygon having ten sides.

defining the variable Choosing a variable and deciding what the variable represents.

dependent variable The variable in a relation with a value that depends on the value of the independent variable.

diameter The distance across a circle through its center.

diameter

datos Información, con frecuencia numérica, que se recoge con fines estadísticos.

decágono Un polígono con diez lados.

definir la variable Elegir una variable y decidir lo que representa.

variable dependiente La variable en una relación cuyo valor depende del valor de la variable independiente.

diámetro La distancia a través de un círculo pasando por el centro.

diámetro

dimensional analysis   The process of including units of measurement when you compute.

análisis dimensional   Proceso que incluye las unidades de medida al hacer cálculos.

distribution   The arrangement of data values.

distributión   El arreglo de valores de datos.

Distributive Property   To multiply a sum by a number, multiply each addend by the number outside the parentheses.

propiedad distributiva   Para multiplicar una suma por un número, multiplica cada sumando por el número fuera de los paréntesis.

Division Property of Equality   If you divide each side of an equation by the same nonzero number, the two sides remain equal.

propiedad de igualdad de la división   Si divides ambos lados de una ecuación entre el mismo número no nulo, los lados permanecen iguales.

dot plot   A diagram that shows the frequency of data on a number line. Also known as a line plot.

diagrama de puntos   Diagrama que muestra la frecuencia de los datos sobre una recta numérica.

Ee

equals sign   A symbol of equality, =.

signo de igualdad   Símbolo que indica igualdad, =.

equation   A mathematical sentence showing two expressions are equal. An equation contains an equals sign, =.

ecuación   Enunciado matemático que muestra que dos expresiones son iguales. Una ecuación contiene el signo de igualdad, =.

equilateral triangle   A triangle having three congruent sides.

triángulo equilátero   Triángulo con tres lados congruentes.

equivalent expressions   Expressions that have the same value.

expresiones equivalentes   Expresiones que poseen el mismo valor, sin importer los valores de la(s) variable(s).

equivalent ratios   Ratios that express the same relationship between two quantities.

razones equivalentes   Razones que expresan la misma relación entre dos cantidades.

evaluate   To find the value of an algebraic expression by replacing variables with numbers.

evaluar   Calcular el valor de una expresión sustituyendo las variables por número.

exponent   In a power, the number that tells how many times the base is used as a factor. In 5^3, the exponent is 3. That is, $5^3 = 5 \times 5 \times 5$.

exponente   En una potencia, el número que indica las veces que la base se usa como factor. En 5^3, el exponente es 3. Es decir, $5^3 = 5 \times 5 \times 5$.

Ff

face   A flat surface.

cara   Una superficie plana.

factor the expression   The process of writing numeric or algebraic expressions as a product of their factors.

factorizar la expresión   El proceso de escribir expresiones numéricas o algebraicas como el producto de sus factores.

first quartile   For a data set with median M, the first quartile is the median of the data values less than M.

primer cuartil   Para un conjunto de datos con la mediana M, el primer cuartil es la mediana de los valores menores que M.

formula An equation that shows the relationship among certain quantities.

fórmula Ecuación que muestra la relación entre ciertas cantidades.

fraction A number that represents part of a whole or part of a set.

$$\frac{1}{2}, \frac{1}{3}, \frac{1}{4}, \frac{3}{4}$$

fracción Número que representa parte de un todo o parte de un conjunto.

$$\frac{1}{2}, \frac{1}{3}, \frac{1}{4}, \frac{3}{4}$$

frequency distribution How many pieces of data are in each interval.

distribución de frecuencias Cantidad de datos asociada con cada intervalo.

frequency table A table that shows the number of pieces of data that fall within the given intervals.

tabla de frecuencias Tabla que muestra el número de datos en cada intervalo.

function A relationship that assigns exactly one output value to one input value.

función Relación que asigna exactamente un valor de salida a un valor de entrada.

function rule An expression that describes the relationship between each input and output.

regla de funciones Expresión que describe la relación entre cada valor de entrada y de salida.

function table A table organizing the input, rule, and output of a function.

tabla de funciones Tabla que organiza las entradas, la regla y las salidas de una función.

Gg

gap An empty space or interval in a set of data.

laguna Espacio o intervalo vacío en un conjunto de datos.

geometric sequence A sequence in which each term is found by multiplying the previous term by the same number.

sucesión geométrica Sucesión en la cual cada término después del primero se determina multiplicando el término anterior por el mismo número.

graph To place a dot at a point named by an ordered pair.

graficar Colocar una marca puntual en el punto que corresponde a un par ordenado.

Greatest Common Factor (GCF) The greatest of the common factors of two or more numbers.

The greatest common factor of 12, 18, and 30 is 6.

máximo común divisor (MCD) El mayor de los factores comunes de dos o más números.

El máximo común divisor de 12, 18 y 30 es 6.

Hh

height The shortest distance from the base of a parallelogram to its opposite side.

altura La distancia más corta desde la base de un paralelogramo hasta su lado opuesto.

height

altura

heptagon A polygon having seven sides.

heptágono Polígono con siete lados.

hexagon   A polygon having six sides.

histogram   A type of bar graph used to display numerical data that have been organized into equal intervals.

hexágono   Polígono con seis lados.

histograma   Tipo de gráfica de barras que se usa para exhibir datos que se han organizado en intervalos iguales.

 Ii

Identity Properties   Properties that state that the sum of any number and 0 equals the number and that the product of any number and 1 equals the number.

independent variable   The variable in a function with a value that is subject to choice.

inequality   A mathematical sentence indicating that two quantities are not equal.

integer   Any number from the set $\{\ldots -4, -3, -2, -1, 0, 1, 2, 3, 4 \ldots\}$ where ... means *continues without end.*

interquartile range   A measure of variation in a set of numerical data, the interquartile range is the distance between the first and third quartiles of the data set.

intersecting lines   *Line*s that meet or cross at a common *point.*

interval   The difference between successive values on a scale.

inverse operations   Operations which *undo* each other. For example, addition and subtraction are inverse operations.

isosceles triangle   A triangle having at least two congruent sides.

propiedades de identidad   Propiedades que establecen que la suma de cualquier número y 0 es igual al número y que el producto de cualquier número y 1 es igual al número.

variable independiente   Variable en una función cuyo valor está sujeto a elección.

desigualdad   Enunciado matemático que indica que dos cantidades no son iguales.

entero   Cualquier número del conjunto $\{\ldots -4, -3, -2, -1, 0, 1, 2, 3, 4 \ldots\}$ donde ... significa que *continúa sin fin.*

rango intercuartil   El rango intercuartil, una medida de la variación en un conjunto de datos numéricos, es la distancia entre el primer y el tercer cuartil del conjunto de datos.

rectas secantes   *Rectas* que se intersectan o se cruzan en un *punto* común.

intervalo   La diferencia entre valores sucesivos de una escala.

operaciones inversas   Operaciones que *se anulan* mutuamente. La adición y la sustracción son operaciones inversas.

triángulo isósceles   Triángulo que tiene por lo menos dos lados congruentes.

lateral face   Any face that is not a base.

cara lateral   Cualquier superficie plana que no sea la base.

least common denominator (LCD)   The least common multiple of the denominators of two or more fractions.

mínimo común denominador (mcd)   El menor múltiplo común de los denominadores de dos o más fracciones.

least common multiple (LCM)   The smallest whole number greater than 0 that is a common multiple of each of two or more numbers.

The LCM of 2 and 3 is 6.

mínimo común múltiplo (mcm)   El menor número entero, mayor que 0, múltiplo común de dos o más números.

El mcm de 2 y 3 es 6.

leaves   The digits of the least place value of data in a stem-and-leaf plot.

hoja   En un diagrama de tallo y hojas, los dígitos del menor valor de posición.

like terms   Terms that contain the same variable(s) to the same power.

términos semejantes   Términos que contienen la misma variable o variables elevadas a la misma potencia.

line   A set of *points* that form a straight path that goes on forever in opposite directions.

recta   Conjunto de *puntos* que forman una trayectoria recta sin fin en direcciones oputestas.

linear function   A function that forms a line when graphed.

función lineal   Función cuya gráfica es una recta.

line graph   A graph used to show how a set of data changes over a period of time.

gráfica lineal   Gráfica que se use para mostrar cómo cambian los valores durange un período de tiempo.

line of symmetry   A line that divides a figure into two halves that are reflections of each other.

eje de simetría   Recta que divide una figura en dos mitades especulares.

line of symmetry

eje de simetría

line plot   A diagram that shows the frequency of data on a number line. Also known as a dot plot.

esquema lineal   Diagrama que muestra la frecuencia de los datos sobre una recta numérica.

line segment   A part of a *line* that connects two points.

segmento de recta   Parte de una *recta* que conecta dos puntos.

line symmetry   Figures that match exactly when folded in half have line symmetry.

simetría lineal   Exhiben simetría lineal las figuras que coinciden exactamente al doblarse una sobre otra.

Mm

mean   The sum of the numbers in a set of data divided by the number of pieces of data.

mean absolute deviation   A measure of variation in a set of numerical data, computed by adding the distances between each data value and the mean, then dividing by the number of data values.

measures of center   Numbers that are used to describe the center of a set of data. These measures include the mean, median, and mode.

measures of variation   A measure used to describe the distribution of data.

median   A measure of center in a set of numerical data. The median of a list of values is the value appearing at the center of a sorted version of the list—or the mean of the two central values, if the list contains an even number of values.

mode   The number(s) or item(s) that appear most often in a set of data.

Multiplication Property of Equality   If you multiply each side of an equation by the same nonzero number, the two sides remain equal.

media   La suma de los números en un conjunto de datos dividida entre el número total de datos.

desviación media absoluta   Una medida de variación en un conjunto de datos numéricos que se calcula sumando las distancias entre el valor de cada dato y la media, y luego dividiendo entre el número de valores.

medidas del centro   Numéros que se usan para describir el centro de un conjunto de datos. Estas medidas incluyen la media, la mediana y la moda.

medidas de variación   Medida usada para describir la distribución de los datos.

mediana   Una medida del centro en un conjunto de datos numéricos. La mediana de una lista de valores es el valor que aparece en el centro de una versión ordenada de la lista, o la media de los dos valores centrales si la lista contiene un número par de valores.

moda   Número(s) de un conjunto de datos que aparece(n) más frecuentemente.

propiedad de multiplicación de la igualdad   Si multiplicas ambos lados de una ecuación por el mismo número no nulo, lo lados permanecen iguales.

Nn

negative integer   A number that is less than zero. It is written with a − sign.

net   A two-dimensional figure that can be used to build a three-dimensional figure.

nonagon   A polygon having nine sides.

numerical expression   A combination of numbers and operations.

entero negativo   Número que es menor que cero y se escribe con el signo −.

red   Figura bidimensional que sirve para hacer una figura tridimensional.

enágono   Polígono que tiene nueve lados.

expresión numérica   Una combinación de números y operaciones.

obtuse angle   Any angle that measures greater than 90° but less than 180°.

ángulo obtuso   Cualquier ángulo que mide más de 90° pero menos de 180°.

obtuse triangle   A triangle having one obtuse angle.

triángulo obtusángulo   Triángulo que tiene un ángulo obtuso.

octagon   A polygon having eight sides.

octágono   Polígono que tiene ocho lados.

opposites   Two integers are opposites if they are represented on the number line by points that are the same distance from zero, but on opposite sides of zero. The sum of two opposites is zero.

opuestos   Dos enteros son opuestos si, en la recta numérica, están representados por puntos que equidistan de cero, pero en direcciones opuestas. La suma de dos opuestos es cero.

ordered pair   A pair of numbers used to locate a point on the coordinate plane. The ordered pair is written in the form (*x*-coordinate, *y*-coordinate).

par ordenado   Par de números que se utiliza para ubicar un punto en un plano de coordenadas. Se escribe de la forma (coordenada *x*, coordenada *y*).

order of operations   The rules that tell which operation to perform first when more than one operation is used.

1. Simplify the expressions inside grouping symbols, like parentheses.
2. Find the value of all powers.
3. Multiply and divide in order from left to right.
4. Add and subtract in order from left to right.

orden de las operaciones   Reglas que establecen cuál operación debes realizar primero, cuando hay más de una operación involucrada.

1. Primero ejecuta todas las operaciones dentro de los símbolos de agrupamiento.
2. Evalúa todas las potencias.
3. Multiplica y divide en orden de izquierda a derecha.
4. Suma y resta en orden de izquierda a derecha.

origin The point of intersection of the *x*-axis and *y*-axis on a coordinate plane.

origen Punto de intersección de los ejes axiales en un plano de coordenadas.

outlier A value that is much greater than or much less than than the other values in a set of data.

valor atípico Dato que se encuentra muy separado de los otros valores en un conjunto de datos.

Pp

parallel lines Lines in a plane that never intersect.

rectas paralelas Rectas en un plano que nunca se intersecan.

parallelogram A quadrilateral with opposite sides parallel and opposite sides congruent.

paralelogramo Cuadrilátero cuyos lados opuestos son paralelos y congruentes.

peak The most frequently occurring value in a line plot.

pico El valor que ocurre con más frecuencia en un diagrama de puntos.

pentagon A polygon having five sides.

pentágono Polígono que tiene cinco lados.

percent A ratio that compares a number to 100.

por ciento Razón en que se compara un número a 100.

percent proportion One ratio or fraction that compares part of a quantity to the whole quantity. The other ratio is the equivalent percent written as a fraction with a denominator of 100.

$$\frac{\text{part}}{\text{whole}} = \frac{\text{percent}}{100}$$

proporción porcentual Razón o fracción que compara parte de una cantidad a toda la cantidad. La otra razón es el porcentaje equivalente escrito como fracción con 100 de denominador.

$$\frac{\text{parte}}{\text{todo}} = \frac{\text{porcentaje}}{100}$$

perfect square Numbers with square roots that are whole numbers. 25 is a perfect square because the square root of 25 is 5.

cuadrados perfectos Números cuya raíz cuadrada es un número entero. 25 es un cuadrado perfecto porque la raíz cuadrada de 25 es 5.

perimeter   The distance around a figure.

$P = 3 + 4 + 5 = 12$ units

perímetro   La distancia alrededor de una figura.

$P = 3 + 4 + 5 = 12$ unidades

pi   The ratio of the circumference of a circle to its diameter. The Greek letter π represents this number. The value of pi is always 3.1415926....

pi   Razón de la circunferencia de un círculo al diámetro del mismo. La letra griega π representa este número. El valor de pi es siempre 3.1415926....

plane   A flat surface that goes on forever in all directions.

plano   Superficie plana que se extiende infinitamente en todas direcciones.

point   An exact location in space that is represented by a dot.

punto   Ubicación exacta en el espacio que se representa con un marca puntual.

polygon   A simple closed figure formed by three or more straight line segments.

polígono   Figura cerrada simple formada por tres o más segmentos de recta.

population   The entire group of items or individuals from which the samples under consideration are taken.

población   El grupo total de individuos o de artículos del cual se toman las muestras bajo estudio.

positive integer   A number that is greater than zero. It can be written with or without a + sign.

entero positivo   Número que es mayor que cero y se puede escribir con o sin el signo +.

powers   Numbers expressed using exponents. The power 3^2 is read *three to the second power,* or *three squared.*

potencias   Números que se expresan usando exponentes. La potencia 3^2 se lee *tres a la segunda potencia* o *tres al cuadrado.*

prism   A three-dimensional figure with at least three rectangular lateral faces and top and bottom faces parallel.

prisma   Figura tridimensional que tiene por lo menos tres caras laterales rectangulares y caras paralelas superior e inferior.

properties   Statements that are true for any number.

propiedades   Enunciados que son verdaderos para cualquier número.

proportion   An equation stating that two ratios or rates are equivalent.

proporción   Ecuación que indica que dos razones o tasas son equivalentes.

pyramid   A three-dimensional figure with at least three triangular sides that meet at a common vertex and only one base that is a polygon.

pirámide   Una figura de tres dimensiones con que es en un un polígono y tres o mas caras triangulares que se encuentran en un vértice común.

Qq

quadrants The four regions in a coordinate plane separated by the x-axis and y-axis.

quadrilateral A closed figure having four sides and four angles.

quartiles Values that divide a data set into four equal parts.

cuadrantes Las cuatro regiones de un plano de coordenadas separadas por el eje x y el eje y.

cuadrilátero Figura cerrada que tiene cuatro lados y cuatro ángulos.

cuartiles Valores que dividen un conjunto de datos en cuatro partes iguales.

Rr

radical sign The symbol used to indicate a nonnegative square root, $\sqrt{\ }$.

radius The distance from the center to any point on the circle.

range The difference between the greatest number and the least number in a set of data.

rate A ratio comparing two quantities with different kinds of units.

rate of change A rate that describes how one quantity changes in relation to another. A rate of change is usually expressed as a unit rate.

ratio A comparison of two quantities by division. The ratio of 2 to 3 can be stated as 2 out of 3, 2 to 3, 2 : 3, or $\frac{2}{3}$.

rational number A number that can be written as a fraction.

ratio table A table with columns filled with pairs of numbers that have the same ratio.

ray A line that has one endpoint and goes on forever in only one direction.

reciprocals Any two numbers that have a product of 1. Since $\frac{5}{6} \times \frac{6}{5} = 1$, $\frac{5}{6}$ and $\frac{6}{5}$ are reciprocals.

rectangle A parallelogram having four right angles.

rectangular prism A prism that has rectangular bases.

signo radical Símbolo que se usa para indicar una raíz cuadrada no negativa, $\sqrt{\ }$.

radio Distancia desde el centro de un círculo hasta cualquier punto del mismo.

rango La diferencia entre el número mayor y el número menor en un conjunto de datos.

tasa Razón que compara dos cantidades que tienen diferentes tipos de unidades.

tasa de cambio Tasa que describe cómo cambia una cantidad con respecto a otra. Por lo general, se expresa como tasa unitaria.

razón Comparación de dos cantidades mediante división. La razón de 2 a 3 puede escribirse como 2 de cada 3, 2 a 3, 2 : 3 ó $\frac{2}{3}$.

número racional Número que se puede expresar como fracción.

tabla de razones Tabla cuyas columnas contienen pares de números que tienen una misma razón.

rayo Recta con un extremo y la cual se extiende infinitamente en una sola dirección.

recíproco Cualquier par de números cuyo producto es 1. Como $\frac{5}{6} \times \frac{6}{5} = 1$, $\frac{5}{6}$ y $\frac{6}{5}$ son recíprocos.

rectángulo Paralelogramo con cuatro ángulos rectos.

prisma rectangular Una prisma que tiene bases rectangulares.

reflection The mirror image produced by flipping a figure over a line.

reflexión Transformación en la cual una figura se voltea sobre una recta. También se conoce como simetría de espejo.

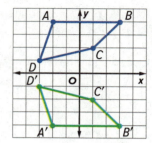

relation A set of ordered pairs such as (1, 3), (2, 4), and (3, 5). A relation can also be shown in a table or a graph.

relación Conjunto de pares ordenados como (1, 3), (2, 4) y (3, 5). Una relación también se puede mostrar en una tabla o una gráfica.

repeating decimal The decimal form of a rational number.

decimal periódico La forma decimal de un número racional.

rhombus A parallelogram having four congruent sides.

rombo Paralelogramo que tiene cuatro lados.

right angle An angle that measures exactly 90°.

ángulo recto Ángulo que mide exactamente 90°.

right triangle A triangle having one right angle.

triángulo rectángulo Triángulo que tiene un ángulo recto.

Ss

sample A randomly selected group chosen for the purpose of collecting data.

muestra Grupo escogido al azar o aleatoriamente que se usa con el propósito de recoger datos.

scale The set of all possible values of a given measurement, including the least and greatest numbers in the set, separated by the intervals used.

escala Conjunto de todos los valores posibles de una medida dada, incluyendo el número menor y el mayor del conjunto, separados por los intervalos usados.

scale   The scale gives the ratio that compares the measurements of a drawing or model to the measurements of the real object.

scale drawing   A drawing that is used to represent objects that are too large or too small to be drawn at actual size.

scalene triangle   A triangle having no congruent sides.

scaling   To multiply or divide two related quantities by the same number.

sequence   A list of numbers in a specific order, such as 0, 1, 2, 3, or 2, 4, 6, 8.

similar figures   Figures that have the same shape but not necessarily the same size.

slant height   The height of each lateral face.

solution   The value of a variable that makes an equation true. The solution of $12 = x + 7$ is 5.

solve   To replace a variable with a value that results in a true sentence.

square   A rectangle having four right angles and four congruent sides.

square root   The factors multiplied to form perfect squares.

statistical question   A question that anticipates and accounts for a variety of answers.

statistics   Collecting, organizing, and interpreting data.

stem-and-leaf plot   A system where data are organized from least to greatest. The digits of the least place value usually form the leaves, and the next place-value digits form the stems.

Stem	Leaf
1	2 4 5
2	
3	1 2 3 3 9
4	0 4 6 7

4 | 7 = 47

escala   Razón que compara las medidas de un dibujo o modelo a las medidas del objeto real.

dibujo a escala   Dibujo que se usa para representar objetos que son demasiado grandes o demasiado pequeños como para dibujarlos de tamaño natural.

triángulo escaleno   Triángulo sin lados congruentes.

homotecia   Multiplicar o dividir dos cantidades relacionadas entre un mismo número.

sucesión   Lista de números en un orden específico como, por ejemplo, 0, 1, 2, 3 ó 2, 4, 6, 8.

figuras semejantes   Figuras que tienen la misma forma, pero no necesariamente el mismo tamaño.

altura oblicua   Altura de cada cara lateral.

solución   Valor de la variable de una ecuación que hace verdadera la ecuación. La solución de $12 = x + 7$ es 5.

resolver   Reemplazar una variable con un valor que resulte en un enunciado verdadero.

cuadrado   Rectángulo con cuatro ángulos rectos y cuatro lados congruentes.

raíz cuadrada   Factores multiplicados para formar cuadrados perfectos.

cuestión estadística   Una pregunta que se anticipa y da cuenta de una variedad de respuestas.

estadística   Recopilar, ordenar e interpretar datos.

diagrama de tallo y hojas   Sistema donde los datos se organizan de menor a mayor. Por lo general, los dígitos de los valores de posición menores forman las hojas y los valores de posición más altos forman los tallos.

Tallo	Hojas
1	2 4 5
2	
3	1 2 3 3 9
4	0 4 6 7

4 | 7 = 47

stems   The digits of the greatest place value of data in a stem-and-leaf plot.

tallo   Los dígitos del mayor valor de posición de los datos en un diagrama de tallo y hojas.

straight angle   An angle that measures exactly 180°.

ángulo llano   Ángulo que mide exactamente 180°.

Subtraction Property of Equality   If you subtract the same number from each side of an equation, the two sides remain equal.

propiedad de sustracción de la igualdad   Si sustraes el mismo número de ambos lados de una ecuación, los dos lados permanecen iguales.

supplementary angles   Two angles are supplementary if the sum of their measures is 180°.

ángulos suplementarios   Dos ángulos son suplementarios si la suma de sus medidas es 180°.

∠1 and ∠2 are supplementary angles.

∠1 y ∠2 son suplementarios.

surface area   The sum of the areas of all the surfaces (faces) of a three-dimensional figure.
$S.A. = 2\ell h + 2\ell w + 2hw$

área de superficie   La suma de las áreas de todas las superficies (caras) de una figura tridimensional.
$S.A. = 2\ell h + 2\ell w + 2hw$

$S.A. = 2(7 \times 3) + 2(7 \times 5) + 2(3 \times 5)$
$= 142$ square feet

$S.A. = 2(7 \times 3) + 2(7 \times 5) + 2(3 \times 5)$
$= 142$ pies cuadrados

survey   A question or set of questions designed to collect data about a specific group of people, or population.

encuesta   Pregunta o conjunto de preguntas diseñadas para recoger datos sobre un grupo específico de personas o población.

symmetric distribution   Data that are evenly distributed.

distribución simétrica   Datos que están distribuidos.

term   Each number in a sequence.

término   Cada uno de los números de una sucesión.

term   Each part of an algebraic expression separated by a plus or minus sign.

término   Cada parte de un expresión algebraica separada por un signo más o un signo menos.

terminating decimal   A decimal is called terminating if its repeating digit is 0.

decimal finito   Un decimal se llama finito si el dígito que se repite es 0.

third quartile   For a data set with median M, the third quartile is the median of the data values greater than M.

tercer cuartil   Para un conjunto de datos con la mediana M, el tercer cuartil es la mediana de los valores mayores que M.

three-dimensional figure   A figure with length, width, and height.

figura tridimensional   Una figura que tiene largo, ancho y alto.

trapezoid   A quadrilateral with one pair of parallel sides.

triangle   A figure with three sides and three angles.

triangular prism   A prism that has triangular bases.

trapecio   Cuadrilátero con un único par de lados paralelos.

triángulo   Figura con tres lados y tres ángulos.

prisma triangular   Prisma con bases triangulares.

Uu

unit price   The cost per unit.

unit rate   A rate that is simplified so that it has a denominator of 1.

unit ratio   A unit rate where the denominator is one unit.

precio unitario   El costo por cada unidad.

tasa unitaria   Tasa simplificada para que tenga un denominador igual a 1.

razón unitaria   Tasa unitaria en que el denominador es la unidad.

Vv

variable   A symbol, usually a letter, used to represent a number.

vertex   The point where three or more faces intersect.

vertex

variable   Un símbolo, por lo general, una letra, que se usa para representar un número.

vértice   El punto en que se intersecan dos o más caras del prisma.

vértice

volume   The amount of space inside a three-dimensional figure. Volume is measured in cubic units.

$V = 10 \times 4 \times 3 = 120$ cubic meters

volumen   Cantidad de espacio dentro de una figura tridimensional. El volumen se mide en unidades cúbicas.

$V = 10 \times 4 \times 3 = 120$ metros cúbicos

x-axis The horizontal line of the two perpendicular number lines in a coordinate plane.

eje x La recta horizontal de las dos rectas numéricas perpendiculares en un plano de coordenadas.

x-coordinate The first number of an ordered pair. The x-coordinate corresponds to a number on the x-axis.

coordenada x El primer número de un par ordenado, el cual corresponde a un número en el eje x.

y-axis The vertical line of the two perpendicular number lines in a coordinate plane.

eje y La recta vertical de las dos rectas numéricas perpendiculares en un plano de coordenadas.

y-coordinate The second number of an ordered pair. The y-coordinate corresponds to a number on the y-axis.

coordenada y El segundo número de un par ordenado, el cual corresponde a un número en el eje y.

Chapter 1 Ratios and Rates

Page 6   Chapter 1   Are You Ready?

1. 29   **3.** 6   **5.** $\frac{1}{4}$ **7.** $\frac{13}{25}$

Pages 11–12   Lesson 1-1   Independent Practice

1. 2   **3.** 7   **5.** 30   **7** 60   **9** 9 pansies   **11.** 30 days
13. Sample answer: A gardener has 27 daisies and 36 marigolds. An equal number of each flower is planted in each row. What is the greatest number of marigolds in each row? 9 marigolds   **15.** 7 and 20, 5 and 8, 4 and 9

Pages 13–14   Lesson 1-1   Extra Practice

17. 5   **19.** 6   **21.** 15   **23.** 30   **25.** 4 baskets
27. 18 months, 36 months   **29.** $\frac{1}{2}$ **31.** $\frac{3}{4}$ **33.** $\frac{2}{8}, \frac{4}{16} \div \frac{2}{2} = \frac{2}{8}$

Pages 23–24   Lesson 1-2   Independent Practice

1. $\frac{2}{1}$; For every 2 flutes, there is 1 drum.   **3.** $\frac{2}{5}$; For every 2 boys, there are 5 girls in the class.   **5.** 12 photos in the first group and 21 photos in the second group   **7a.** $\frac{2}{5}$, 2 to 5, or 2:5; From 1968–2012, for every 2 championships won by Australia, the United States has won 5.   **7b.** $\frac{6}{41}$, 6 to 41, or 6:41; Between 1968 and 2012, the Australians have won 6 of the 41 championships.   **9.** 1,440; The ratios are 1:2, 1:3, 1:4, and 1:5.

Pages 25–26   Lesson 1-2   Extra Practice

11. $\frac{1}{4}$; for every 1 triangle there are 4 rectangles.
13. $\frac{1}{3}$; For every 1 puppy, there are 3 kittens available for adoption.   **15.** $\frac{2}{7}$, 2 to 7, or 2:7; Two out of every 7 food items donated were cans of fruit.   **17.** $\frac{1}{3}$; 1:3; or 1 to 3; Sample answer: If 6 students own a cell phone, 24 − 6 or 18 do not. The ratio is $\frac{6}{18}$ or $\frac{1}{3}$.   **19.** $\frac{7}{50}$ **21.** 4
23. 195 miles   **25.** 15 girls

Pages 35–36   Lesson 1-3   Independent Practice

1. $\frac{12 \text{ oz}}{1 \text{ steak}}$ **3.** $\frac{5.1 \text{ gal}}{1 \text{ container}}$ **5.** Divide the time by the number of laps. Evans drove the fastest at 2.3 minutes per lap.
7. $4 per mile   **9a.** 268 miles   **9b.** about 2 h   **11.** A unit rate has a denominator of 1. $\frac{\$108}{6 \text{ weeks}} = \frac{\$18}{1 \text{ week}}$ **13.** $1\frac{1}{3}$ min, or about 1 min, 20 s; 45 mph = $\frac{45}{60}$ mi per s, so $\frac{60}{45}$ gives the seconds per mile

Pages 37–38   Lesson 1-3   Extra Practice

15. 4 tulips per minute   **17.** 15 miles per hour   **19.** $63 per ticket   **21a.** 10.4 m per s   **21b.** 9.3 m per s
21c. 10.3 m per s   **23.** the ones that are 4 for $6; $4   **25.** $\frac{1}{5}$
27. $\frac{6}{25}$

Pages 43–44   Lesson 1-4   Independent Practice

1.

Number of Pies	5	10	20	8 pounds
Pounds of Apples	2	4	8	

3.

American Dollars	270	27	9	$9
Mexican Pesos	3,000	300	100	

5a.

People Served	24
Liters of Soda	4
Pints of Sherbet	2
Cups of Ice	6

5b. 2 L soda, 1 pt sherbet, 3 c ice; 6 L soda, 3 pt sherbet, 9 c ice   **5c.** 3 L soda, 1.5 pt sherbet, 4.5 c ice; Since 18 is half of 36, half the recipe that serves 36 people will serve 18 people. 6 L ÷ 2 = 3 L, 3 pt ÷ 2 = 1.5 pt, and 9 c ÷ 2 = 4.5 c.
7. Larger Quantity: ×; larger batches, Smaller Quantity: ÷; unit rate

9.

Bulls	18	2	22
Cows	45	5	55

No; if 4 bulls and 4 cows are added, there would be 22 bulls and 49 cows on the ranch. Using the ratio table, there should be 55 cows for 22 bulls.

Pages 45–46   Lesson 1-4   Extra Practice

11.

Number of Adults	1	2	3	4	4 adults
Number of Students	7	14	21	28	

13.

Ounces of Nectar	16	2	12	60 birds
Number of Birds Fed	80	10	60	

15. 160 mi   **17.** 7.5 c   **19.** (4, 5)

21.

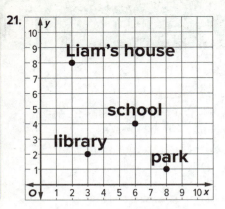

Pages 51–52   Lesson 1-5   Independent Practice

1

3.

Ken's Home Supply		
Fencing (ft), x	Cost ($), y	(x, y)
1	5	(1, 5)
2	10	(2, 10)
3	15	(3, 15)
4	20	(4, 20)

Wayne's Warehouse		
Fencing (ft), x	Cost ($), y	(x, y)
1	6	(1, 6)
2	12	(2, 12)
3	18	(3, 18)
4	24	(4, 24)

5 Sample answer: As the number of feet of fencing increases, the cost at Wayne's Warehouse increases at a faster rate than the cost at Ken's Home Supply. The cost at Wayne's Warehouse is shown on the graph as a steeper line.

7a.

Length (x)	Width (y)	(x, y)
1.618	1	(1.618, 1)
3.236	2	(3.236, 2)
4.854	3	(4.854, 3)
6.472	4	(6.472, 4)

7b.

7c. The area of the first rectangle in the table is 1.618 square units. The areas increase to 6.472, 14.562, and 25.888.
9. (2, 2.5)

Pages 53–54   Lesson 1-5   Extra Practice

11.

13.

Tiger Exhibit		
Animals, x	Employees, y	(x, y)
1	2	(1, 2)
2	4	(2, 4)
3	6	(3, 6)
4	8	(4, 8)

Elephant Exhibit		
Animals, x	Employees, y	(x, y)
1	4	(1, 4)
2	8	(2, 8)
3	12	(3, 12)
4	16	(4, 16)

15. Sample answer: The number of employees for the elephant exhibit increases at a faster rate than the number of employees for the tiger exhibit. The line representing the elephant exhibit is a steeper line.   **17.** She needs to mow 8 lawns. She'll earn $120 from mowing 8 lawns. This is more than she needs, but if she mowed only 7 lawns she would get $105 and that wouldn't be enough.   **19.** $\frac{1}{5}$ **21.** 6 students

Page 57   Problem-Solving Investigation   The Four-Step Plan

Case 3. 2.86 mi   **Case 5.** $70

Pages 63–64   Lesson 1-6   Independent Practice

1 No; Since the unit rates, $\frac{\$0.50}{1\text{ bagel}}$ and $\frac{\$0.38}{1\text{ bagel}}$, are not the same, the rates are not equivalent.   **3** Yes; Since $\frac{3\text{ h} \times 3}{\$12 \times 3} = \frac{9\text{ h}}{\$36}$, the fractions are equivalent; $\frac{3\text{ h}}{\$12} = \frac{9\text{ h}}{\$36}$.   **5.** No; Sample answer: since $\frac{8\text{ pairs}}{\$12} \neq \frac{3\text{ pairs}}{\$6}$, the ratios are not equivalent.   **7** No; Sample answer: Kiera did $\frac{6\text{ problems}}{30\text{ minutes}}$ or $\frac{1\text{ problem}}{5\text{ minutes}}$, and Heath did $\frac{18\text{ problems}}{40\text{ minutes}}$ or $\frac{9\text{ problems}}{20\text{ minutes}}$. So, the ratios are not equivalent.   **9.** 86.4 feet in 18 seconds; Sample answer: The other three are equivalent rates.   **11.** 6 girls

Pages 65–66   Lesson 1-6   Extra Practice

13. Yes; Since the unit rates are the same, $\frac{32\text{ words}}{1\text{ minute}}$, the rates are equivalent; $\frac{96\text{ words}}{3\text{ minutes}} = \frac{160\text{ words}}{5\text{ minutes}}$.   **15.** No; Since $\frac{16\text{ students}}{28\text{ students}} \neq \frac{240\text{ students}}{560\text{ students}}$, the ratios are not equivalent.   **17.** yes; The length to width ratio for the model and sofa form equivalent fractions.   **19.** Yes; sample answer: $\frac{\$35}{5\text{ weeks}} = \frac{\$7}{1\text{ week}}$ and $\frac{\$56}{56\text{ days}} = \frac{\$56}{8\text{ weeks}}$ or $\frac{\$7}{1\text{ week}}$.   **21.** $30.50   **23.** 64   **25.** $10

Pages 75–76   Lesson 1-7   Independent Practice

1. 90 cookies   **3** 840 gal   **5** 60 students   **9.** Elisa did not set up the equivalent ratios in the correct order. She should have set it up as $\frac{1}{12} = \frac{\blacksquare}{276}$. There are 23 teachers at the preschool.   **11.** 15 people; The unit rate of people that said they like to play disc golf is $\frac{1}{7}$. In a group of 252 people, 252 ÷ 7 or 36 people would like to play disc golf. Using equivalent ratios, $\frac{5}{12} = \frac{\blacksquare}{36}$. So, 15 people would have a personalized flying disc.

Pages 77–78   Lesson 1-7   Extra Practice

13. 54 teenagers   **15.** 3 baseballs   **17.** 28 DVDs   **19.** 18   **21.** $\frac{1}{7}$ **23.** $\frac{1}{6}$ **25.** 32 cars

Page 81   Chapter Review   Vocabulary Check

Across
3. ordered pair   **9.** graph   **11.** y coordinate
Down
1. y axis   **5.** scaling   **7.** ratio table

Page 82   Chapter Review   Key Concept Check

1. d   **3.** a   **5.** b

Chapter 2 Fractions, Decimals, and Percents

Page 88   Chapter 2   Are You Ready?

1. 4   **3.** 18   **5.** 60   **7.** 3 rotations

Pages 93–94   Lesson 2-1   Independent Practice

1. $\frac{1}{2}$ **3.** $\frac{33}{100}$ **5** 0.385   **7.** 0.16   **9** Mercury: 87.96; Venus: 224.7; Mars: 686.98   **11a.** meat: $\frac{7}{20}$; vegetables: $\frac{3}{20}$; sauce: $\frac{1}{20}$; bread: $\frac{1}{20}$ **11b.** $\frac{1}{5}$ lb   **11c.** $\frac{3}{5}$ lb   **13** Sample answer: $\frac{1}{5}$ in. and $\frac{7}{20}$ in.   **15.** Always; a decimal that ends in the thousandths place can have a denominator of 1,000. Since 1,000 is divisible by 2 and 5, the denominator of every such terminating decimal is divisible by 2 and 5.   **17.** yes; Sample answer: The fabric costs about $10, so $15 is enough.

Pages 95–96   Lesson 2-1   Extra Practice

19. $\frac{13}{20}$ **21.** $9\frac{7}{20}$ **23.** 0.622   **25.** 14.6   **27.** 23.375   **29.** 0.6   **31.** $\frac{4}{5}$ mi   **33.** $\frac{1}{5}$ **35.** $\frac{18}{25}$ **37.** 115 students

Pages 105–106   Lesson 2-2   Independent Practice

1 $\frac{1}{50}$ **3.** $\frac{17}{20}$ **5.** 20%   **7.** 35%   **9.** $\frac{7}{25}$ **11** $\frac{9}{50}$ **13.** Do not prefer: 80%, prefer: 20%; the sum of the percents is 100%.   **15.** Sample answer: $\frac{11}{20} = \frac{55}{100}$ or 55%, $\frac{3}{5} = \frac{60}{100}$ or 60%, $\frac{7}{10} = \frac{70}{100}$ or 70%   **17.** $\frac{8}{45}$; The other numbers are equivalent to $\frac{9}{20}$.   **19.** Sample answer: When written as a fraction, $33\frac{1}{3}\%$ is $\frac{1}{3}$ and 33% is $\frac{33}{100}$, which does not simplify.

Pages 107–108   Lesson 2-2   Extra Practice

21. $\frac{47}{100}$ **23.** $\frac{22}{25}$ **25.** 84%   **27.** 72%   **29.** 95%   **31a.** 44%   **31b.** 16%   **31c.** 60%   **31d.** 40%   **33.** 19 students   **35.** 0.9   **37.** Lap 2

Pages 113–114   Lesson 2-3   Independent Practice

1 0.35   **3.** 0.31   **5.** 22%   **7.** 10%   **9.** 0.04   **11** 12%   **13.** C: $59.50, A: $70, B: $87.50   **15.** 0.88, 0.90, 0.92   **17.** Sample answer: Since $\frac{3}{4}$ is equal to 0.75, write $43\frac{3}{4}\%$ as 43.75%. Then change 43.75% to the decimal 0.4375.   **19.** Sample answer: Every percent can be written as a fraction with a denominator of 100, and since every fraction is a rational number, every percent is a rational number.   **21.** Square D

Pages 115–116   Lesson 2-3   Extra Practice

23. 0.03   **25.** 0.11   **27.** 62%   **29.** 87%   **31.** 0.65   **33.** 82%   **35.** 0.8, 80%   **37.** =   **39.** >   **41.** Aliah's brother

Pages 121–122 Lesson 2-4 Independent Practice

1. 3.5; $3\frac{1}{2}$ 0.0015; $\frac{3}{2,000}$ **5.** 250% **7.** 420% **9.** 850%
11. 0.9% **13** 140% **15.** 0.003; $\frac{3}{1,000}$; 3 out of every 1,000 people are Japanese. **17a.** 0.0005 **17b.** sulfur
19. 30 mph **21.** Sample answer: Since 135% > 100%, two 10-by-10 grids will be used. The first will be completely shaded and the second will have 35 of the sections shaded.

Pages 123–124 Lesson 2-4 Extra Practice

23. 4; 4 **25.** 0.0004; $\frac{1}{2,500}$ **27.** 3,500% **29.** 0.77%
31. 9.8% **33.** 0.0012 **35.** 125% **37.** $\frac{18}{50,000}$, $\frac{27}{75,000}$, $\frac{9}{25,000}$ **39.** <
41.

Page 127 Problem-Solving Investigation Solve a Simpler Problem

Case 3. 1,440 days **Case 5.** 420 pieces

Pages 133–134 Lesson 2-5 Independent Practice

1. < **3** < **5.** $\frac{1}{4}$, $\frac{1}{2}$, $\frac{2}{3}$, $\frac{5}{6}$ **7.** Alex; 0.35 < 0.40
9a. 0.20, 0.25, 0.20 **9b.** Two are the same; 0.25 is the greatest score. **11** 8%, 17%, 0.2, $\frac{11}{20}$ **13.** $\frac{3}{9}$, $\frac{3}{8}$, and $\frac{3}{7}$; Because the numerators are the same, the larger the denominator, the smaller the fraction. **15.** less than; Sample answer: 0.4 is equivalent to 0.40, and 44% is equivalent to 0.44. Zero is less than 4 when you compare the hundredths.

Pages 135–136 Lesson 2-5 Extra Practice

17. = **19.** > **21.** $\frac{1}{2}$, $\frac{9}{16}$, $\frac{5}{8}$, $\frac{3}{4}$ **23.** 3%, $\frac{2}{50}$, 0.08
25. $\frac{1}{2}$, 0.55, $\frac{5}{7}$ **27.** $\frac{13}{64}$ inch **29.** 0.62 **31.** 5.11 **33.** $24.15

Pages 141–142 Lesson 2-6 Independent Practice

1 Sample answer: $\frac{1}{2}$ of $120 is $60. **3** Sample answer: $\frac{2}{5}$ of 15 is 6. **5.** Sample answer: 24 + 24 + 24 = 72 **7.** Sample answer: $\frac{3}{4}$ of 20 yr is 15 yr. **9a.** No; 40% is $\frac{2}{5}$, and $\frac{2}{5}$ of 15 is 6. He needs 7 baskets to win a prize. **9b.** about 50%
11 Sample answer: 71 − 37 = 34 missed shots and $\frac{34}{71}$ is about $\frac{35}{70}$ or $\frac{1}{2}$. Since $\frac{1}{2}$ = 50%, he missed about 50% of his shots. **13.** about 75% **15.** more; Rachel rounded $32 down to $30, so the actual amount she will save will be more than $12. **17.** Sample answer: First, round 42% to 40%, and $122

to $125. Next, rewrite 40% as $\frac{2}{5}$. Then find $\frac{1}{5}$ of $125. Finally, multiply this result by 2 to find $\frac{2}{5}$ of $125.

Pages 143–144 Lesson 2-6 Extra Practice

19. Sample answer: $\frac{1}{2}$ of 60 is 30. **21.** Sample answer: $\frac{1}{4}$ of 120 is 30. **23.** Sample answer: 19 + 19 + 19 = 57
25. Sample answer: 61 + 61 + 61 + 61 = 244 **27.** Sample answer: $\frac{3}{4}$ of 8 h is 6 h. **29.** about 423; 47 × 9 = 423
31. about 75% **33.** 200 teens **35.** 0.22 **37.** 0.67 **39.** 0.12
41. $29.91

Pages 151–152 Lesson 2-7 Independent Practice

1. 46 **3.** 92 **5** 9 **7.** 336 **9** $8.40 **11.** no; 70% of 30 is 21, not 24. 80% of 30 is 24. **13.** Sample answers given for examples. Percent: equal to; 100%; less than; 25%; greater than; 125%. Fraction: equal to; $\frac{3}{3}$; less than; $\frac{1}{3}$; greater than; $\frac{4}{3}$
15. yes; 16% of 40 is 6.4 and 40% of 16 is 6.4 **17.** Sample answer: If a number n is 25% of a and 35% of b, it is a greater part of b than it is of a. So, $a > b$.

Pages 153–154 Lesson 2-7 Extra Practice

19. 16.5 **21.** 161 **23.** 159.84 **25.** 0.45 **27.** 18 ounces of bleach **29.** 398.24 **31.** 212.85 **33.** 0.3 × 42, $\frac{3}{10}$ × 42, 3 × 4.2 **35.** 32.6 **37.** 9 **39.** 0.32

39.

Trip	Part of Club
Carnegie Museum of Art	0.32
Fallingwater	0.20
Westemoreland Museum of American Art	0.48

Pages 159–160 Lesson 2-8 Independent Practice

1. 70

3 $\frac{22}{\blacksquare} = \frac{44}{100}$; 50 **5** $50 **7.** 15 cups **9.** 20 cups
11. Sample answer: $\frac{21}{25} = \frac{\blacksquare}{100}$; 84 **13.** 18 karats; 24 is the whole and 75 is the percent, so $\frac{18}{24} = \frac{75}{100}$. **15.** 5%

Pages 161–162   Lesson 2-8   Extra Practice

17. 400

19. $\frac{270}{\blacksquare} = \frac{90}{100}$; 300   **21.** 24 cars   **23.** 300 students
25. 240 ounces   **27.** 30 students   **29.** 49   **31.** 123   **33.** 16
35. 4.5 miles

Page 165   Chapter Review   Vocabulary Check

1. percent   **3.** percent proportion   **5.** least common
denominator

Page 166   Chapter Review   Key Concept Check

1. not correct;

$$\frac{4}{5} \rightarrow 5\overline{)4.0} \quad \begin{array}{r} 0.8 \\ -40 \\ \hline 0 \end{array}$$

3. not correct;

$$120\% = \frac{120}{100}$$
$$= \frac{6}{5}$$
$$= 1\frac{1}{5}$$

Chapter 3 Compute with Multi-Digit Numbers

Page 176   Chapter 3   Are You Ready?

1. 300   **3.** 1,078   **5.** 49   **7.** 4 million albums

Pages 181–182   Lesson 3-1   Independent Practice

1. 16.7   **3** 103.01   **5.** 80.02   **7** 1.73 s   **9.** $24.85
11. Luis did not annex a zero before subtracting. 8.9 − 3.72 =
5.18   **13.** Sample answer: When you add the whole numbers
the sum is 20. The sum of the decimals will be added on,
which will make the sum greater than 20.   **15.** 6; A zero is
annexed to subtract from 8.5. So, you will subtract 10 − 4 to
find the value in the hundredths place.

Pages 183–184   Lesson 3-1   Extra Practice

17. 7.9   **19.** 39.99   **21.** 14.82   **23.** $207.85   **25.** 34.15
degrees Celsius   **27.** 166.85 mi   **29.** 5   **31.** 22
33. Thursday

Pages 189–190   Lesson 3-2   Independent Practice

1. Sample answer: 30; 10 × 3 = 30   **3** Sample answer: 160;
20 × 8 = 160   **5.** Sample answer: 240; 30 × 8 = 240
7. about 80 million tons   **9** More; her wage and hours
worked were rounded up, so the actual total is less than the
estimate.   **11a.** Raj needs to save $132 more.   **11b.** No; $6 ×
20 = $120   **13.** about $12   **15.** between $5 and $6; Sample
answer: The best estimate for the cost of the peppers is $3 × 2
or $6. Since both values were rounded up, the cost will be less
than the estimate.

Pages 191–192   Lesson 3-2   Extra Practice

17. Sample answer: 30 × 80 = 2,400   **19.** about 90   **21.** yes
23. no; 32   **25.** Calories: 400; Vitamin C: 400 mg;
carbohydrates: 100 g; calcium: 80 mg   **27.** $50   **29.** 345
31. $4.37

Pages 197–198   Lesson 3-3   Independent Practice

1. 8.4   **3.** 2.6   **5.** 0.06   **7** $215.27   **9** 134.6°F; Sample
answer: 1,346 × 10 = 13,460. Since 13.46 has 2 decimal places,
13.46 × 10 = 134.60.   **11.** Sample answer: 32 mm; 20 × 1.95 =
39; 4 × 1.75 = 7; 39 − 7 = 32   **13.** Sample answer: First
evaluate 1.17 × 100 to be 117. Then, multiply 117 by 5.4 to get
the answer of 631.8. Or first evaluate 5.4 × 100 to be 540.
Then multiply 540 by 1.17 to get the answer of 631.8. Or first
evaluate 5.4 × 10 to be 54 and 1.17 × 10 to be 11.7. Then
multiply 54 by 11.7 to get the answer of 631.8.   **15.** No; zero
represents the number of hundredths and should be counted.

Pages 199–200   Lesson 3-3   Extra Practice

17. 8.5   **19.** 19.2   **21.** 0.084   **23.** 2.24 g   **25.** 93.5 in^2
27. 7.44 miles   **29.** $687.50   **31.** 6   **33.** 13

Pages 205–206   Lesson 3-4   Independent Practice

1. 0.28   **3** 1.092   **5.** 167.0067   **7** 84.474 ft; 46.93 × 1.8
≈ 45 × 2 = 90; 84.474 ≈ 90   **9** $5.76; Each price is about $1.
He bought about 6 pounds of fruit. 6 × 1 = 6 ≈ $5.76
11. 1.03515   **13a.** 209.6 mi   **13b.** 413,170.3 mi
15. 32.013341…; Sample answer: 3.9853 × 8.032856 rounds
to 4 × 8 = 32, so the answer must be about 32.   **17.** greater
than 0.4; It is being multiplied by a decimal greater than 1.
19. Sample answer: Diego is growing plants from seeds. Each
day, the plant grows 0.5 inch. How many inches tall is the plant
after 1.5 days? 0.75 in.

Pages 207–208   Lesson 3-4   Extra Practice

21. 2.48   **23.** 16.128   **25.** 0.02255   **29a.** Junnie
29b. 0.62 mile farther   **31.** 32.4   **33.** 12   **35.** 7 boxes

Page 213   Problem-Solving Investigation   Look for a Pattern

Case 3. $95.40   **Case 5.** 383.5, 1,153, 3,461.5; multiply by 3,
then add 2.5

Selected Answers

Pages 219–220   Lesson 3-5   Independent Practice

1. 29   **3.** 15   **5** 130 R30   **7.** 170   **9** 60 mi
11. 175 names   **13.** 144 cups   **15.** Sample answer: Mary
saved $2,400 in 12 months. What was the average amount
she saved each month?; $200   **17.** No; Sample answer: if the
remainder equals the divisor, then the quotient should be
increased by 1.   **19.** Sample answer: Find the sum of the
distances that she drove each day and divide the total by the
number of hours; 60 mph

Pages 221–222   Lesson 3-5   Extra Practice

21. 57 R3   **23.** 10 R21   **25.** 166 R24   **27.** 224 R1
29. 845 cups   **31.** 1,652 castles; 59 employees   **33.** 60
35. 700

Pages 227–228   Lesson 3-6   Independent Practice

1. Sample answer: $33 \div 3 = 11$ **3.** Sample answer: $36 \div 12 =$
3   **5** Sample answer: about 3   **7** about 6 gal; $53 \div 8.5$
$\approx 54 \div 9 = 6$ **9.** 1; 2; 4; 8   **11.** Sample answer: 160.23 ÷
6.54   **13.** Sample answer: Look for multiplication or division
math facts containing numbers close to the decimal dividend
and divisor that give the whole number quotient. For example,
you can estimate $13.8 \div 7.1$ by finding $14 \div 7$.

Pages 229–230   Lesson 3-6   Extra Practice

15. Sample answer: $46 \div 23 = 2$ **17.** Sample answer: about
7 inches; $45.9 \div 7 \approx 49 \div 7 = 7$ **19** Sample answer:
$474.72 \div 12 \approx \$480 \div 12 = \$40$ **21a.** 5; $100 \times 5 = 500$
21b. 10   **23.** about 10 locks   **25.** 0.147   **27.** 7.3456; 0.73456;
0.073456; Sample answer: Divide by ten to move the decimal
point one place value to the left.

Pages 235–236   Lesson 3-7   Independent Practice

1. 13.1   **3** 23.7   **5.** 1.2   **7.** $770.56   **9** 22.8 ft; Area of a
rectangle is length times width, so divide the area by the
length to find the width. $752.4 \div 33 = 22.8$ **11.** Brand B; The
cost of each bottled water for Brand B is about $0.44. For
Brand A the cost is about $0.58 and for Brand C the cost is
about $0.46. So Brand B has the best cost per bottle.
13. Amanda placed the decimal point in the wrong place of
the quotient. $11.2 \div 14 = 0.8$ **15.** Since $40 \div 20 = 2$, the
answer is about 2, so the answer should have one number to
the left of the decimal point.

Pages 237–238   Lesson 3-7   Extra Practice

17. 18.4   **19.** 1.6   **21.** 1.9   **23.** $3.75   **25.** Dominoes; The
cost of each domino set is about $0.66. The cost of each peg
game is about $0.83, and the cost of each mini football is
about $0.75. So, the domino set has the best cost per toy.
27. 11.8 points   **29.** 17.2   **31.** Sample answer: $34.3 < 34.32$
33. 4 m

Pages 243–244   Lesson 3-8   Independent Practice

1. 3.6   **3** 250   **5.** 450   **7.** 20 steps   **9a.** 24 h
9b. 21.12 h   **11** **a.** 2.2 times   **b.** 3.8 times   **13.** $49 \div 7$; the
quotient is 7 and all of the other problems have a quotient of
0.7.

Pages 245–246   Lesson 3-8   Extra Practice

15. 0.2   **17.** 0.0492   **19.** 420   **21.** 6 pieces   **23.** about
4.4 times   **25.** about 41.3 times   **27.** <   **29.** >   **31.** $\frac{7}{12}$ of
her free time

Page 249   Chapter Review   Vocabulary Check

Across
1. compatible numbers   **5.** decimal
Down
3. multidigit   **5.** dividend

Page 249   Chapter Review   Key Concept Check

Across
1. 483   **3.** 178   **5.** 21   **9.** 4930   **13.** 203
Down
1. 40   **3.** 108   **5.** 239   **7.** 72   **9.** 463   **11.** 880

Chapter 4 Multiply and Divide Fractions

Page 256   Chapter 4   Are You Ready?

1. 12   **3.** 6   **5.** $4\frac{11}{21}$ **7.** $6\frac{7}{8}$ in.

Pages 261–262   Lesson 4-1   Independent Practice

1. Sample answer: $\frac{1}{4} \times 20 = 5$

3. Sample answer: $1 \times 0 = 0$ **5** Sample answer: $12 \times \frac{1}{4} =$
3 pizzas   **7a.** about 20 lb   **7b.** about $75   **9** Sample
answer: $10 \times 3 = 30$ in^2 **11.** Sample answer: $\frac{5}{9}$; $8\frac{1}{2}$ is about $\frac{5}{9}$
$\times 9 = 5$.   **13.** 6 square yards; Sample answer: $2\frac{2}{3}$ is closer to 3
than 2 and $1\frac{5}{6}$ is closer to 2 than 1. $3 \times 2 = 6$

Pages 263–264   Lesson 4-1   Extra Practice

15. Sample answer: 14   **17.** Sample answer: $\frac{2}{9} \times 90 = 20$
19. Sample answer: $\frac{1}{2} \times 4 = 2$ **21.** Sample answer: $8 \times 1 =$
8 cm^2 **23.** 23 movies   **25.** between 70 and 100   **27.** $\frac{1}{2}$
29. 0
31.

33. 120 square feet

Pages 269–270 Lesson 4-2 Independent Practice

1. 15 **3.** 2 **5.** $\frac{22}{5}$ or $4\frac{2}{5}$ **7.** $2\frac{2}{5}$ in. **9.** neither;

$\frac{4}{5} \times 30 = 24$ and $\frac{2}{3} \times 36 = 24$. So, 24 = 24. **11.** seventh

13. Sample answer: Melinda baked a dozen cookies. Three-fourths of the cookies were oatmeal raisin. How many cookies were oatmeal raisin cookies?; $\frac{3}{4} \times 12 = 9$; 9 cookies

15. Sample answer: $\frac{5}{3} \times 2$

Pages 271–272 Lesson 4-2 Extra Practice

17. 4 **19.** 2 **21.** $\frac{24}{7}$ or $3\frac{3}{7}$ **23.** 16,800 votes **25.** Sample

answer: Joshua has 24 classmates and $\frac{3}{8}$ of them play soccer.

How many of his classmates play soccer?; 9 students;

$24 \times \frac{1}{2} = 12$ **27.** Sample answer: David will need 8 cups of

milk for 12 batches, so he has enough milk. **29.** 198

31. 45 minutes

Pages 277–278 Lesson 4-3 Independent Practice

1. $\frac{2}{15}$ $2\frac{2}{3}$ **5.** $\frac{1}{6}$ **7.** $\frac{3}{8}$ **9.** $\frac{3}{10}$

11a. Sample answer: Olivia withdrew $\frac{3}{4}$ of her savings. She

used $\frac{1}{5}$ of what was left to buy a book. If she had $100 in

savings, how much did she spend on the book?

11b.

11c. Sample answer: Multiply $\frac{1}{5} \times \frac{1}{4}$. Multiply the product, $\frac{1}{20}$

by $100. She spent $5 on a book. **13.** Sample answer: $a = \frac{3}{8}$

and $b = \frac{5}{7}$; $a = \frac{5}{8}$ and $b = \frac{3}{7}$; $a = \frac{5}{14}$ and $b = \frac{3}{4}$ **15.** Sample

answer: Makayla gives her cat $\frac{3}{4}$ cup of cat food each day. How

much cat food will she have given her cat after 14 days? Since

$\frac{3}{4}$ is close to 1, Makayla will have given her cat about 1 × 14 or

14 cups of cat food. $\frac{3}{4} \times 14 = 10\frac{1}{2}$ which is close to 14.

Pages 279–280 Lesson 4-3 Extra Practice

17. $\frac{6}{35}$ **19.** $4\frac{1}{8}$ **21.** $\frac{1}{3}$ **23.** Nyemi: 138; Luke: 69; Natalie: 23

25. $\frac{1}{2}$ c; $\frac{3}{4} \times \frac{2}{3} = \frac{3 \times 2}{4 \times 3} = \frac{1}{2}$ **27.** 48 text messages **29.** 330

31. 4 square feet

Pages 285–286 Lesson 4-4 Independent Practice

1. $1\frac{1}{6}$ **3.** $2\frac{27}{32}$ **5.** 9 **7.** $9\frac{1}{4}$ mi **9.** $3\frac{3}{8}$ c

11a. about $69\frac{27}{40}$ million mi **11b.** about $139\frac{7}{40}$ million mi

11c. about $487\frac{29}{40}$ million mi **11d.** about $882\frac{11}{20}$ million mi

13. $1\frac{1}{24}$ **15.** B; the product must be greater than $\frac{2}{3}$ and

less than $2\frac{1}{2}$.

Pages 287–288 Lesson 4-4 Extra Practice

17. $2\frac{1}{8}$ **19.** $\frac{17}{20}$ **21.** $12\frac{3}{4}$ **23.** $1,259\frac{1}{4}$ in^2 **25.** 31 inches

27. $28\frac{1}{8}$ in. × $16\frac{7}{8}$ in. **29.** 12 **31.** 3

Pages 293–294 Lesson 4-5 Independent Practice

1. 6 **3.** 52 **5.** $2\frac{1}{2}$ **7.** 1,500 lb **9.** $4\frac{4}{5}$ oz

11a. The x-value represents the number of quarts and the

y-value represents the equivalent number of gallons.

11b. Sample answer: The point on the line whose y-value is

equal to 2.5 is (10, 2.5), so 10 qt = 2.5 gal. **11c.** $\frac{4 \text{ qt}}{1 \text{ gal}}$

11d. 2.5 gal × $\frac{4 \text{ qt}}{1 \text{ gal}}$ **13.** <; 16 in. is equivalent to 1 ft 4 in.;

$1\frac{1}{2}$ ft is equivalent to 1 ft 6 in.; So, 16 in. < $1\frac{1}{2}$ ft. **15.** Sample

answer: 5 pt; 80 fl oz

Pages 295–296 Lesson 4-5 Extra Practice

17. $4\frac{1}{2}$ **19.** 24 **21.** $6\frac{1}{2}$ **23.** 3,520 ft

25. No; 15 in. + $4\frac{1}{2}$ in. + $6\frac{3}{4}$ in. = $26\frac{1}{4}$ in.; $2\frac{1}{2}$ ft = 30 in.; So,

$26\frac{1}{4}$ in. < 30 in. **27.** $\frac{1}{2}$ **29a.** False **29b.** False **29c.** True

29d. False **31.** 39 **33.** 7.5 **35.** $41\frac{2}{3}$ feet

Page 299 Problem-Solving Investigation Draw a Diagram

Case 3.

Case 5. 100 miles

Pages 309–310 Lesson 4-6 Independent Practice

1. $\frac{5}{3}$ **3.** 1 **5.** $6\frac{2}{3}$ **7.** 10 **9.** $4\frac{1}{2}$ **11.** 110 horses

13. 6 activities; $4 \div \frac{2}{3} = 4 \times \frac{3}{2} = \frac{12}{2} = 6$ **15.** Daniella

did not multiply by the reciprocal of 4, which is $\frac{1}{4}$.

$\frac{8}{9} \div 4 = \frac{8}{9} \times \frac{1}{4} = \frac{8}{36}$ or $\frac{2}{9}$ **17a.** $1.40 **17b.** $\frac{1}{2}$ pound

17c. 6 bags **17d.** 4 bags; $25 \times \frac{3}{4} = 18\frac{3}{4}$; $18\frac{3}{4} \div 5 = 3\frac{3}{4}$; Since

you can't buy part of a bag, they would need to buy 4 bags.

Pages 311–312 Lesson 4-6 Extra Practice

19. $\frac{9}{7}$ **21.** $3\frac{1}{3}$ **23.** $3\frac{3}{5}$ **25.** $7\frac{1}{5}$ **27.** 8 dinners **29.** $6\frac{2}{3}$ c;

5 c **31.** 6 **33.** 6 **35.** 24 **37.** $2\frac{2}{3}$ miles

Pages 321–322 Lesson 4-7 Independent Practice

1. $\frac{1}{4}$ **3.** $\frac{1}{12}$ **5.** $\frac{1}{24}$ **7.** Sample answer: David has $\frac{5}{6}$ foot of

tape. He uses $\frac{1}{12}$ foot of tape to hang each photo on the

bulletin board. How many photos can he hang on the bulletin

board? 10 photos

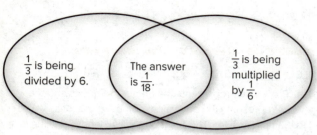 $\frac{3}{4} \div \frac{3}{8} = 2$; 2 T-shirts

11.

Left oval: $\frac{1}{3}$ is being divided by 6.

Center overlap: The answer is $\frac{1}{18}$.

Right oval: $\frac{1}{3}$ is being multiplied by $\frac{1}{6}$.

13. greater than 1; the dividend is greater than the divisor; less than 1; the dividend is less than the divisor   **15.** $\frac{2}{3} \div \frac{1}{6}$

Pages 323–324   Lesson 4-7   Extra Practice

17. $\frac{1}{20}$ **19.** $\frac{1}{7}$ **21.** $\frac{5}{12}$ **23.** $\frac{1}{10} \div 4 = \frac{1}{40}$; $\frac{1}{40}$ kilometer
25. $\frac{7}{8} \div \frac{1}{24} = 21$; 21 cycles   **27.** Yes; Sample answer: the shelf
can hold $24 \div \frac{3}{8}$ or 64 cases.   **29.** 4   **31.** 4   **33.** 4
35. 2 feet

Pages 329–330   Lesson 4-8   Independent Practice

1. $\frac{5}{12}$ **3.** $\frac{2}{3}$ **5.** 28 slices   **7.** fraction; improper;
reciprocal; fractions   **9.** less than; Sample answer: The
expression $5\frac{1}{6} \div 3\frac{5}{8}$ represents $5\frac{1}{6}$ being divided into a
greater number of parts than the expression $5\frac{1}{6} \div 2\frac{2}{5}$. If $5\frac{1}{6}$ is
divided into a greater number of parts, each part will be
smaller. So, $5\frac{1}{6} \div 3\frac{5}{8} < 5\frac{1}{6} \div 2\frac{2}{5}$.

Pages 331–332   Lesson 4-8   Extra Practice

11. $2\frac{3}{4}$ **13.** $2\frac{2}{3}$ **15.** 2   **17.** 2   **19.** $2\frac{1}{10}$ **21.** $6\frac{3}{8} \div \frac{3}{8}$;
17 bags   **23.** 20 books; Sample answer: Twenty books weigh
25 pounds and twenty-one books weigh $26\frac{1}{4}$ pounds, so the
shelf can hold 20 books.   **25.** $\frac{3}{4}$ **27.** $\frac{1}{4}$ **29.** $\frac{1}{10}$
31. $\frac{5}{12}$ square mile

Page 335   Chapter Review   Vocabulary Check

1. mixed number   **3.** reciprocal   **5.** denominator
7. unit fraction   **9.** Compatible numbers

Page 336   Chapter Review   Key Concept Check

1. not correct; $13 \times \frac{1}{3} = \frac{13}{5}$ or $4\frac{1}{3}$ **3.** correct   **5.** correct

Chapter 5 Integers and the Coordinate Plane

Page 342   Chapter 5   Are You Ready?

1. =   **3.** <   **5.** >   **7.** peanuts

Pages 349–350   Lesson 5-1   Independent Practice

1. −3; The integer 0 represents at sea level.
3. −5; The integer 0 represents neither moving backward
nor moving forward.

5.

7.

9. Sample answers are given.

Positive Integer	Negative Integer
• gain	• lose
• above	• below
• earn	• spend
• +	• −

11. Negative; Sample answer: A drop of 15° would result in a
temperature of 0°F. Since the drop of 20° is greater than 15°,
the temperature is below zero and will be represented by a
negative integer.   **13.** Sample answer: Locate −2 and 3 on a
number line. Count the number of units between each integer
and 0. −2 is 2 units to the left of zero and 3 is 3 units to the
right. So, the number of units between −2 and 3 is 2 + 3 or
5 units.

Pages 351–352   Lesson 5-1   Extra Practice

15. −25; The integer 0 represents neither spending nor
earning.

17.

19.   **21.**

23.

25.

27. <   **29.** >   **31.** <   **33.** 104 raffle tickets

Pages 359–360   Lesson 5-2   Independent Practice

1. −6   **3.** 0   **5.** −9   **7.** 14   **9** 21   **11.** 4 feet
13 70°F   **15.** 14   **17.** Absolute value cannot be a negative number. So, the absolute value of −14 is 14, not −14.
19. Never; distance cannot be negative.   **21.** Absolute value is distance and distance cannot be negative.   **23.** sometimes; Sample answer: If *n* is positive, then −*n* is negative. If *n* is negative, then −*n* is positive.

Pages 361–362   Lesson 5-2   Extra Practice

25. −15   **27.** 9   **29.** −8   **31.** 0   **33.** 15   **35.** 11   **37.** 9 yards
39. −10   **41.** acetone   **43.** <   **45.** =   **47.** 45 minutes

Pages 367–368   Lesson 5-3   Independent Practice

1. >   **3.** >   **5** −9 < 26; The temperature in Flagstaff, Arizona, was warmer.   **7** −79, −55, 18, 44, 101, 143
9a. Sun   **9b.** Sun, 100-Watt Bulb, Full Moon, Venus, Andromeda Galaxy, Alpha Centauri   **9c.** −27   **11.** Sample answer: Elise owes her brother $15. Jacob has $7. Elise has less money than Jacob.   **13.** $-\frac{12}{4}, -\frac{1}{2}, \frac{1}{6}, \frac{7}{8}, \frac{5}{2}$

Pages 369–370   Lesson 5-3   Extra Practice

15. <   **17.** <   **19.** −20 > −25; Michael owes less money than Yvonne.   **21.** −221, −89, −71, −10, 54, 63

23a.

B   C   A   D
−5 −4 −3 −2 −1 0 1 2 3 4 5 6 7

23b. D; Since −4 < −3 < 2 < 6, player D had the most strokes over par.

25.

	Player
Least Shots	Cristian
	Liam
	Bailey
Most Shots	Marisol

27. 0.2   **29.** Julio; $5\frac{5}{6} > 5\frac{1}{4}$

Page 373   Problem-Solving Investigation   Work Backward

Case 3. 488 meters   **Case 5.** 3

Pages 383–384   Lesson 5-4   Independent Practice

1 0.4$\overline{6}$ **3.** −0.$\overline{6}$ **5.** 3.34$\overline{09}$ **7.** 0.$\overline{34}$ **9.** $-\frac{9}{10}$ **11.** $-3\frac{4}{5}$
13. $\frac{4}{13}$ **15** 0.$\overline{7}$ **17a.** 43   **17b.** $\frac{24}{43}$; 0.558   **19.** $\frac{17}{36}$ is not a terminating decimal since decimals are based on powers of 10 and 36 is not a factor of any power of 10.   **21.** $\frac{1}{11} = 0.\overline{09}$, $\frac{2}{11} = 0.\overline{18}$, $\frac{3}{11} = 0.\overline{27}$; The digits that are repeated are equal to the numerator times 9. So, $\frac{7}{11} = 0.\overline{63}$ and $\frac{8}{11} = 0.\overline{72}$.

Pages 385–386   Lesson 5-4   Extra Practice

23. 0.$\overline{27}$ **25.** −0.7   **27.** −1.$\overline{80}$ **29.** $-\frac{3}{20}$ **31.** $-12\frac{27}{50}$
33. 0.1$\overline{6}$ **35.** <   **37.** >   **39.** <   **41.** 5.691 < 5.78

Pages 391–392   Lesson 5-5   Independent Practice

1. >   **3** =   **5** $-2\frac{3}{4}$, −2.$\overline{2}$, 2.8, $3\frac{1}{8}$ **7.** $-4\frac{1}{2}$, $-2\frac{3}{8}$, 1.35, 5.6   **9** −10.8, −9.7, 9.0, 11.4   **11.** always; The greater a number is, the farther away from zero. Therefore, its opposite will also be farther from zero.   **13.** The first decimal is a terminating decimal, so its thousandths place is zero. The second decimal has a repeating digit of 3, so its thousandths place is 3. −0.330 > −0.$\overline{333}$ **15.** Sample answer: The temperature of a freezer changed throughout a day as the door was opened and shut. The temperatures were −11°F, 13°F, −12°F, and 15°F. Order the set of temperatures from least to greatest.; −12°F, −11°F, 13°F, 15°F

Pages 393–394   Lesson 5-5   Extra Practice

17. >   **19.** >   **21.** <   **23.** $1\frac{1}{5}$, 1.25, 1.2$\overline{5}$, $1\frac{3}{4}$ **25.** <   **27.** >
29a. False   **29b.** True   **29c.** False   **29d.** True

31–37.

39.

A
0.0 0.1 0.2 0.3 0.4 0.5 0.6 0.7 0.8 0.9 1.0

Pages 399–400   Lesson 5-6   Independent Practice

1. (2, 2); I   **3.** (−4, 2); II   **5** (5, 0); none   **7** *Z*; II   **9.** *A*; IV
11. *N*; none   **13a.** The Clock   **13b.** the Wonder Wheel; (2, −4)   **13c.** the Big Coaster; (−3, 1)   **13d.** (−1, −2)
15. Quadrants I and III; Sample answer: In Quadrant I, both coordinates are positive and in Quadrant III, both coordinates are negative.   **17.** Sample answer: The first coordinate corresponds to a number on the *x*-axis. The second coordinate corresponds to a number on the *y*-axis. A point is defined by only one ordered pair.

Pages 401–402   Lesson 5-6   Extra Practice

19. (1, 3); I   **21.** (−2, 1); II   **23.** (−4, −5); III   **25.** *L*; II   **27.** *S*; IV
29. *B*; III   **31a.** (4, 2)   **31b.** (4, 4)   **31c.** (−1, −4)   **33.** (4, 2)

35.

2.7   2.8   2.9   3.0   3.1   3.2   3.3

37. 19 magazines

Pages 407–408   Lesson 5-7   Independent Practice

1–7.

9–11.

13. (4.25, −1.75)

15.

6 sq. units

17. Sample answer: (7, 2), (−5, 2)   **19.** always; The y-coordinate will be the opposite of the original following the reflection across the x-axis. The x-coordinate will be the opposite of the original following the reflection across the y-axis.   **21.** never; The x-coordinate of any point on the y-axis is always zero.

Pages 409–410   Lesson 5-7   Extra Practice

23–29.

31.

33.

rectangle

35. 7 units   **37.** (2.5, −3.25)   **39.** 27   **41.** 25

Page 417   Chapter Review   Vocabulary Check

1. rational number   **3.** positive integer
5. terminating decimal

Page 418   Chapter Review   Key Concept Check

1. 4   **3.** x-coordinate   **5.** 6.543

Index

Index

Kk

Ll

Mm

Index

0 1 2 3 4 5 6 7 8 9

−11 −10 −9 −8 −7 −6 −5 −4 −3 −2 −1 0 1 2 3 4 5 6 7 8 9 10 11

What Are Foldables and How Do I Create Them?

Foldables are three-dimensional graphic organizers that help you create study guides for each chapter in your book.

Step 1 Go to the back of your book to find the Foldable for the chapter you are currently studying. Follow the cutting and assembly instructions at the top of the page.

Step 2 Go to the Key Concept Check at the end of the chapter you are currently studying. Match up the tabs and attach your Foldable to this page. Dotted tabs show where to place your Foldable. Striped tabs indicate where to tape the Foldable.

How Will I Know When to Use My Foldable?

When it's time to work on your Foldable, you will see a Foldables logo at the bottom of the **Rate Yourself!** box on the Guided Practice pages. This lets you know that it is time to update it with concepts from that lesson. Once you've completed your Foldable, use it to study for the chapter test.

How Do I Complete My Foldable?

No two Foldables in your book will look alike. However, some will ask you to fill in similar information. Below are some of the instructions you'll see as you complete your Foldable. **HAVE FUN** learning math using Foldables!

Instructions and what they mean

Best Used to...	Complete the sentence explaining when the concept should be used.
Definition	Write a definition in your own words.
Description	Describe the concept using words.
Equation	Write an equation that uses the concept. You may use one already in the text or you can make up your own.
Example	Write an example about the concept. You may use one already in the text or you can make up your own.
Formulas	Write a formula that uses the concept. You may use one already in the text.
How do I ...?	Explain the steps involved in the concept.
Models	Draw a model to illustrate the concept.
Picture	Draw a picture to illustrate the concept.
Solve Algebraically	Write and solve an equation that uses the concept.
Symbols	Write or use the symbols that pertain to the concept.
Write About It	Write a definition or description in your own words.
Words	Write the words that pertain to the concept.

Meet Foldables Author Dinah Zike

Dinah Zike is known for designing hands-on manipulatives that are used nationally and internationally by teachers and parents. Dinah is an explosion of energy and ideas. Her excitement and joy for learning inspires everyone she touches.

numbers

diagram

table

graph

Foldables

Write About It

Write About It

Write About It

Write About It

page 82

page 82

Tab 2

Tab 1

Fractions, Decimals, and Percents

percents and fractions

percents and decimals

percent of a number

Write About It

Write About It

Write About It

page 166

Divide with Decimals

decimal
÷
whole number

decimal
÷
decimal

Foldables

page 250

How do I divide a decimal by a decimal?

How do I divide a decimal by a whole number?

Multiply and Divide Fractions

multiply

divide

Example

fraction × whole number

Example

whole number ÷ fraction

Example

fraction × fraction

Example

fraction ÷ fraction

page 336   **Tab 3**

How do I divide a whole number by a fraction?

How do I multiply a fraction by a whole number?

page 336   **Tab 2**

How do I divide a fraction by a fraction?

How do I multiply a fraction by a fraction?

page 336   **Tab 1**

How do I divide a mixed number by a fraction?

How do I multiply a fraction by a mixed number?

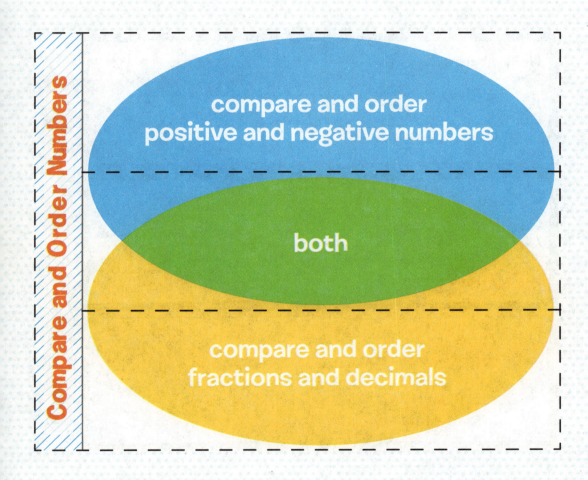

Compare and Order Numbers

compare and order
positive and negative numbers

both

compare and order
fractions and decimals

✂ cut on all dashed lines   ▭ fold on all solid lines   tape to page 418   **FOLDABLES**

Write About It

Write About It

Write About It

page 418